Texts in Computational Science and Engineering

T0237976

Editors

Timothy J. Barth
Michael Griebel
David E. Keyes
Risto M. Nieminen
Dirk Roose
Tamar Schlick

For further volumes:
http://www.springer.com/series/5151

Aslak Tveito • Hans Petter Langtangen
Bjørn Frederik Nielsen • Xing Cai

Elements of Scientific Computing

With 88 Figures and 18 Tables

 Springer

Aslak Tveito
Hans Petter Langtangen
Xing Cai
Center for Biomedical Computing
Simula Research Laboratory
P.O. Box 134
1325 Lysaker
Norway
aslak@simula.no
hpl@simula.no
xingca@simula.no

and

Department of Informatics
University of Oslo
P.O. Box 1080 Blindern
0316 Oslo
Norway

Bjørn Frederik Nielsen
Department of Mathematical
Sciences and Technology
Norwegian University of Life Sciences
P.O. Box 5003
1432 Ås
Norway
bjorn.f.nielsen@umb.no

and

Center for Biomedical Computing
Simula Research Laboratory
P.O. Box 134
1325 Lysaker
Norway

Aslak Tveito has received financial support from the Norwegian Non-fiction Literature Fund.

ISSN 1611-0994
ISBN 978-3-642-26519-8   ISBN 978-3-642-11299-7 (eBook)
DOI 10.1007/978-3-642-11299-7
Springer Heidelberg Dordrecht London New York

Mathematics Subject Classification (2010): 97-01, 97M10, 97N80

Cover design: deblik, Berlin

Printed on acid-free paper

Springer is part of Springer Science+Business Media (www.springer.com)

Preface

Science used to be experiments and theory; now it is experiments, theory, and computations. The computational approach to understanding nature and technology is currently expanding in many fields, such as physics, mechanics, geophysics, astrophysics, chemistry, biology, and most engineering disciplines. The computational methods used in these branches are very similar, and this book is a first introduction to such methods. Many books have been written on the subject. The present text aims to provide a gentle introduction, explaining the methods through examples taken from various fields of science.

As a computational scientist, you will work with other applications, models, and methods than those covered herein. The field is vast and it is impossible to capture more than a small fraction of it in a reasonably sized text. Therefore, we will teach principles and ideas. We believe that principles and ideas carry over from field to field, whereas particularly clever tricks tend to be application specific. We urge you to focus on the ideas and not to get too concerned about the context in which the models appear. We describe the context and provide examples merely to simplify the setting and make the text easier to read.

To read this text, you must know calculus (functions, differentiation, integration etc.) and the basics of linear algebra (vectors, matrices etc.), and you should be familiar with elementary programming. This book is just a gentle start to show what scientific computing is about and present some background that will simplify your future study of more advanced texts on numerical methods and their applications in science and engineering.

All the problems in the text have been solved, and the solutions are provided at

http://www.ifi.uio.no/cs/

where you can also find lecture slides for all the chapters.

We hope you enjoy reading this book as much as we have enjoyed writing it.

Fornebu
September 2010

Aslak Tveito
Hans Petter Langtangen
Bjørn Fredrik Nielsen
Xing Cai

Contents

Chapter 1
Computing Integrals

1.1 Background

In Oslo, there is a chain of small cafés called Bagel and Juice that serve fresh bagels and tasty juice. We know of such a café on Hegdehaugsveien, fairly close to the University of Oslo. The owner of this café, as well as all the other owners, faces one particular problem each night: She has to determine how many bagels to order for the next day. Obviously, on the one hand, she wants to have a sufficient supply for the customers. However, on the other hand, she does not want to order more than she will be able to sell, because the surplus has to be discarded or sold elsewhere at a loss.

A reasonable approach to this problem is to try to ensure that on, for instance, 95% of the days, the owner has enough bagels, but in the remaining 5% she has to disappoint the last few customers. The problem of determining how many bagels are needed in order to fulfill the need on 95% of the days is a problem of statistics. Here, we will just accept the statistical approach to this problem and focus on the computational problem that it generates. This situation is rather common in scientific computing: You are not the one to formulate the problem – you are the one to solve it. Often, solving the problem is only possible when you understand the background of the problem.

A standard statistical approach to this problem is to use a probability density function to model the number of bagels sold per day. By integrating this function from a value a to another value b, we get a figure representing the probability that the number of bagels sold on one particular day is larger than a and smaller than b. But how should we choose an appropriate probability density function? In the present problem, we can easily compute the average number of bagels sold per day. It is more likely that the number of bagels sold on an arbitrary day is close to this average rather than far away. In such probability problems, the so-called normal, or Gaussian, probability density function is a good candidate. This probability density function is uniquely determined by two parameters: the average and the standard deviation. Suppose we have counted the number of bagels sold each day over a·

A. Tveito et al., *Elements of Scientific Computing*, Texts in Computational Science and Engineering 7, DOI 10.1007/978-3-642-11299-7_1,
© Springer-Verlag Berlin Heidelberg 2010

Fig. 1.1 The figure illustrates the normal probability distribution in the case of $\bar{x} = 300$ and $s = 20$

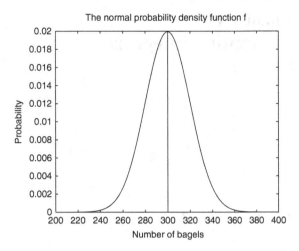

The normal probability density function f

long period. From these observations, we can compute[1] the sample mean $\bar{x}$ and the sample standard deviation s. Using these two parameters, the normal probability function is given by

$$f(x) = \frac{1}{\sqrt{2\pi}s}e^{-\frac{(x-\bar{x})^2}{2s^2}}.$$

Let us assume that $\bar{x} = 300$ and $s = 20$. Then we have

$$f(x) = \frac{1}{\sqrt{2\pi}20}e^{-\frac{(x-300)^2}{2\cdot20^2}}, \tag{1.1}$$

which is plotted in Fig. 1.1.

Figure 1.1 shows that in most cases, the demand of bagels is around 300. It is very unlikely that it is above 370 or below 230. More precisely, if we consider any one particular day, the probability that the demand of bagels is less than b is given by

$$p = \int_{-\infty}^{b} f(x)dx, \tag{1.2}$$

---

[1] Suppose we have n measurements resulting in the numbers $x_1, x_2 \ldots, x_n$. Then the sample mean is given by

$$\bar{x} = \frac{1}{n}\sum_{j=1}^{n} x_j$$

and the sample standard deviation is given by

$$s = \sqrt{\frac{\sum_{j=1}^{n}(x_j - \bar{x})^2}{n-1}}.$$

For a discussion of these concepts, see any introductory book in Statistics, e.g., Johnson and Bhattacharyya [18].

where f is given by (1.1). And this is exactly the problem we want to solve in this chapter: How do we obtain a numerical value for p, given a value of the number of bagels b? The reason for us to seek a numerical solution is that (1.2) cannot be integrated analytically when f is of the form (1.1).

1.2 The Computational Problem

As stated above, the problem we want to solve is to compute the probability p given by (1.2) for given values of the number of bagels b. Before we introduce a numerical method for this problem, we can do some observations that will simplify our task.

Our problem is to estimate the number of bagels b such that the owner can meet the demand on at least 95% of the days. Thus we want to find b such that

$$p \geqslant 0.95. \tag{1.3}$$

In order to do this, we have to be able to compute p given by (1.2), for any relevant value of the upper limit b. We will devise a numerical algorithm for computing such an approximation. One main step in deriving this algorithm is to divide the interval into a number of small subintervals and then approximate the integral on each of these smaller domains. However, negative infinity $-\infty$ as the lower limit is difficult to deal with, because it will result in an infinite number of subintervals. We therefore want to rephrase the problem in terms of a definite integral on a bounded domain. In order to do this, we start by recalling that the average number of bagels sold each day is 300. Thus, if we have chosen $b = 300$, we would only have a sufficient supply on 50% of the days. Consequently, we only consider the case of

$$b > 300. \tag{1.4}$$

Now, we can divide the integral of (1.2) into two parts:

$$p = \int_{-\infty}^{300} f(x)dx + \int_{300}^{b} f(x)dx. \tag{1.5}$$

Here we can compute the first term analytically. To do this, we first note that f is symmetric, in the sense that

$$f(300 + x) = f(300 - x)$$

for all x. It follows from this property that

$$\int_{-\infty}^{300} f(x)dx = \int_{300}^{\infty} f(x)dx,$$

and consequently we have

$$\int_{-\infty}^{300} f(x)dx = \frac{1}{2} \int_{-\infty}^{\infty} f(x)dx.$$

Since f is a probability density function, we must have[2]

$$\int_{-\infty}^{\infty} f(x)dx = 1,$$

and therefore

$$\int_{-\infty}^{300} f(x)dx = \frac{1}{2},$$

which indeed is consistent with the argument indicated above, that if $b = 300$, we would have a sufficient supply on 50% of the days.

By using this result, we can reformulate the problem stated in (1.5) as

$$p = \frac{1}{2} + \int_{300}^{b} f(x)dx, \tag{1.6}$$

where

$$f(x) = \frac{1}{\sqrt{2\pi}20} e^{-\frac{(x-300)^2}{2\cdot20^2}}. \tag{1.7}$$

We want to be able to compute the value of p for b larger than 300. By the arguments given above, we will be interested in p for b ranging from 300 to 370. Thus, the computational problem is now reduced to that of computing a definite integral on a bounded domain.

1.3   The Trapezoidal Method

We will derive a numerical algorithm based on dividing the interval of integration into very small intervals, where the integral can be approximated by simple formulas. Then, we add all the contributions from the small intervals to yield an approximation of the entire integral. We will start this process by considering only one interval – the original interval – and then proceed by using two intervals and finally we will divide the original interval into an arbitrarily large number of n subintervals.

---

[2] This property can be derived in a straightforward manner. By a change of variable, we have

$$\frac{1}{\sqrt{2\pi}20} \int_{-\infty}^{\infty} e^{-\frac{(x-300)^2}{2\cdot20^2}} dx = \frac{1}{\sqrt{\pi}} \int_{-\infty}^{\infty} e^{-y^2} dy = 1.$$

The latter integral can be found in any table of definite integrals, see, e.g., [24].

1.3.1 Approximating the Integral Using One Trapezoid

The general problem motivated by (1.6) above, is to compute numerical approximations of definite integrals, i.e., approximations of

$$\int_a^b f(x)dx$$

for fairly general choices of functions f. Let us start by considering

$$f(x) = e^x$$

which we want to integrate from 1 to 2, i.e., we want to compute

$$\int_1^2 e^x dx.$$

In Fig. 1.2, we have graphed f and we note that we want to compute the area under the curve. We have also graphed the straight line interpolating f in the endpoints $x = 1$ and $x = 2$. This straight line is given by

$$y(x) = e\left[1 + (e - 1)(x - 1)\right],$$

which is verified by noting that

$$y(1) = e$$

and

$$y(2) = e^2.$$

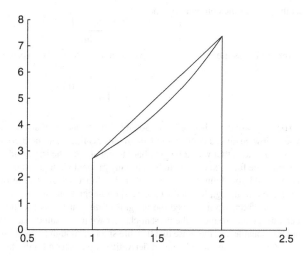

Fig. 1.2 The figure illustrates how the integral of $f(x) = e^x$ (*lower curve*) can be approximated by a trapezoid on a given interval

Since

$$y(x) \approx f(x)$$

for $1 \leq x \leq 2$, it follows that also

$$\int_1^2 e^x dx \approx \int_1^2 y(x)dx.$$

In this case, we can compute both integrals and compare the result. We have

$$\int_1^2 e^x dx = e(e-1) \approx 4.6708$$

and

$$\int_1^2 y(x)dx = \int_1^2 e\left[1 + (e-1)(x-1)\right] dx = \frac{1}{2}e + \frac{1}{2}e^2 \approx 5.0537,$$

so the relative error[3] is

$$\frac{|4.6708 - 5.0537|}{4.6708} \cdot 100\% \approx 7.4\% .$$

We can generalize this approximation to any reasonable[4] function f and any finite limits a and b. Again, we let $y = y(x)$ be a linear function interpolating f in

---

[3] Here we need to clarify our concepts. Let u be the exact number that we want to compute, and let v be an approximation generated by some kind of method. Then we refer to

$$|u - v|$$

as the error of the computation and

$$\frac{|u - v|}{|u|}$$

is referred to as the relative error. If we want the latter to be in percentages, we compute

$$\frac{|u - v|}{|u|} \cdot 100\%.$$

[4] You may wonder whether "unreasonable" functions exist and we can assure you that they do. If you follow advanced courses in mathematics dealing with the theory of integration, you will learn about functions that would be very hard to get at with the trapezoidal method. To illustrate this, we mention the function defined to be 1 for any rational value of x and 0 for any irrational values of x. How would you define the integral of such a function? If you are interested in this topic, you can read about it in, e.g., Royden's book [25] on integration theory.

Less subtle, we may need the integral of discontinuous functions, or functions with jumps in derivatives, and so on. In the present chapter we just assume that all functions under consideration are "reasonable", in the sense that they are smooth enough to justify the techniques we are studying. Typically, functions with bounded derivatives up to second order are allowed in this chapter.

the endpoints a and b. Hence, y is defined by

$$y(x) = f(a) + \frac{f(b) - f(a)}{b - a}(x - a), \tag{1.8}$$

and we have the following trapezoid approximation:

$$\int_a^b f(x)dx \approx \int_a^b y(x)dx. \tag{1.9}$$

Since y is linear, it is easy to compute the integral analytically:

$$\int_a^b y(x)dx = \int_a^b \left[f(a) + \frac{f(b) - f(a)}{b - a}(x - a) \right] dx = (b - a)\frac{1}{2}(f(a) + f(b)).$$

The trapezoidal rule, using a single trapezoid, is therefore given by

$$\boxed{\int_a^b f(x)dx \approx (b - a)\frac{1}{2}(f(a) + f(b)).} \tag{1.10}$$

Example 1.1. Let

$$f(x) = \sin(x), \quad a = 1, \quad b = 1.5.$$

Then, the trapezoidal method gives

$$\int_1^{1.5} f(x)dx \approx (1.5 - 1)\frac{1}{2}(\sin(1) + \sin(1.5)) \approx 0.4597,$$

whereas the exact value is

$$\int_1^{1.5} f(x)dx = -[\cos(x)]_1^{1.5} = -(\cos(1.5) - \cos(1)) \approx 0.4696.$$

Thus, the relative error is

$$\frac{0.4696 - 0.4597}{0.4696} \cdot 100\% \approx 2.11\%.$$

$\blacksquare$

1.3.2 Approximating the Integral Using Two Trapezoids

The reasoning behind the trapezoidal method and the examples given above clearly indicate that, in general, the error is smaller when the length of the interval, i.e., $b - a$, is smaller. This observation can be utilized to derive a composite scheme just by observing that the integral can be split into several parts. Let $c = (a+b)/2$, i.e., the midpoint between a and b. Then we have

$$\int_a^b f(x)dx = \int_a^c f(x)dx + \int_c^b f(x)dx,$$

and thus, using (1.10) on each interval, we have

$$\int_a^b f(x)dx \approx \left[(c-a)\frac{1}{2}\left(f(a) + f(c)\right) \right] + \left[(b-c)\frac{1}{2}\left(f(c) + f(b)\right) \right].$$

By observing that

$$c - a = b - c = \frac{1}{2}(b-a),$$

we can simplify this expression to obtain

$$\int_a^b f(x)dx \approx \frac{1}{4}(b-a)\left[f(a) + 2f(c) + f(b)\right]. \qquad (1.11)$$

This method is illustrated in Fig. 1.3.

Example 1.2. Let us go back to the example above and use this slightly refined method to compute an approximation of

$$\int_1^{1.5} \sin(x)\,dx. \qquad (1.12)$$

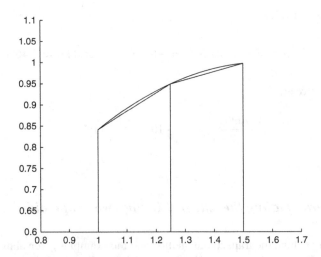

Fig. 1.3 The figure illustrates how the integral of $f(x) = \sin(x)$ can be approximated by two trapezoids on a given interval

By (1.11), using $c = (1 + 1.5)/2 = 1.25$, we have

$$\int_1^{1.5} \sin(x)\,dx \approx \frac{1}{4} \cdot \frac{1}{2} [\sin(1) + 2\sin(1.25) + \sin(1.5)] \approx 0.4671.$$

The relative error of this approximation is

$$\frac{0.4696 - 0.4671}{0.4696} \cdot 100\% \approx 0.53\%,$$

which, as we expected, is significantly smaller than the error obtained by using only one trapezoid.

∎

1.3.3 Approximating the Integral Using n Trapezoids

It is not hard to realize that we can proceed further and split the interval into n parts, where $n \geq 1$ is an integer. Let

$$h = \frac{b - a}{n}$$

and define

$$x_i = a + ih$$

for $i = 0, 1, \ldots n$. These points

$$a = x_0 < x_1 < \cdots < x_{n-1} < x_n = b$$

divide the interval from a to b into n subintervals, see Fig. 1.4.

Due to the additive property of the integral, we have

$$\int_a^b f(x)dx = \int_{x_0}^{x_1} f(x)dx + \int_{x_1}^{x_2} f(x)dx + \cdots + \int_{x_{n-1}}^{x_n} f(x)dx$$

$$= \sum_{i=0}^{n-1} \int_{x_i}^{x_{i+1}} f(x)dx. \tag{1.13}$$

Let us consider one of these intervals and use the approximation derived above. If we use x_i and x_{i+1} as the two end-points in (1.10), we get

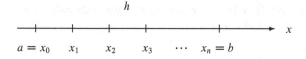

Fig. 1.4 The interval from $x = a$ to $x = b$ is divided into sub-intervals of length $x_{i+1} - x_i = h$

$$\int_{x_i}^{x_{i+1}} f(x)dx \approx (x_{i+1} - x_i)\frac{1}{2}[f(x_i) + f(x_{i+1})].$$

Since

$$x_{i+1} - x_i = h,$$

we have the approximation

$$\int_{x_i}^{x_{i+1}} f(x)dx \approx \frac{h}{2}[f(x_i) + f(x_{i+1})]. \tag{1.14}$$

Next, we apply this approximation to each of the terms in (1.13). This leads to

$$\int_a^b f(x)dx = \sum_{i=0}^{n-1}\int_{x_i}^{x_{i+1}} f(x)dx \approx \frac{h}{2}\sum_{i=0}^{n-1}[f(x_i) + f(x_{i+1})]. \tag{1.15}$$

This expression can be simplified slightly. By expanding the sum, we observe that

$$\sum_{i=0}^{n-1}[f(x_i) + f(x_{i+1})] = [f(x_0)+f(x_1)] + [f(x_1) + f(x_2)] + [f(x_2) + f(x_3)]$$

$$+ \cdots + [f(x_{n-2}) + f(x_{n-1})] + [f(x_{n-1}) + f(x_n)].$$

Here we note that, except for $i = 0$ and $i = n$, all the $f(x_i)$ terms appear twice in this sum. Hence, we have

$$\sum_{i=0}^{n-1}[f(x_i) + f(x_{i+1})] = f(x_0) + 2[f(x_1) + f(x_2) + \cdots + f(x_{n-1})] + f(x_n).$$

By using this observation in (1.15), we get

$$\int_a^b f(x)dx \approx h\left[\frac{1}{2}f(x_0) + f(x_1) + f(x_2) + \cdots + f(x_{n-1}) + \frac{1}{2}f(x_n)\right],$$

which can be written more compactly as

$$\boxed{\int_a^b f(x)dx \approx h\left[\frac{1}{2}f(x_0) + \sum_{i=1}^{n-1}f(x_i) + \frac{1}{2}f(x_n)\right].} \tag{1.16}$$

This approximation is referred to as the *composite trapezoidal rule*[5] of integration. A simple example is illustrated in Fig. 1.5.

---

[5] In the rest of the text we often use the term trapezoidal rule as shorthand for the composite trapezoidal rule.

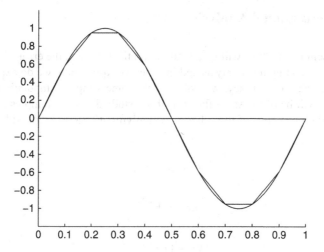

Fig. 1.5 The figure shows how $f(x) = \sin(2\pi x)$ graphed on the unit interval can be approximated by linear functions over $n = 10$ subintervals

Example 1.3. We return again to the problem studied in the examples above, i.e., we want to compute numerical approximations of

$$\int_1^{1.5} \sin(x)\,dx. \tag{1.17}$$

In the composite trapezoidal rule, we set $a = 1, b = 1.5$ and $n = 100$. This gives

$$h = \frac{b - a}{n} = \frac{0.5}{100} = 0.005,$$

and then, by (1.16), we get

$$\int_a^b f(x)dx \approx 0.469564,$$

which should be compared with the exact solution given by

$$\int_1^{1.5} f(x)dx = -[\cos(x)]_1^{1.5} = -(\cos(1.5) - \cos(1)) \approx 0.469565.$$

The relative error is now

$$\frac{0.469565 - 0.469564}{0.469565} \cdot 100\% = 0.0002\%.$$

1.4 Computational Analysis

For the present method, we will give a full theoretical study of the error,[6] but for the purpose of motivating this analysis and in order to explain how we do experiments suggesting a certain accuracy, we will present some simple computations. These experiments will be of the same flavor as those studied in the examples above; we investigate the properties of the scheme by applying it to a problem with a known solution.

Let

$$F(x) = xe^x,$$

and define

$$f(x) = F'(x),$$

so

$$f(x) = (1 + x)e^x. \tag{1.18}$$

Consequently, we have

$$\int_0^1 f(x)dx = [F(x)]_0^1 = e.$$

Since we have the analytical solution, this is a good test problem for the method outlined above. Let T_h denote the approximation of the integral computed by the trapezoidal method (1.16), i.e.,

$$T_h = h \left[\frac{1}{2}f(x_0) + \sum_{i=1}^{n-1} f(x_i) + \frac{1}{2}f(x_n) \right],$$

where f is given by (1.18), and where

$$x_i = ih,$$

for $i = 0, \ldots, n$, and

$$h = \frac{1}{n}. \tag{1.19}$$

We want to study the error defined by

$$E_h = |e - T_h|.$$

In Table 1.1, we present numerical values of this error for decreasing values of h. We observe from this table that

---

[6] See Project 1.7.1.

Table 1.1 The table shows the number of intervals, n, the length of the intervals, h, the error, E_h, and finally E_h/h^2. The latter term seems to converge toward a constant

n	h	E_h	E_h/h^2
1	1.0000	0.5000	0.5000
2	0.5000	0.1274	0.5096
4	0.2500	0.0320	0.5121
8	0.1250	0.0080	0.5127
16	0.0625	0.0020	0.5129
32	0.0313	0.0005	0.5129
64	0.0156	0.0001	0.5129

$$E_h \approx 0.5129h^2, \tag{1.20}$$

which means that we can get as accurate an approximation as we want just by choosing h sufficiently small which is done by increasing the number of intervals n. Say, we want

$$E_h \leqslant 10^{-5}. \tag{1.21}$$

Then it follows from (1.20) that h must satisfy

$$0.5129h^2 \leqslant 10^{-5},$$

or

$$h \leqslant 0.0044.$$

Using (1.19), we have

$$n = 1/h \geqslant 226.47.$$

Since n has to be an integer, we conclude that any $n \geqslant 227$ will yield an error less than 10^{-5}.

We have seen that the trapezoidal method gives good results for the sine-function and for the function given by (1.18). Let us now challenge the method by applying it to a few other functions. In Fig. 1.6 we have plotted the functions x^4, x^{20}, and $\sqrt{x}$, and in Table 1.2 we have given numerical results obtained by applying the trapezoidal method to these functions. We observe from Table 1.2 that for x^4 and for x^{20}, the error is about $0.33h^2$ and $1.67h^2$, respectively. So for these functions, the trapezoidal method gives an error of basically the same form as we observed above. We note that the error is about five times larger for x^{20} compared to that of x^4. If you consider the plot above and recall the derivation of the trapezoidal method, it is not hard to realize that the lower curve is more difficult, since the second-order derivative is much larger. Compare this with Fig. 1.2 where it is apparent that large second-order derivatives may cause large errors. Now, this observation also applies to the $\sqrt{x}$ function. In this case the second derivative is infinite as x approaches zero, and we see from Table 1.2 that E_h/h^2 does not converge toward any constant. These issues will be given further consideration in Project 1.7.1 below.

Fig. 1.6 The figure shows the graphs of $\sqrt{x}$ (*upper*), x^4 (*middle*), and x^{20} (*lower*) on the unit interval

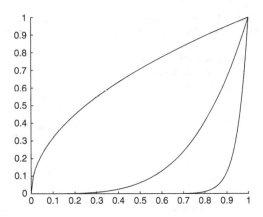

Table 1.2 The table shows how accurate the trapezoidal method is for three definite integrals where the exact solutions are known. Note that convergence seems to be obtained for all three cases. For the two first cases, we note that E_h/h^2 is constant as h is reduced. This is, however, not the case for the third function

h	$\int_0^1 x^4 dx = \frac{1}{5}$		$\int_0^1 x^{20} dx = \frac{1}{21}$		$\int_0^1 \sqrt{x} dx = \frac{2}{3}$	
	$10^5 E_h$	E_h/h^2	$10^5 E_h$	E_h/h^2	$10^5 E_h$	E_h/h^2
0.01	3.33	0.33	16.66	1.67	20.37	2.04
0.005	0.83	0.33	4.17	1.67	7.25	2.90
0.0025	0.21	0.33	1.04	1.67	2.57	4.17
0.00125	0.05	0.33	0.26	1.67	0.91	5.84

1.5 Back to the Bagels

Let us recall that our problem is to compute

$$p = p(b) = \frac{1}{2} + \int_{300}^{b} f(x)dx, \qquad (1.22)$$

where

$$f(x) = \frac{1}{\sqrt{2\pi}20} e^{-\frac{(x-300)^2}{2 \cdot 20^2}}. \qquad (1.23)$$

See the discussion leading to (1.6). Recall also that for a given value of b, the function $p = p(b)$ denotes the probability that the number of bagels sold on one particular day is less than or equal to b. In particular, we want to find the smallest possible value of $b = b^*$ such that $p = p(b^*) \geqslant 0.95$. When the owner orders b^* bagels, she knows that this will be sufficient on 95% of the days.

In the section above, we derived a method for computing integrals of the form encountered in (1.22). Here we will apply this method to compute $p(b)$ for integer values of b ranging from 300 to 370. Note that since

$$p(b) = \frac{1}{2} + \int_{300}^{b} f(x)dx,$$

we have

$$p(b + 1) = \frac{1}{2} + \int_{300}^{b+1} f(x)dx$$

$$= \frac{1}{2} + \int_{300}^{b} f(x)dx + \int_{b}^{b+1} f(x)dx$$

$$= p(b) + \int_{b}^{b+1} f(x)dx.$$

Since

$$p(300) = \frac{1}{2},$$

we find the value of $p(b)$ for any integer value of b larger than 300 by performing the following iterative procedure

$$p(b + 1) = p(b) + \int_{b}^{b+1} f(x)dx, \qquad b = 300, 301, \ldots$$

Hence, we only have to apply the trapezoidal rule to intervals of length one. In Fig. 1.7, we have plotted $p = p(b)$ using this procedure with[7] $n = 4$ in the trapezoidal rule at each step from b to $b + 1$. We observe from the figure that the desired value of b ensuring a sufficient supply on at least 95% of the days is between 330 and 340. In Table 1.3 we have listed the values of $p(b)$ for some relevant values of b and we observe that choosing $b^* = 333$ will give a sufficient supply on 95.1% of the days.

1.6 Exercises

Exercise 1.1. Let

$$f(x) = \frac{1}{x + 1}.$$

(a) Use the trapezoidal method (1.10) to estimate

$$\int_{0}^{1} f(x)\,dx. \qquad (1.24)$$

---

[7] We have also tried $n = 6$ and $n = 8$ with the same results; in fact, the difference between the associated p functions is less than $2 \cdot 10^{-6}$.

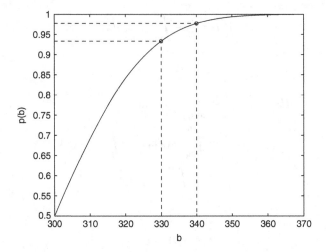

Fig. 1.7 The graph of $p(b)$, where b is the number of bagels and $p(b)$ is the probability that the demand on one given day is less than or equal to b

Table 1.3 The table shows the probability $p(b)$ for the demand of bagels on one particular day to be less than or equal to b

b	$p(b)$
331	0.939
332	0.945
333	0.951
334	0.955

(b) Compute the exact integral above and compute the relative error.
(c) Use the composite trapezoidal rule (1.16) with $n = 2$ to estimate the integral in (1.24). Compute the relative error.
(d) Repeat (c) using $n = 3$.
(e) Use the results for $n = 1, 2,$ and 3 to estimate how large n has to be in order for the relative error to be less than $1/1,000$.

◇

Exercise 1.2. Let
$$f(x) = \sin(x)$$
and
$$g(x) = \sin(5x).$$

(a) Compute the exact values
$$\int_0^{1/2} f(x)\,dx \tag{1.25}$$

and

$$\int_0^{1/2} g(x)\,dx.\tag{1.26}$$

(b) Compute approximations to (1.25) and (1.26) using the composite trapezoidal rule (1.16) with $n = 2$. Compute the relative error.

(c) Graph f and g on the interval $[0, 1/2]$ and draw the two linear interpolants as in Fig. 1.3. Use these graphs to explain why the error of the approximation of (1.26) is larger than the error of the approximation of (1.25).

◇

Exercise 1.3. Which of the following integrals would you expect to be difficult to approximate using the composite trapezoidal rule?

(a) $\displaystyle\int_0^1 x^3\,dx$

(b) $\displaystyle\int_{-1}^1 |x|\,dx$

(c) $\displaystyle\int_{-1}^1 \sin(\pi x)\,dx$

(d) $\displaystyle\int_0^1 1/\sqrt{x}\,dx$

◇

Exercise 1.4. Make a computer code that implements the composite trapezoidal method (1.16). The code should take a, b, and n as input parameters for a given function $f = f(x)$.

(a) Use the code to compute $\int_0^1 x^4 dx$ and compare your results with the results given in Table 1.2 on page 14.

(b) Use your code to check the validity of your results in Exercise 1.3 above.

(c) Use your code to investigate the rate of convergence for the examples in Exercise 1.3. Use the method applied in Table 1.2.

◇

1.7 Projects

1.7.1 Show that the Error is $O(h^2)$

We have seen some computations above that clearly indicate that the error for the composite trapezoidal rule is[8] $O(h^2)$. More precisely, the computations indicate that there exists a constant c independent of h such that the error is bounded by ch^2. In

---

[8] In numerical analysis, it is very common to use this O terminology. If the error is bounded by some constant c times h^2, we say that the error is $O(h^2)$. Generally, if the error is bounded by ch^β, the error is said to be $O(h^\beta)$. Note that it is crucial that the constant c be independent of h.

the particular example we considered above, we found that c could be chosen to be 0.5129, cf. (1.20). The aim of this project is to show that this result is fairly general and holds for all functions f having a bounded second-order derivative. In this project, we will use some results from Calculus. They are listed below, see page 20.

We start by recalling that the problem at hand is to compute an approximation to the definite integral

$$\int_a^b f(x)dx, \tag{1.27}$$

and we first consider the trapezoidal method using one single interval. We recall that this method was derived using a linear function given by

$$y(x) = f(a) + \frac{f(b) - f(a)}{b - a}(x - a), \tag{1.28}$$

which approximates the function f, cf. Fig. 1.2. Note, in particular, that $y(a) = f(a)$ and $y(b) = f(b)$. By integrating this linear function, we get the trapezoidal approximation given by[9]

$$T(f, a, b) = \int_a^b y(x)dx = (b - a)\frac{1}{2}(f(a) + f(b)). \tag{1.29}$$

(a) The error of the trapezoidal method is defined by

$$E = \left| \int_a^b f(x)dx - T(f, a, b) \right|.$$

Show that

$$E = \left| \int_a^b (f(x) - y(x))dx \right|.$$

(b) Use (1.34) below to show that

$$E \leq \int_a^b |f(x) - y(x)| \, dx.$$

(c) Use (1.35) below to show that

$$E \leq (b - a) \max_{a \leq x \leq b} |f(x) - y(x)|.$$

Furthermore, if $\beta = 1$, the convergence is said to be *linear*. If $\beta = 2$, the convergence is *quadratic*, if $\beta = 3$, the convergence is *cubic*, and so on.

Now this is probably a bit more than you would like to know, but, nevertheless; if the error is ch and c depends on h and goes to zero as h goes to zero, the convergence is said to be *superlinear*.

[9] We add the explicit dependence of T on f, a, and b for future use.

(d) Note that the problem of bounding the error of the trapezoidal method is now reduced to bounding the difference between the function $f = f(x)$ and the linear function $y = y(x)$, which coincides with f in the endpoints $x = a$ and $x = b$. We will bound this difference by subsequent use of the Taylor series given in (1.41) below.

Use the Taylor series with $n = 1$ to show that there is a value ξ between a and x such that

$$f(x) = f(a) + (x - a)f'(a) + \frac{1}{2}(x - a)^2 f''(\xi), \qquad (1.30)$$

and use this expansion to argue that

$$f(x) - y(x) = \left[f'(a) - \frac{f(b) - f(a)}{b - a} \right](x - a) + \frac{1}{2}(x - a)^2 f''(\xi).$$

(e) Use the Taylor expansion with $n = 1$ once more to show that

$$\frac{f(b) - f(a)}{b - a} = f'(a) + \frac{1}{2}(b - a)f''(\eta)$$

for some η in the interval from a to b. Apply this relation to show that

$$f(x) - y(x) = -\frac{1}{2}(x - a)(b - a)f''(\eta) + \frac{1}{2}(x - a)^2 f''(\xi).$$

(f) Let

$$M = \max_{a \le x \le b} |f''(x)|, \qquad (1.31)$$

and use the triangle inequality (1.37) to show that

$$|f(x) - y(x)| \le M(b - a)^2.$$

(g) Show that

$$E \le M(b - a)^3.$$

(h) Next we consider the composite trapezoidal rule. Recall that the composite trapezoidal rule is defined as follows:

$$T_n = h\left[\frac{1}{2}f(x_0) + \sum_{i=1}^{n-1} f(x_i) + \frac{1}{2}f(x_n) \right],$$

as in (1.16). Show that

$$T_n = \sum_{i=1}^{n} T(f, x_{i-1}, x_i)$$

where T is defined in (1.29) and

$$x_i = a + ih = a + i\frac{b-a}{n}.$$

(i) Define

$$E_n = \left| \int_a^b f(x)dx - T_n \right|,$$

and show that

$$E_n \leq \sum_{i=1}^n \left| \int_{x_{i-1}}^{x_i} f(x)dx - T(f, x_{i-1}, x_i) \right|.$$

(j) Show that

$$E_n \leq M \sum_{i=1}^n (x_i - x_{i-1})^3.$$

(k) Use the fact that $nh = b - a$ to show that

$$E_n \leq M(b-a)h^2. \tag{1.32}$$

(l) In comparison with (1.32), there exists a sharper error estimate of the form:

$$E_n \leq \frac{h^2}{12}(b-a) \max_{a \leq x \leq b} |f''(x)|. \tag{1.33}$$

The above error estimate is obtained by using a refined representation of the error. You will find the argument in the book of Conte and de Boor [10]. Discuss the quality of the estimate (1.33) in light of the experiments of Sect. 1.4.

Useful Results from Calculus

In the project above, you will need some results from calculus.

– *An Integral Inequality.* Let $g = g(x)$ be a bounded function defined on the interval $[a, b]$. Then,

$$\left| \int_a^b g(x)dx \right| \leq \int_a^b |g(x)| \, dx. \tag{1.34}$$

– *Another Integral Inequality.* Let $g = g(x)$ be a bounded function defined on the interval $[a, b]$. Then,

$$\int_a^b |g(x)|\,dx \le (b-a)\max_{a\le x\le b}|g(x)|. \tag{1.35}$$

− *Yet Another Integral Inequality.* A more general result can also be useful; Let $f = f(x)$ and $g = g(x)$ be bounded functions defined on the interval $[a, b]$. Then,

$$\int_a^b |f(x)g(x)|\,dx \le \max_{a\le x\le b}|g(x)|\int_a^b |f(x)|\,dx. \tag{1.36}$$

− *The Triangle Inequality.* For any real numbers x and y, we have

$$|x + y| \le |x| + |y|. \tag{1.37}$$

Generally, for real numbers $x_1, x_2, \ldots, x_n$, we have

$$\left|\sum_{i=1}^n x_i\right| \le \sum_{i=1}^n |x_i|. \tag{1.38}$$

− *The Taylor Series.* It is hard to find a more important tool for the development and analysis of numerical methods than the Taylor series. For future reference, we shall state it here in several versions. Let $g = g(x)$ be a function with $n + 1$ continuous derivatives. Then we have the following expansion[10]:

$$g(x + \alpha) = g(x) + \alpha g'(x) + \frac{1}{2}\alpha^2 g''(x) + \cdots \frac{1}{n!}\alpha^n g^{(n)}(x) + R_{n+1}, \tag{1.39}$$

where

$$R_{n+1} = \frac{1}{(n+1)!}\alpha^{n+1} g^{(n+1)}(x + \eta),$$

for some η in the interval bounded by x and $x + \alpha$. This can be written more compactly as (recall that $0! = 1$)

$$g(x + \alpha) = \sum_{m=0}^n \frac{\alpha^m}{m!} g^{(m)}(x) + R_{n+1}. \tag{1.40}$$

The Taylor series expansion can also be written as

---

[10] The notation $g^{(n)}$ may seem a bit unfamiliar, but it simply means

$$\frac{d^n g}{dx^n}$$

i.e., $g^{(n)}$ is the nth order derivative of the function g with respect to x.

$$g(x) = g(a) + (x - a)g'(a) + \frac{1}{2}(x - a)^2 g''(a) \qquad (1.41)$$

$$+ \cdots + \frac{1}{n!}(x - a)^n g^{(n)}(a) + Q_{n+1}$$

where

$$Q_{n+1} = \frac{1}{(n + 1)!}(x - a)^{n+1} g^{(n+1)}(\xi)$$

for some ξ in the interval bounded by x and a. This can again be written more compactly as

$$g(x) = \sum_{m=0}^{n} \frac{(x - a)^m}{m!} g^{(m)}(a) + Q_{n+1}. \qquad (1.42)$$

1.7.2 Derive Other Methods for Numerical Integration

The purpose of this project is to derive other methods of numerical integration. Since we argued above that the error of the trapezoidal scheme is $O(h^2)$, it is reasonable to ask why we should need any other schemes. It is correct that the trapezoidal rule is accurate and a good approximation can be achieved if we choose sufficiently many grid points. Essentially, the trapezoidal scheme requires about n arithmetic operations and n function evaluations. For any reasonable function, we can evaluate the integral using, say, $n = 10^6$ in less than a second with a fairly modern computer. So why do we need anything else? There are several reasons. First, there are situations where we need to compute millions of integrals and we need them fast. One million seconds is more than 11 days and that's a long time to wait for the result of a computation. So even if the trapezoidal scheme is accurate, there are situations where we want to be able to compute an accurate solution more quickly. Second, the methods discussed here also have multidimensional counterparts. In three dimensions, using $n = 10^6$ grid points in each coordinate direction would lead to 10^{18} function evaluations for a method similar to the trapezoidal scheme. On a very fast computer, one arithmetic operation takes about 10^{-9} s and thus such a grid is not feasible with today's technology. We may circumvent this difficulty by using more accurate methods or parallel computers. In this project we will concentrate on developing two methods that are more accurate than the trapezoidal scheme. But more important is that by completing this project, you will understand the principle of how such a method is derived.

The problem is to compute an approximation of the integral

$$\int_a^b f(x)\, dx.$$

One way of doing this is to find a function $p_0(x)$ such that

$$p_0(x) \approx f(x)$$

for all $x \in [a, b]$. If we are able to integrate $p_0(x)$, we have an estimate

$$\int_a^b f(x)\, dx \approx \int_a^b p_0(x)\, dx.$$

We used this strategy above to derive the trapezoidal rule and in this project we will pursue this a little further.

(a) The midpoint rule.
   Let

$$p_0(x) = f\left(\frac{a+b}{2}\right), \quad x \in [a, b]$$

and show that

$$\int_a^b p_0(x)dx = (b-a)f\left(\frac{a+b}{2}\right). \tag{1.43}$$

The midpoint rule is given by

$$\int_a^b f(x)dx \approx (b-a)f\left(\frac{a+b}{2}\right). \tag{1.44}$$

(b) Note that

$$\int_0^1 \frac{x}{(4-x^2)^2}\, dx = \frac{1}{24}. \tag{1.45}$$

Use the midpoint rule to approximate the integral (1.45) and compute the relative error.

(c) Let

$$h = \frac{b-a}{n}$$

where $n > 0$ is an integer, and define

$$x_{i+1/2} = a + (i + 1/2)h$$

for $i = 0, 1, 2, \ldots, n-1$. Show that $x_{1/2} = a+h/2$ and that $x_{n-1/2} = b-h/2$. Derive the composite midpoint rule

$$\int_a^b f(x)\, dx \approx h \sum_{i=0}^{n-1} f(x_{i+1/2}). \tag{1.46}$$

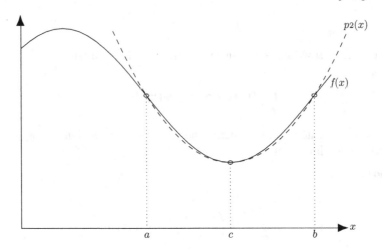

Fig. 1.8 A polynomial of degree 2 that approximates a function $f(x)$

Set $n = 2$ and use (1.46) to approximate the integral in (1.45). Compute the relative error.

(d) Implement the composite midpoint rule (1.46) and use your program to try to estimate c and α such that the error of the scheme is approximately ch^α. You should probably consider Sect. 1.4 first.

(e) In the midpoint rule we used the constant $f\left(\frac{a+b}{2}\right)$ to approximate f on the interval $[a, b]$. Of course, the constant is trivial to integrate. The next level of complexity is to approximate f by a straight line. Read the derivation given in Sect. 1.3.1 again and use the trapezoidal rule (1.10) to approximate the integral in (1.45). Compute the relative error and compare the result with the one you obtained in (b) above.

(f) Simpson's[11] rule.
Define

$$c = \frac{1}{2}(a + b) \qquad (1.47)$$

and let $p_2(x)$ be a polynomial of degree ≤ 2 that interpolates f at the points a, c and b, see Fig. 1.8.
More precisely, we want to find $p_2 = p_2(x)$ such that

$$p_2(x) = A + Bx + Cx^2, \qquad (1.48)$$

where the constants A, B and C are chosen such that

---

[11] This is probably not the Simpson you have heard of. Thomas Simpson, 1710–1761, worked on interpolation and numerical methods for integration.

$$p_2(a) = f(a),$$
$$p_2(c) = f(c),$$
$$p_2(b) = f(b).$$

(1.49)

In order to determine A, B and C such that these conditions hold, it is common to introduce so-called "divided differences". By using these, A, B and C can be expressed uniquely in terms of $f(a)$, $f(c)$ and $f(b)$. That is, however, a bit technical, and we refer the interested reader to, e.g., Conte and de Boor [10]. When A, B and C are determined, $p_2(x)$ can be integrated and it turns out that we get

$$\int_a^b p_2(x)\, dx = \frac{b-a}{6} [f(a) + 4f(c) + f(b)]$$

and hence Simpson's rule is

$$\boxed{\int_a^b f(x)\, dx \approx \frac{b-a}{6} [f(a) + 4f(c) + f(b)].}$$

(1.50)

Redo (b) above using (1.50). Discuss the relative error for the midpoint rule, the trapezoidal rule, and Simpson's rule for this problem.

(g) Let

$$h = \frac{b-a}{n}$$

for an integer $n > 0$ and define

$$x_i = a + ih$$

for $i = 0, 1, 2, \ldots, n$.

Use (1.50) to approximate $\int_{x_{i-1}}^{x_i} f(x)dx$ and use this result to derive the composite Simpson's rule

$$\int_a^b f(x)\, dx \approx \frac{h}{6} \sum_{i=1}^n [f(x_{i-1}) + 4f(x_{i-1/2}) + f(x_i)],$$

where

$$x_{i-1/2} = \frac{1}{2}(x_{i-1} + x_i).$$

(h) Redo (d) for the composite Simpson's rule.

(i) You work at Numerical Integrators, Inc. A customer has sent you a mail and asked whether you can write a program that computes

$$a(t) = \int_0^1 \sin(\pi^4 t x^2) \ln\left(\sqrt{x+t}\right) dx$$

for any value of t in the interval $[1, 2]$. She wants a program where t is input, $a(t)$ is output, where the relative error is less $1/1,000$ (no percent sign here), and the computation must take less than 1 s of CPU time on a typical desk-top computer.

Write a project proposal for the customer. Mention what kinds of methods you would like to try, how you would ensure the accuracy, what kind of computer you would use, and how many hours you need to do the job in, e.g., Java.[12]

1.7.3 Compute the Length of a Cable

In Fig. 1.9 you see the beautiful Storebælt bridge in Denmark connecting Sjælland and Fyn. The bridge was opened in 1998 and is now used by more than 20,000 cars every day. In the year 2000, the Øresund bridge connecting Denmark and Sweden was also opened. These two bridges, together with the smaller Lillebælt bridge, now tie Norway and Sweden directly to the European mainland. Building bridges is a major challenge for engineers and applied mathematicians. Most bridges are safe, but there have been disastrous designs. Probably the most infamous disaster is the break-down of the Tacoma bridge.[13]

In this project we will consider one of the simpler problems connected to building bridges. Take a look at Fig. 1.9. Suppose we want to know the length of the cables that are needed. For the purpose of illustration, we consider a simplified design, see Fig. 1.10.

Our problem is to compute the length of a graph. To solve this problem, we start by considering the graph of some smooth function

$$y = y(x)$$

for x ranging from a to b. In particular, we consider a small interval ranging from x to $x + \Delta x$, where $0 < \Delta x \ll 1$. In Fig. 1.11 we have depicted y on this interval. We want to estimate the length s of the graph of y from x to $x + \Delta x$. If Δx is sufficiently small, we see from Fig. 1.11 that s can be approximated by the length of the straight line connecting the two points $(x, y(x))$ and $(x + \Delta x, y(x + \Delta x))$.

---

[12] You may specify the programming environment but nothing that requires a license for the customer.

[13] See http://www.ifi.uio.no/~scicomp/tacoma.mpg.

Fig. 1.9 The Storebælt bridge in Denmark
Source: Wikimedia Commons, which is a media repository making available public domain and freely-licensed educational media content to all; Photo by Tone V. V. Rosbach Jensen.

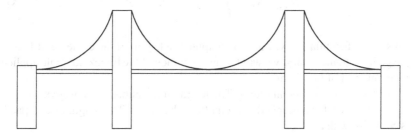

Fig. 1.10 A simplified sketch of a bridge similar to the Storebælt bridge in Denmark

Fig. 1.11 If Δx is small enough, the length s can be approximated by the length of a straight line

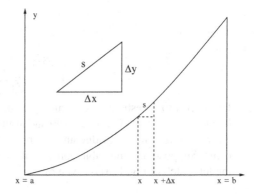

(a) Use the theorem of Pythagoras to show that

$$s^2 = (\Delta x)^2 + (y(x + \Delta x) - y(x))^2.$$

(b) Use Taylor's theorem – see (1.39) – to show that

$$s^2 \approx (\Delta x)^2(1 + (y'(x))^2).$$

(c) Divide the interval $[a, b]$ into n subintervals and let s_i denote the length of the ith interval. Give an argument for the approximation

$$l(y, a, b) \approx \sum_{i=1}^{n} \Delta x \sqrt{1 + (y'(x_i))^2},$$

where $l(y, a, b)$ denotes the length of the graph of y from $x = a$ to $x = b$, and where x_i denotes the midpoint of the ith interval.

(d) Let $\Delta x \to 0$ and argue by Riemann's integration theory that the limit is given by

$$l(y, a, b) = \int_a^b \sqrt{1 + (y'(x))^2}dx. \tag{1.51}$$

This is the formula we will use to compute the length of the cables. But before we go back to the cables, we will test the accuracy of the trapezoidal rule applied to an integral of this form.

(e) Consider a circle with unit radius. The length of the entire circle is given by 2π. In Fig. 1.12 we have graphed one-eighth of the circle. The length of this patch is $\pi/4$. Show that

$$\frac{\pi}{4} = \int_0^{\sqrt{2}/2} \sqrt{1 + (y'(x))^2} \, dx,$$

where $y(x) = \sqrt{1 - x^2}$.

(f) Show that

$$\frac{\pi}{4} = \int_0^{\sqrt{2}/2} \frac{1}{\sqrt{1 - x^2}}dx. \tag{1.52}$$

(g) Use (1.52) to investigate the accuracy of the composite trapezoidal rule given by (1.16). Use $n = 10, 20, \ldots, 100$ and argue that the error is $\approx 0.1667h^2$.

(h) Now we return to the bridge and the problem of computing the length of the cable. Suppose that the cable consists of four similar patches and suppose that the left patch of Fig. 1.10 is given by

Fig. 1.12 The figure illustrates a *circle* with unit radius. The marked segment has length $\pi/4$

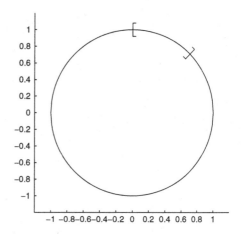

Fig. 1.13 One patch of the cable

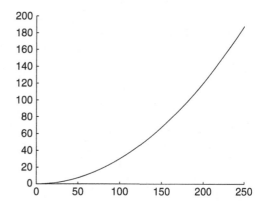

$$y(x) = 0.003x^2,$$

where x goes from $x = 0$ to $x = 250$ meters, see Fig. 1.13. Use formula (1.51) to show that the total length of the entire cable, counting both sides of the bridge, is given by

$$L = 4 \int_0^{250} \sqrt{1 + (0.006x)^2}\, dx.$$

(i) Use the composite trapezoidal scheme to estimate[14] L.

---

[14] Using $n = 100$, we get $L \approx 4 \cdot 324.91 = 1299.6$ m.

Chapter 2
Differential Equations: The First Steps

2.1 Modeling Changes

If you know the characteristics of something today, and you know the laws of change, then you can figure out what the characteristics will be tomorrow. This is the basic idea of modeling lots of natural processes. Since we know the weather today and we know the equations modeling the changes of the weather, we can predict it some days ahead. We know the heat of an object now and we know the equations describing how heat changes; thus we can predict how warm an object will be later on. This is really at the heart of science and has been so for quite a while. But how do we express change? How do we make these vague statements precise and suitable for computer simulations? We do so by expressing the laws of change in terms of differential equations.[1] The laws of planetary motions, the laws of fluid flow, the laws of electric and magnetic fields, the laws of mechanics, the laws of population growth, and so forth are all formulated in terms of differential equations. Major progress in science and technology will come about because the new generation of scientists will be better at solving differential equations than the previous ones. And the most important reason for this is the computer. In fact, the computer was invented for solving differential equations and weather forecasting was early on one of the prime applications.[2] Because of these facts, computer solutions of differential equations are the core subject of this text; it is the most important problem in scientific computing. We will teach you how to solve these equations on a computer by describing the basic principles as they apply to some examples.

---

[1] Of course, this is a simplification. There are other ways of expressing the laws of nature in a systematic manner. One approach to this has been presented by S. Wolfram [30] in an attempt to use programs as building blocks for models of nature.

[2] Weather forecasting is still a major challenge for computers. Moreover, it is not clear that it is possible to predict the weather with reasonable accuracy more than a few days – possibly a couple of weeks – ahead. You can check http://weather.yahoo.com to see whether their predictions 5 days ahead are accurate. Store their predictions and wait 5 days and then evaluate their correctness.

A. Tveito et al., *Elements of Scientific Computing*, Texts in Computational Science and Engineering 7, DOI 10.1007/978-3-642-11299-7_2,
© Springer-Verlag Berlin Heidelberg 2010

2.1.1 The Cultivation of Rabbits

We are entering the business of rabbit farming. Let us assume that we place a number
of rabbits on an isolated island with a perfect environment for them. How will the
number of rabbits grow? We cannot solve this problem just based on mathematical
reasoning. Some fundamental data have to be provided. But by using mathematical
models, we can figure out exactly what we need to know in order to make realistic
predictions, and we can figure out some interesting facts about the growth by doing
some very simple assumptions.

2.1.2 The Simplest Possible Case

Let $r = r(t)$ be the number of rabbits on the island at time t. We assume that at
time $t = 0$, the number is given by r_0, so

$$r(0) = r_0. \tag{2.1}$$

That is all we know initially. That is the basic state. Next we consider the change.
Let $\Delta t > 0$ be a small period of time. If the change of rabbits per time is given by
$f(t)$, we have

$$\frac{r(t + \Delta t) - r(t)}{\Delta t} \approx f(t). \tag{2.2}$$

Next we assume that the number of rabbits is large and that it can be modeled as a
continuous function of time. This is, of course, wrong since the number of rabbits
has to change discontinuously when one single rabbit is born. But when the number
is large, a continuous r can be a fair approximation. If we assume that the number
of rabbits is continuous and differentiable, we obtain the model

$$r'(t) = f(t) \tag{2.3}$$

by letting Δt go to zero.[3] If we assume that (2.1) and (2.3) hold, we get

$$r(t) = r(0) + \int_0^t f(s)ds \tag{2.4}$$

----

[3] Recall the definition of the derivative from calculus:

$$y'(x) = \lim_{\Delta x \to 0} \frac{y(x + \Delta x) - y(x)}{\Delta x}.$$

from the fundamental theorem of Calculus.[4] By the methods introduced in the previous chapter, we know that $r = r(t)$ can be computed as accurately as we want for any reasonable function f.

Equation (2.3) is a mathematical model of rabbit growth, but is it useful? No, not really. It simply states that if you know the number at $t = 0$ and each day you add the net number of new rabbits, you will have r each time. This is somewhat trivial. Let us turn to something slightly more advanced.

2.1.3 Exponential Growth

How does the number of rabbits really change? What characterizes the growth? It is reasonable to assume that safe sex is not a big issue among rabbits, and it is also fairly well known that they do *it* very often. We find it reasonable to assume that the rate of change is proportional to the number of rabbits. More specifically, we assume that

$$\frac{r(t + \Delta t) - r(t)}{\Delta t} = ar(t), \tag{2.5}$$

where a is a positive constant. Another way to put this is to state that the relative change is constant:

$$\frac{r(t + \Delta t) - r(t)}{r(t)\Delta t} = a. \tag{2.6}$$

In order to phrase this as a differential equation, we assume that r is differentiable and let Δt go to zero. This gives the exponential growth equation[5]

$$r'(t) = ar(t). \tag{2.7}$$

Let us assume that the growth rate a is a given positive constant. Of course, in a real-life application, this number has to be measured and we will return to this issue later. But let us just assume that it is given.

---

[4] The fundamental theorem of Calculus states that if $f = f(x)$ is continuous and $F'(t) = f(t)$, then $\int_a^b f(t)dt = F(b) - F(a)$.

[5] We can state this slightly more precisely: Let $b = b(t)$ be births and $d = d(t)$ be deaths. Then

$$r'(t) = b(t) - d(t).$$

Now we assume that the births are proportional to the population, i.e., $b(t) = \beta r(t)$ and similarly $d(t) = \delta r(t)$. Thus,

$$r'(t) = (\beta - \delta)r(t),$$

which is (2.7) with $a = \beta - \delta$.

Analytical Solution

Model (2.7) is a differential equation. If we write this equation and the associated initial equation together, we have the following initial value problem:

$$r'(t) = ar(t), \qquad (2.8)$$
$$r(0) = r_0.$$

The amazing thing is that these two simple equations determine r uniquely for all time. So if we are able to estimate the initial number of rabbits r_0 and the growth rate a, we get an estimate for the number of rabbits in the future. Since

$$\frac{dr}{dt} = ar,$$

we have[6]

$$\frac{1}{r} dr = a \, dt,$$

and then integration

$$\int \frac{1}{r} dr = \int a \, dt$$

gives

$$\ln(r) = at + c \qquad (2.9)$$

where c is a constant of integration. By setting $t = 0$, we have

$$c = \ln(r_0),$$

where we have used the initial condition $r(0) = r_0$. It now follows from (2.9) that

$$\ln(r(t)) - \ln(r_0) = at,$$

---

[6] It is perfectly ok to multiply $\frac{dr}{dt} = ar$ by dt to obtain $\frac{1}{r} dr = a \, dt$ and then integrate. But if you are unfamiliar with that way of manipulating differentials, you may also observe directly that since $\frac{1}{r} r'(t) = a$, we can integrate in time to get

$$\int \frac{1}{r} r'(t) dt = \int a \, dt.$$

If we now substitute $r = r(t)$, we have $dr = r'(t)dt$ and thus the left-hand side is

$$\int \frac{1}{r} dr,$$

and we arrive at

$$\ln(r) = at + \text{const.}$$

so

$$\ln(\frac{r(t)}{r_0}) = at$$

and thus

$$r(t) = r_0 e^{at}. \tag{2.10}$$

From this model it follows that the number of rabbits increases exponentially in time. Note that this is a highly nontrivial observation; we only assume that the rate of change is proportional to the number of rabbits and then it follows that the growth has to be exponential. And, in fact, this is often a fairly accurate description.

Uniqueness

We mentioned above that the function r is uniquely determined by the two equations

$$r'(t) = ar(t), \tag{2.11}$$
$$r(0) = r_0.$$

Then we constructed the solution

$$r(t) = r_0 e^{at},$$

but we did not ask whether other solutions were possible. This is a general issue; you cannot claim that a solution is unique simply by constructing one. So let us assume that there are two solutions, r and q, where r satisfies (2.11) and q satisfies the same two conditions, i.e.,

$$q'(t) = aq(t), \tag{2.12}$$
$$q(0) = q_0.$$

Define the difference

$$E(t) = r(t) - q(t)$$

and assume that $q_0 = r_0$. Then

$$E_0 = E(0) = r(0) - q(0) = 0.$$

Moreover,

$$E'(t) = r'(t) - q'(t)$$
$$= a(r(t) - q(t))$$
$$= aE(t).$$

Hence E satisfies the following two conditions:

$$E'(t) = aE(t),$$ (2.13)
$$E(0) = 0.$$

If we multiply both sides of (2.13) by e^{-at}, we observe that

$$e^{-at} E'(t) - a e^{-at} E(t) = 0,$$

and thus

$$(e^{-at} E(t))' = 0,$$

and so we have

$$e^{-at} E(t) = E(0)$$

for all time. Since $E(0) = 0$, we have

$$E(t) = 0$$

for all time, and therefore

$$q(t) = r(t)$$

for all time, and hence problem (2.11) has only one solution.

Stability

Counting rabbits sounds simple, but it is not![7] That means that we cannot be absolutely sure about the initial state. What happens if our initial estimate is wrong? Suppose the number of rabbits are modeled by

$$r'(t) = ar(t),$$ (2.14)
$$r(0) = r_0.$$

Let us also consider the following problem:

$$q'(t) = aq(t),$$ (2.15)
$$q(0) = q_0.$$

---

[7] Just think about it: They all look almost the same and they run around like crazy. If we are absolutely positive that the island do not have any rabbits before $t = 0$, we could, of course, count how many we introduce, but that is an unlikely situation. Probably, we would only be able to get a rough estimate of the number of rabbits at $t = 0$. And this is the typical case for many parameters introduced in differential equations: They are based on some sort of measurements and therefore their values are uncertain. The issue of stability is about how this uncertainty affects the solution. If small perturbations produce gigantic changes, we have an unstable problem. But if small perturbations in the data produce small changes in the solution, the solution is stable with respect to perturbations of these data.

If we let
$$d(t) = r(t) - q(t),$$

we find that d solves the problem

$$d'(t) = ad(t), \tag{2.16}$$
$$d(0) = d_0 = r_0 - q_0.$$

It follows that
$$d(t) = d_0 e^{at},$$

so
$$r(t) - q(t) = (r_0 - q_0)e^{at},$$

or
$$|r(t) - q(t)| = |r_0 - q_0| e^{at}. \tag{2.17}$$

This equation states that the difference is magnified by a factor of e^{at}. To understand what this means, we divide the left-hand side of (2.17) by $r(t)$ and the right-hand side by $r_0 e^{at}$. This is fine since

$$r(t) = r_0 e^{at}.$$

We find that
$$\frac{|r(t) - q(t)|}{r(t)} = \frac{|r_0 - q_0|}{r_0}, \tag{2.18}$$

which means that the relative difference between the two solutions remains constant. We also note from (2.18) that if we let q_0 tend toward r_0, then $q(t)$ will also tend toward $r(t)$. Since the relative difference between these two solutions does not blow up, we refer to this problem as *stable* with respect to changes in the initial data.

More generally, we refer to a problem as stable with respect to perturbations in one parameter if such changes do not lead to very different solutions. With the rabbit counting in mind, we hope you appreciate the need for some sort of stability. Usually, the parameters that we use in such equations are based on some sort of measurements, and measurements often involve errors.

2.1.4 Logistic Growth

As mentioned above, model (2.8) is an interesting model for the growth of rabbits on an island, but it is not perfect. Mathematical models are not perfect; we can never incorporate every tiny effect in a model, and therefore it will always remain a *model*. However, there is one particular feature that is unrealistic about model (2.8), and that is that the number of rabbits goes to infinity as time increases. This is unrealistic, because each rabbit needs sufficient food and some space to stay alive. Therefore, we need to refine the model with an additional term. Let R denote the *carrying capacity* of the island. This number is the maximum number of rabbits that the island can supply with food, space, and so forth. We introduce the so called *logistic model*

$$r'(t) = ar(t)\left(1 - \frac{r(t)}{R}\right),\tag{2.19}$$

where $a > 0$ is the growth rate and R is the carrying capacity. We assume that R is a very big number and, in particular, we assume that $r_0 << R$. So for t close to zero, we have

$$\frac{r(t)}{R} \approx 0,$$

and thus

$$r'(t) \approx ar(t),$$

which means that the logistic model and the exponential model give similar predictions for small values of t. But as t increases, the two models provide very different predictions. By looking at (2.19), you can notice that the right hand side is positive at time $t = 0$. In fact, since $r_0 < R$, we have

$$r'(0) = ar_0\left(1 - \frac{r_0}{R}\right) > 0.$$

This means that r is increasing. And we observe that r will continue to increase as long as

$$\left(1 - \frac{r(t)}{R}\right) > 0,$$

that is, as long as

$$r(t) < R.$$

Thus, as time increases, r will increase, but as it approaches R, the growth rate will become smaller and smaller, and if we reach $r(t) = R$ at, say, $t = t^*$, we will have

$$r'(t^*) = 0.$$

This means that the logistic model predicts that the number of rabbits will steadily increase but never exceed the carrying capacity. Note, however, that we have not proved that we will actually reach the carrying capacity; more on this later.

Exceeding the Carrying Capacity

Suppose that we place lots of rabbits on the island initially, such that

$$r_0 > R,$$

that is, the initial number of rabbits exceeds the carrying capacity of the island. Then it follows from (2.19) that

$$r'(0) = ar_0 \left(1 - \frac{r_0}{R}\right) < 0,$$

and consequently the number of rabbits will decrease. In fact, by (2.19), it will continue to decrease until we reach $r = R$.

To summarize, we have observed that if the initial number of rabbits is smaller than the carrying capacity, then the number of rabbits will be monotonically increasing but will never become larger than the carrying capacity. Similarly, if $r_0 > R$, then $r(t)$ will be monotonically decreasing but will always satisfy $r(t) \geq R$. The carrying capacity is referred to as an *equilibrium point* because if $r(t^*) = R$, then $r'(t^*) = 0$, and the solution will remain equal to R for all $t \geq t^*$.

Analytical Solution

To gain further insight about the rabbit population as predicted by (2.19), we can solve the initial value problem

$$r'(t) = ar(t) \left(1 - \frac{r(t)}{R}\right), \tag{2.20}$$

$$r(0) = r_0,$$

analytically. Since

$$\frac{dr}{dt} = ar\left(1 - \frac{r}{R}\right),$$

we have

$$\frac{dr}{r\left(1 - \frac{r}{R}\right)} = a\,dt,$$

and by integration we get

$$\ln\left(\frac{r}{R - r}\right) = at + c,$$

where c is a constant that we have to determine using the initial condition. At $t = 0$, we have

$$\ln\left(\frac{r_0}{R - r_0}\right) = c$$

and thus

$$\ln\left(\frac{\frac{r}{R-r}}{\frac{r_0}{R-r_0}}\right) = at,$$

or

$$\frac{r}{R - r} = \frac{r_0}{R - r_0}e^{at}.$$

By solving this equation with respect to r, we get

$$r(t) = \frac{r_0}{r_0 + e^{-at}(R - r_0)}R. \tag{2.21}$$

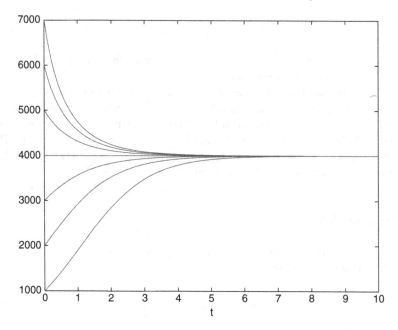

Fig. 2.1 Different solutions of (2.21) using different values of r_0

In Fig. 2.1, we show seven different solutions of (2.21) using $a = 1$, $R = 4,000$, and seven different initial values $r_0 = k \cdot 1,000$, for $k = 1, \ldots, 7$.

From Fig. 2.1 we note that whatever positive initial condition r_0 we give, the solution ends up as $r = R$ as time goes to infinity. This is also easy to see from the analytical solution (2.21): Let t go to infinity; then the term e^{-at} goes to zero,[8] and consequently $r(t)$ approaches R.

2.2   Numerical Solution

We have seen that it is possible to derive models of population growth using differential equations. The models we derived are so simple that analytical arguments are indeed sufficient to tell us all we ever want to know about the solutions. Unfortunately, this is very rare in real-life applications. Analytical tools can solve very few realistic models of nature. When analytical methods are inadequate, we have to rely on numerical computations performed on computers. Now, do not let this lead you to believe that finding analytical solutions of differential equations is a useless

---

[8] Recall that $a > 0$.

topic. That is not so at all. In fact, very often successful computer solutions of differential equations rely heavily on analytical insight. In many cases we know a lot about a solution without actually having computed it – see our arguments above about increasing and decreasing the number of rabbits. We were able to figure out quite a bit without solving anything. All such knowledge can be utilized either in computing the solution or checking that the solution that we have computed is in fact a reasonable approximation of the exact solution. Computer codes for nontrivial problems almost always contain bugs, and methods for checking the validity of a computed solution are therefore very important.

2.2.1 The Simplest Possible Model

In the simplest model above, the change in the rabbit population was given by an explicit function that was independent of the population. More precisely, the model reads

$$r'(t) = f(t), \tag{2.22}$$

with $r(0) = r_0$ and where $f = f(t)$ is assumed to be a given function. Although this model can be integrated directly, we will start by viewing it as a differential equation. Suppose we want to find an approximation of $r(t)$ for t ranging from[9] $t = 0$ to $t = 1$. We discretize this problem by picking an integer $N \geq 1$ and define the time step

$$\Delta t = 1/N$$

and time levels

$$t_n = n\Delta t$$

for $n = 0, 1, \ldots, N$. Note, in particular, that $t_0 = 0$ and $t_N = 1$. Furthermore, we let r_n denote an approximation[10] of $r(t_n)$. In order to derive an approximation of the solution of (2.22), we will need the Taylor series, see Project 1.7.1 from Chap. 1. For a sufficiently smooth function $r = r(t)$, we have

$$r(t + \Delta t) = r(t) + \Delta t r'(t) + O(\Delta t^2).$$

Consequently,

---

[9] Suppose time t is measured in years. Then it makes sense to consider how the number of rabbits changes from $t = 0$ to $t = 1$.

[10] It is important that you get this right: $r(t)$ is the correct solution, $r(t_n)$ is the correct solution at time $t = t_n$, and r_n is the approximate solution at time t_n. More specifically, we have

$$r_n \approx r(t_n).$$

Note that r_0 denotes both the correct and approximate solution at $t = 0$. This is ok since the solution is given at $t = 0$.

$$r'(t) = \frac{r(t + \Delta t) - r(t)}{\Delta t} + O(\Delta t).$$

If we apply this observation in the case of $t = t_n$, we have

$$r'(t_n) = \frac{r(t_{n+1}) - r(t_n)}{\Delta t} + O(\Delta t).$$

Thus, it follows from (2.22) that it is reasonable[11] to require that

$$\frac{r_{n+1} - r_n}{\Delta t} = f(t_n),$$

and then we have

$$r_{n+1} = r_n + \Delta t f(t_n).$$

Since r_0 is known by the initial condition, we can compute

$$r_1 = r_0 + \Delta t f(t_0).$$

And since, by now, r_1 is known, we can compute

$$r_2 = r_1 + \Delta t f(t_1) = r_0 + \Delta t \left(f(t_0) + f(t_1) \right),$$

and so on. We get

$$r_N = r_0 + \Delta t \sum_{n=0}^{N-1} f(t_n), \tag{2.23}$$

which is the Riemann sum[12] approximation of the integral.

Example 2.1. Let us, just for the purpose of illustration, consider the case of $f(t) = t^2$ and $r(0) = 0$. Then

---

[11] Why is this reasonable? The idea is that $r'(t)$ can be approximated by a finite difference, i.e.,

$$r'(t_n) \approx \frac{r(t_{n+1}) - r(t_n)}{\Delta t}.$$

Now since

$$r'(t_n) = f(t_n),$$

we can *define* the numbers $\{r_n\}$ by requiring that

$$\frac{r_{n+1} - r_n}{\Delta t} = f(t_n).$$

There is nothing mysterious about this; it is simply a reasonable way of putting up a condition that is sufficient to compute the numbers $\{r_n\}$.

[12] Consult your calculus book.

$$r(1) = r(0) + \int_0^1 t^2 dt = 1/3.$$

By using (2.23) with $N = 10$, we get[13]

$$
\begin{aligned}
r(1) &\approx r_{10} \\
&= 0 + \Delta t \left(\Delta t^2 + (2\Delta t)^2 + \cdots + (9\Delta t)^2 \right) \\
&= \Delta t^3 (1 + 2^2 + \cdots + 9^2) \\
&= \frac{1}{6} \frac{9 \cdot 10 \cdot 19}{10^3} \\
&= 0.285,
\end{aligned}
$$

which is a reasonable approximation to the exact solution. Let us also try $N = 100$. Then

$$
\begin{aligned}
r(1) &\approx r_{100} \\
&= 0 + \Delta t \left(\Delta t^2 + (2\Delta t)^2 + \cdots + (99\Delta t)^2 \right) \\
&= \Delta t^3 (1 + 2^2 + \cdots + 99^2) \\
&= \frac{1}{6} \frac{99 \cdot 100 \cdot 199}{100^3} \\
&= 0.3285,
\end{aligned}
$$

which is an even better approximation.

■

Numerical Integration

Since the simple model (2.22) can be integrated directly, we can also use the trapezoidal method derived in Chap. 1. By integration of (2.22), we get

$$r(t) = r_0 + \int_0^t f(s)ds. \tag{2.24}$$

In order to apply the trapezoidal method, we introduce

$$\Delta t = t/N$$

where $N \geq 1$ is an integer. As above, we define

---

[13] Recall that

$$\sum_{x=1}^{m} x^2 = \frac{1}{6}m(m+1)(2m+1).$$

$$t_n = n \Delta t$$

for $n = 0, 1, \ldots N$. The trapezoidal approximation then gives

$$r(t) \approx r_0 + \Delta t \left(\frac{1}{2} f(0) + \sum_{n=1}^{N-1} f(t_n) + \frac{1}{2} f(t) \right). \qquad (2.25)$$

Example 2.2. Let us again consider the case of

$$f(t) = t^2,$$
$$r(0) = 0,$$

where we want an approximation of $r(1)$. We choose $N = 10$ and get

$$r(1) \approx \Delta t \left(\frac{1}{2} f(0) + \sum_{n=1}^{9} f(t_n) + \frac{1}{2} f(t) \right)$$

$$= \Delta t^3 \left(\sum_{n=1}^{9} n^2 + \frac{1}{2} 10^2 \right)$$

$$= \frac{1}{10^3} \left(\frac{9 \cdot 10 \cdot 19}{6} + 50 \right)$$

$$= 0.335,$$

which is a very good approximation to the correct value, $1/3$. If we choose $N = 100$, we get

$$r(1) \approx \Delta t \left(\frac{1}{2} f(0) + \sum_{n=1}^{99} f(t_n) + \frac{1}{2} f(t) \right)$$

$$= \Delta t^3 \left(\sum_{n=1}^{99} n^2 + \frac{1}{2} 100^2 \right)$$

$$= \frac{1}{100^3} \left(\frac{99 \cdot 100 \cdot 199}{6} + 5000 \right)$$

$$= 0.33335.$$

∎

2.2.2 Numerical Approximation of Exponential Growth

Our next goal is to derive a numerical method for the initial value problem

$$r'(t) = ar(t), \qquad t \in (0, T), \qquad (2.26)$$
$$r(0) = r_0,$$

where a is a given constant. We want to compute an approximate solution of r in the time interval ranging from $t = 0$ to $t = T$. Recall that r_n denotes an approximation of $r(t_n)$, where

$$t_n = n \Delta t$$

and

$$\Delta t = T/N$$

denotes the time step. Since

$$r'(t_n) \approx \frac{r(t_{n+1}) - r(t_n)}{\Delta t},$$

we define the scheme

$$\frac{r_{n+1} - r_n}{\Delta t} = a r_n \tag{2.27}$$

for $n \geqslant 0$, where we recall that r_0 is given. We get

$$r_{n+1} = (1 + a \Delta t) r_n, \tag{2.28}$$

so

$$r_1 = (1 + a \Delta t) r_0,$$
$$r_2 = (1 + a \Delta t) r_1 = (1 + a \Delta t)^2 r_0,$$

and so on. In general, we have

$$r_n = (1 + a \Delta t)^n r_0. \tag{2.29}$$

Example 2.3. Let us assume that $a = 1$, $r_0 = 1$, $T = 1$, and $N = 10$. Then

$$r(1) \approx r_{10} = (1 + \frac{1}{10})^{10} \approx 2.594.$$

The exact solution of this problem is

$$r(t) = e^t,$$

and thus

$$r(1) = e \approx 2.718.$$

By choosing $N = 100$, we get

$$r_{100} = (1 + \frac{1}{100})^{100} \approx 2.705.$$

■

Convergence

We want, of course, the numerical scheme to be convergent in the sense that the numerical solution converges toward the analytical solution as Δt approaches zero, or, equivalently, as N goes to infinity. For the simple model considered here, we can prove convergence in a direct manner. With $a = 1$ and $T = 1$, we have

$$r(1) \approx r_N = (1 + \frac{1}{N})^N,$$

and since[14]

$$\lim_{N \to \infty} (1 + \frac{1}{N})^N = e = r(1),$$

it follows that r_N converges to $r(1)$ as N goes to infinity.

2.2.3 Numerical Stability

Let us now consider the initial value problem

$$y'(t) = -100y(t), \tag{2.30}$$
$$y(0) = 1,$$

which has the analytical solution

$$y(t) = e^{-100t}.$$

Let us try to solve this problem numerically from $t = 0$ to $t = 1$ by the method introduced above. We let y_n denote an approximation to $y(t_n)$ where, as usual, $t_n = n\Delta t$ and $\Delta t = 1/N$. The numerical approximation is defined by the finite difference scheme

$$\frac{y_{n+1} - y_n}{\Delta t} = -100y_n,$$

so

$$y_{n+1} = (1 - 100\Delta t)y_n. \tag{2.31}$$

By reasoning as above, we get

$$y_n = \left(1 - \frac{100}{N}\right)^n.$$

---

[14] See your calculus textbook.

Note that the analytical solution e^{-100t} is equal to one initially and decreases rapidly and monotonically toward zero. Let us set $N = 10$ in our numerical scheme. Then we have

$$y_0 = 1,$$

$$y_1 = \left(1 - \frac{100}{10}\right) = -9,$$

$$y_2 = \left(1 - \frac{100}{10}\right)^2 = 18,$$

$$y_3 = \left(1 - \frac{100}{10}\right)^3 = -729,$$

and so on. We note that the values oscillates between positive and negative values and that their absolute values increase very quickly. Now, these values have nothing whatsoever to do with the correct solution. The approximation fails completely. This phenomenon is a very unfortunate and common problem in the numerical solution of differential equations. It is referred to as *numerical instability* and leads to erroneous solutions. This example is very simple and therefore we are able to fully understand what is going on and how to deal with it. But for complicated models, stability problems can be very difficult.

Let us reconsider the scheme (2.31) and require that the solution be positive. For a given n, we assume that $y_n > 0$ and we want to derive a condition on Δt that also ensures that y_{n+1} is positive. It follows from scheme (2.31) that we must have

$$1 - 100\Delta t > 0$$

or

$$\Delta t < \frac{1}{100}, \tag{2.32}$$

which means that the number of time steps N must satisfy

$$N \geqslant 101.$$

If we now choose $N = 101$, we get

$$y_n = \left(1 - \frac{100}{101}\right)^n = \frac{1}{101^n},$$

where we note that all the values are strictly positive, and in particular we get

$$y(1) \approx y_{101} = \frac{1}{101^{101}} \approx 0.$$

We refer to conditions of the type in (2.32) as *stability conditions*. Numerical schemes that behave well for any positive value Δt are commonly called *unconditionally stable*, whereas schemes of the type encountered here are referred to as *conditionally stable*.

2.2.4 An Implicit Scheme

The scheme we derived above was based on the observation that

$$r'(t_n) = \frac{r(t_{n+1}) - r(t_n)}{\Delta t} + O(\Delta t).$$

Since

$$r'(t_n) = ar(t_n),$$

we get

$$\frac{r_{n+1} - r_n}{\Delta t} = ar(t_n).$$

But using the Taylor series,[15] we also see that

$$r'(t_{n+1}) = \frac{r(t_{n+1}) - r(t_n)}{\Delta t} + O(\Delta t),$$

and thus we get the scheme

$$\frac{r_{n+1} - r_n}{\Delta t} = ar_{n+1},$$

or

$$r_{n+1} = \frac{1}{1 - \Delta t a} r_n,$$

which leads to

$$r_n = \left(\frac{1}{1 - \Delta t a} \right)^n.$$

---

[15] The Taylor series states that

$$r(t + \varepsilon) = r(t) + \varepsilon r'(t) + O(\varepsilon^2).$$

Setting $\varepsilon = -\Delta t$ and $t = t_{n+1}$, we have

$$r(t_n) = r(t_{n+1}) - \Delta t \, r'(t_{n+1}) + O(\Delta t^2)$$

and therefore

$$r'(t_{n+1}) = \frac{r(t_{n+1}) - r(t_n)}{\Delta t} + O(\Delta t).$$

Let us consider the initial value problem

$$y'(t) = -100y(t),$$
$$y(0) = 1,$$

which gave us some difficulties above. Since $a = -100$, we have

$$y_n = \left(\frac{1}{1 + 100\Delta t}\right)^n$$
$$= \left(\frac{N}{N + 100}\right)^n,$$

which is easily seen to yield uniformly positive solutions; the oscillatory behavior observed above is not present here. Note, in particular, that at time $t_N = T = 1$, we have

$$y_N = \left(\frac{N}{N + 100}\right)^N.$$

From the table below, we observe that this formula provides a fair approximation of the exact solution.

N	y_N
10^1	$3.85 \cdot 10^{-11}$
10^2	$7.89 \cdot 10^{-31}$
10^3	$4.05 \cdot 10^{-42}$
10^7	$3.72 \cdot 10^{-44}$

Note that the exact solution is $e^{-100} \approx 3.72 \cdot 10^{-44}$.

2.2.5 Explicit and Implicit Schemes

The scheme obtained here is *implicit* and it is also *unconditionally stable*. So what is the difference between an explicit and an implicit scheme? Suppose we have an equation of the form

$$v'(t) = something, \tag{2.33}$$

and suppose also that we derive the numerical method by replacing the term $v'(t)$ with a term of the form

$$\frac{v_{n+1} - v_n}{\Delta t}.$$

Now, if we replace the right hand side of (2.33) by *something* evaluated at time $t = t_n$, then the scheme is called explicit and the reason is simply that we get an explicit formula for v_{n+1}; in fact,

$$v_{n+1} = v_n + \Delta t \, something(t_n).$$

But, on the other hand, if we evaluate *something* at time $t = t_{n+1}$, we get the implicit definition of v_{n+1}:

$$v_{n+1} = v_n + \Delta t \; something(t_{n+1}),$$

which cannot be evaluated directly, because *something* may depend on v_{n+1}. So this latter type is referred to as an implicit scheme. Generally speaking, implicit schemes are often unconditionally stable but suffer from the fact that an equation has to be solved, whereas explicit schemes are usually only conditionally stable but are very simple to use.

Example 2.4. Let us consider the initial value problem

$$y'(t) = y^2(t), \qquad\qquad (2.34)$$
$$y(0) = 1,$$

with the analytical solution[16]

$$y(t) = \frac{1}{1-t}.$$

An explicit scheme for this problem reads

$$\frac{y_{n+1} - y_n}{\Delta t} = y_n^2,$$

so we get the explicit definition

$$y_{n+1} = y_n + \Delta t y_n^2, \qquad\qquad (2.35)$$

and thus we can easily compute the values of y_n for $n = 1, 2, \ldots$.
    Similarly, an implicit scheme reads

$$\frac{z_{n+1} - z_n}{\Delta t} = z_{n+1}^2,$$

which can be rewritten as

$$z_{n+1} - \Delta t \, z_{n+1}^2 = z_n.$$

Hence, if z_n is already computed, we can find z_{n+1} by solving the nonlinear equation

$$x - \Delta t x^2 = z_n \qquad\qquad (2.36)$$

and then setting $z_{n+1} = x$. Since this is a simple second order polynomial equation, we can solve it analytically and use one of the two solutions,

---

[16] See Exercise 2.8.

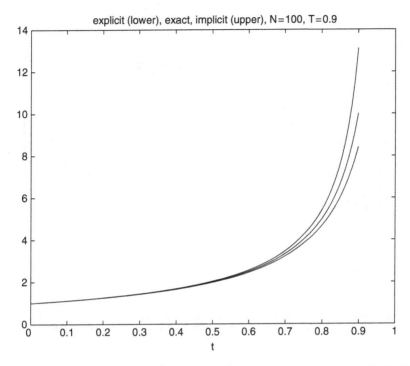

Fig. 2.2 The figure shows numerical approximations computed by an explicit and an implicit scheme. The *upper curve* is the solution generated by the implicit scheme and the *lower curve* is computed by the explicit scheme; the analytical solution is in between

$$z_{n+1} = \frac{1}{2\Delta t} \left(1 - \sqrt{1 - 4\Delta t \, z_n}\right). \tag{2.37}$$

In Fig. 2.2 we have plotted the two numerical solutions and the exact solution for t ranging from 0 to 0.9. In the numerical computations we used $N = 100$. We note that both numerical solutions behave well.

■

Example 2.5. We consider the problem above but with a negative initial condition:

$$y'(t) = y^2(t), \tag{2.38}$$
$$y(0) = -10.$$

The solution of this problem is given by[17]

$$y(t) = \frac{-10}{1 + 10t}.$$

---

[17] See Exercise 2.8.

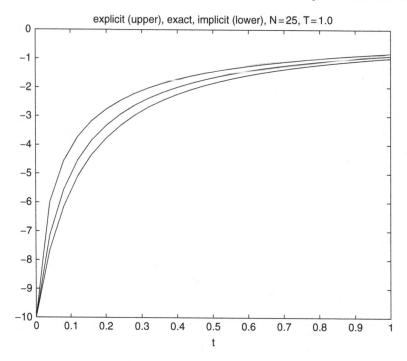

Fig. 2.3 The figure shows the numerical solution generated by an explicit scheme (*upper curve*), the analytical solution (*middle*), and the solution generated by an implicit scheme (*lower*). All the solutions are plotted by drawing a *straight line* between the computed values. This is also done for the analytical solution, which is therefore correct only at $t_0, t_1 \cdots t_N$

In Fig. 2.3 we have plotted the two numerical solutions and the exact solution for t ranging from 0 to 1. In the numerical computations we used $N = 25$. We note that both numerical solutions behave well.

But let us look a bit closer at what happens when we reduce the number of time steps, i.e., we increase the size of Δt. In Table 2.1, we compare the numerical solution generated by the implicit and explicit scheme at time $t = 1$.

We see from Table 2.1 that the implicit scheme gives reasonable solutions for any Δt, whereas the explicit scheme runs into serious trouble as N becomes smaller than 11.

The Explicit Scheme

We can see this effect directly from the explicit scheme. Suppose $y_n < 0$, and recall that

$$y_{n+1} = y_n + \Delta t y_n^2.$$

Hence, in order for y_{n+1} to be negative, we must have

$$y_n + \Delta t y_n^2 < 0,$$

Table 2.1 The table shows the analytical, explicit and implicit solutions at time $t = 1$ for problem (2.38). Note that the implicit scheme provides reasonable approximations for any values of Δt, whereas the explicit scheme requires Δt to be small in order to give solutions less than zero

N	Δt	$y(1)$	Explicit at $t = 1$	Implicit at $t = 1$
1,000	$\frac{1}{1,000}$	$-\frac{10}{11} \approx -0.9091$	-0.9071	-0.9111
100	$\frac{1}{100}$	$-\frac{10}{11} \approx -0.9091$	-0.8891	-0.9288
25	$\frac{1}{25}$	$-\frac{10}{11} \approx -0.9091$	-0.9871	-0.8256
12	$\frac{1}{12}$	$-\frac{10}{11} \approx -0.9091$	-0.6239	-1.0703
11	$\frac{1}{11}$	$-\frac{10}{11} \approx -0.9091$	-0.4835	-1.0848
10	$\frac{1}{10}$	$-\frac{10}{11} \approx -0.9091$	0.0	-1.1022
9	$\frac{1}{9}$	$-\frac{10}{11} \approx -0.9091$	5.7500	-1.1235
8	$\frac{1}{8}$	$-\frac{10}{11} \approx -0.9091$	$6.4 * 10^3$	-1.1501
7	$\frac{1}{7}$	$-\frac{10}{11} \approx -0.9091$	$1.8014 * 10^7$	-1.1843
5	$\frac{1}{5}$	$-\frac{10}{11} \approx -0.9091$	$1.6317 * 10^7$	-1.2936
2	$\frac{1}{2}$	$-\frac{10}{11} \approx -0.9091$	840	-1.8575

and consequently we must require that

$$1 + \Delta t y_n > 0.$$

If we set $n = 0$, we have $y_0 = -10$, and hence we require

$$1 - 10\Delta t > 0$$

which implies that

$$\Delta t < \frac{1}{10},$$

or

$$N > 10.$$

The computations presented in the table shows that breaking this criterion for the explicit scheme leads to completely erroneous numerical solutions.

The Implicit Scheme

For completeness, we also look a bit closer at the implicit scheme. We observed above that the implicit scheme can be formulated as follows: Suppose z_n is given. Then solve the equation

$$x - \Delta t x^2 = z_n \qquad (2.39)$$

and set $z_{n+1} = x$. So in order to prove that z_n remains negative, we have to show that (2.39) has a unique negative solution. To this end we define the auxiliary function

$$f(x) = x - \Delta t x^2 - z_n,$$

where we assume that

$$z_n < 0.$$

First we note that

$$f'(x) = 1 - 2\Delta t x$$

and thus $f'(x) > 0$ for all $x < 0$. Also, $f(0) = -z_n > 0$, and we note that f tends to minus infinity when x tends to minus infinity for any positive value of Δt. Hence, as x goes from minus infinity to zero, f is monotonically increasing from minus infinity to $f(0) > 0$, and then it follows that there must be a unique negative value x^* such that

$$f(x^*) = 0.$$

This is also illustrated in Fig. 2.4, where we see that the function $x - \Delta t x^2$ intersects the straight line z_n for a uniquely determined negative value of x. It now follows by induction on n that all values generated by the implicit scheme are negative.

∎

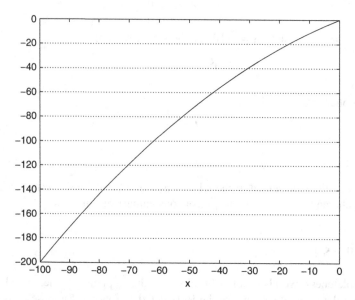

Fig. 2.4 The figure depicts the function $f(x) = x - \Delta t x^2$ with $\Delta t = 1/100$ for the negative values $-100 \le x \le 0$. It illustrates that for any given *horizontal dotted line* $f = z_n$, where $z_n < 0$, the $f = f(x)$ curve has a single intersection point with the line for $x < 0$

2.2.6 Numerical Solution of the Logistic Equation

The Explicit Scheme

Certainly, the methods introduced to solve the equations above can also be used to solve the logistic equation. Recall that this initial value problem is given by

$$r'(t) = ar(t)\left(1 - \frac{r(t)}{R}\right),$$ (2.40)

$$r(0) = r_0,$$ (2.41)

where $a > 0$ is the growth rate and R is the carrying capacity. An explicit scheme for this model is given by

$$\frac{r_{n+1} - r_n}{\Delta t} = ar_n(1 - \frac{r_n}{R}),$$

or

$$r_{n+1} = r_n + ar_n\Delta t(1 - \frac{r_n}{R}).$$ (2.42)

Properties

In our discussions of the differential equation (2.40) above, we made the following observations just based on the form of the equation:

- If $R \gg r_0$, then for small t, we have $r'(t) \approx ar(t)$, and thus approximately exponential growth.
- If $0 < r_0 < R$, then the solution satisfies $r_0 \leqslant r(t) \leqslant R$ and $r'(t) \geqslant 0$ for all time.
- If $r_0 > R$, then the solution satisfies $R \leqslant r(t) \leqslant r_0$ and $r'(t) \leqslant 0$ for all time.

We can make exactly the same observations for the numerical solution generated by the scheme (2.42). In order to do so, we have to assume that the time step is not too big. More precisely, we assume that

$$\Delta t < 1/a.$$ (2.43)

Numerical solutions generated by the explicit scheme (2.42) have the following properties:

- If $R \gg r_0$, we have $r_{n+1} \approx r_n + a\Delta t r_n$, for small values of n. This is the explicit scheme for the exponential growth model; see (2.28).
- Assume that $0 < r_0 < R$. If, for some value of n, we have that $r_n \leqslant R$, then

$$(1 - \frac{r_n}{R}) \geqslant 0,$$

and consequently

$$a r_n \Delta t \left(1 - \frac{r_n}{R}\right) \geq 0,$$

and thus

$$r_{n+1} = r_n + a r_n \Delta t \left(1 - \frac{r_n}{R}\right) \geq r_n, \qquad (2.44)$$

which resembles the monotonicity property of the analytical solution.
In order to show that

$$r_0 \leq r_n \leq R$$

for all values of n, we will study the function

$$g(x) = x + a x \Delta t \left(1 - \frac{x}{R}\right)$$

for x in the interval from 0 to R. Observe that

$$g'(x) = 1 + a \Delta t - \frac{2 a \Delta t}{R} x,$$

so, for x in $[0, R]$, we have

$$g'(x) \geq 1 + a \Delta t - \frac{2 a \Delta t}{R} R$$
$$= 1 - a \Delta t$$
$$> 0,$$

where we used the assumption that $\Delta t < 1/a$ (see (2.43)). Note that

$$r_{n+1} = g(r_n),$$

and assume that $0 \leq r_n \leq R$. Then, since $g'(x) > 0$ for $0 \leq x \leq R$, we have

$$r_{n+1} = g(r_n) \leq g(R) = R,$$

and

$$r_{n+1} = g(r_n) \geq g(0) = 0.$$

Hence, if $0 \leq r_n \leq R$, then also $0 \leq r_{n+1} \leq R$, and then it follows by induction on n that $0 \leq r_n \leq R$ holds for all $n \geq 0$, provided that $0 < r_0 < R$. Moreover, since we have already seen that $r_{n+1} \geq r_n$, it follows that

$$r_0 \leq r_n \leq R, \qquad (2.45)$$

for all $n \geq 0$.

- In the same way, we can show that if $r_0 > R$, then $r_{n+1} \leqslant r_n$ and

$$r_0 \geqslant r_n \geqslant R. \tag{2.46}$$

- Finally, we also note that if $r_0 = R$, then

$$r_1 = R + a R\Delta t \left(1 - \frac{R}{R}\right) = R,$$

and we can easily prove by induction that

$$r_n = R \tag{2.47}$$

for all $n \geqslant 0$.

The Implicit Scheme

By proceeding as explained above, we can derive an implicit scheme[18] for the logistic initial value problem. As usual, we replace the left hand side of

$$r'(t) = ar(t) \left(1 - \frac{r(t)}{R}\right)$$

with a finite difference approximation, and we evaluate the right-hand side at time t_{n+1}. This gives the scheme

$$\frac{r_{n+1} - r_n}{\Delta t} = ar_{n+1}(1 - \frac{r_{n+1}}{R}),$$

which can be rewritten in the form

$$r_{n+1} - \Delta t \, ar_{n+1}(1 - \frac{r_{n+1}}{R}) = r_n.$$

---

[18] As discussed earlier, the prime reason for introducing implicit schemes is to get rid of a numerical stability condition. We will see examples later on illustrating that this may be very important and definitely worthwhile. However, for the logistic model this is not really a big issue, since the condition

$$\Delta t \leqslant 1/a$$

is not very strict for reasonable values of a. In fact, for practical computations we would probably use such small time steps anyway. Since implicit schemes are harder to implement and analyze, you may ask why we bother. The idea is this: By teaching both explicit and implicit schemes for these simple models, it will be easier for you to understand implicit schemes for complicated models when they are really needed. And when are they really needed? They are needed when the restriction on the time step is so severe that we have to do millions, or perhaps even billions or trillions, of time steps in order to compute an approximation of the solution. Then we have to consider alternatives, and such alternatives almost always involve some sort of implicit procedure.

We note that in order to compute r_{n+1} for a given value of r_n, we have to solve a second-order polynomial equation. In Exercise 2.6 below, see page 64, we show that this scheme is unconditionally stable and that it mimics the properties of the exact solution in the same manner as we showed for the explicit scheme above.

Numerical Experiments

Since we have the exact solution of the logistic model, we are able to investigate the accuracy of the numerical schemes by experiments. But let us first simplify the equation a bit. Let s be the scaled number of rabbits defined as

$$s(t) = \frac{r(t)}{R}.$$

Recall that

$$r'(t) = ar(t)\left(1 - \frac{r(t)}{R}\right).$$

Since

$$r(t) = Rs(t),$$

it follows that

$$r'(t) = Rs'(t),$$

and consequently

$$Rs'(t) = aRs(t)\left(1 - \frac{Rs(t)}{R}\right)$$

or

$$s'(t) = as(t)(1 - s(t)).$$

For simplicity,[19] we also choose $a = 1$, and thus consider the model

$$s'(t) = s(t)(1 - s(t)), \qquad (2.48)$$
$$s(0) = s_0.$$

For this model, we have the explicit scheme

$$y_{n+1} = y_n + \Delta t y_n (1 - y_n) \qquad (2.49)$$

---

[19] We can also get rid of a by a scaling of time. Set $\tau = at$ and $u(\tau) = s(t)$. Then

$$\frac{du(\tau)}{d\tau} = \frac{ds(t)}{dt}\frac{dt}{d\tau} = \frac{1}{a}\frac{ds(t)}{dt} = \frac{1}{a}(as(t)(1 - s(t))),$$

so we get the equation

$$u'(\tau) = u(\tau)(1 - u(\tau)).$$

and the implicit scheme

$$z_{n+1} - \Delta t z_{n+1}(1 - z_{n+1}) = z_n. \tag{2.50}$$

Both schemes are started using the initial condition

$$y_0 = z_0 = s_0$$

where s_0 is given. By solving the algebraic equation (2.50), we find that the implicit scheme can be written as

$$z_{n+1} = \frac{1}{2\Delta t}\left(-1 + \Delta t + \sqrt{(1 - \Delta t)^2 + 4\Delta t z_n}\right). \tag{2.51}$$

In Fig. 2.5 we show the numerical solutions as time ranges from 0 to 10. We have used $N = 100$, so $\Delta t = 1/10$. We have also plotted the analytical solution given by

$$s(t) = \frac{s_0}{s_0 + e^{-t}(1 - s_0)}, \tag{2.52}$$

with $s_0 = 0.2$. Note that the three solutions are more or less indistinguishable. We also recognize the properties of the logistic model discussed above. For a short

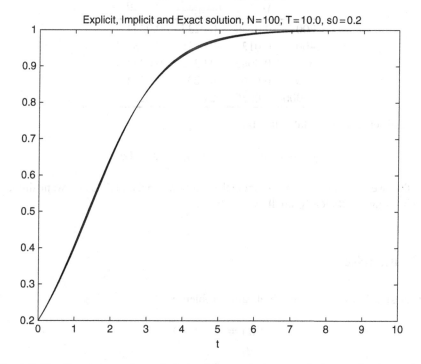

Fig. 2.5 The figure shows the analytical solution and two numerical approximations using $\Delta t = 1/20$. Both numerical solutions provide excellent approximations to the analytical solution

time, the solution looks like an exponential curve, but as time increases the growth is reduced and the equilibrium point of $s = 1$ acts as a barrier for the solution. The monotonicity explained above is also observed.

The Error is $O(\Delta t)$

Let us look a bit closer at the error of the numerical solutions. We consider the problem

$$s'(t) = s(t)(1 - s(t)),$$
$$s(0) = 0.2,$$

with t ranging from 0 to 5. We want to compare the analytical solution and the two numerical solutions at time $t = 5$. The analytical solution is

$$s(5) = \frac{0.2}{0.2 + 0.8e^{-5}} \approx 0.9737555,$$

see (2.52). In the table below we present the error for the numerical solution generated by the explicit scheme, y_N, and the explicit scheme, z_N.

N	$\Delta t = \frac{5}{N}$	$\frac{|y_N - s(5)|}{\Delta t}$	$\frac{|z_N - s(5)|}{\Delta t}$
200	0.0250	0.0235	0.0234
400	0.0125	0.0235	0.0234
600	0.0083	0.0235	0.0234
800	0.0063	0.0235	0.0234
1,000	0.0050	0.0234	0.0234

We observe from the table that both

$$|y_N - s(5)| \approx |z_N - s(5)| \approx 0.0234 \Delta t$$

and thus we know that we can compute the solution as accurately as we want simply by choosing a sufficiently small time step Δt.

2.3 Exercises

Exercise 2.1. Consider the initial value problem

$$r'(t) = 1 + 4r(t), \tag{2.53}$$
$$r(0) = 0.$$

(a) Verify by direct differentiation that the analytical solution of this problem is
given by

$$r(t) = \frac{1}{4}(e^{4t} - 1).$$

(b) An explicit scheme for the initial value problem (2.53) can be written in the
form

$$y_{n+1} = y_n + \Delta t(1 + 4y_n).$$

Explain the derivation of this scheme.

(c) Similarly, derive the following implicit[20] scheme:

$$z_{n+1} = \frac{z_n + \Delta t}{1 - 4\Delta t}.$$

(d) Set $\Delta t = 1/10$ and compute, by hand, y_1, y_2, y_3 and z_1, z_2, z_3. Compare these
values with the analytical values given by $r(\Delta t), r(2\Delta t)$, and $r(3\Delta t)$.

(e) Write a computer program for the implicit and the explicit scheme. The program
should accept N and T as input data and make a graph of the explicit,[21] implicit,
and exact solutions for t ranging from 0 to T. The numerical schemes should
use $\Delta t = T/N$.

(f) Test the program for $T = 1$. Evaluate the error for various values of Δt. Do the
computations indicate that the error is $O(\Delta t)$?

◇

Exercise 2.2. Consider the initial value problem

$$r'(t) = r(t) + 2t - t^2, \qquad (2.54)$$
$$r(0) = 1,$$

for t ranging from $t = 0$ to $t = T$.

(a) Verify that the analytical solution of this problem is given by

$$r(t) = e^t + t^2.$$

(b) Derive the explicit scheme

$$y_{n+1} = y_n + \Delta t(y_n + 2t_n - t_n^2),$$

---

[20] Note that setting $\Delta t = 1/4$ or larger is not a good idea. This also illustrates that implicit schemes
can impose stability restrictions on time stepping. But usually, implicit schemes allow much longer
time steps than explicit schemes.

[21] From time to time we use the terms "explicit solution" and "implicit solution". This is not
entirely accurate; we should use instead the term numerical solution generated by the explicit
scheme, which is the precise statement, but it is too lengthy to use all the time, so now and then we
cheat a little bit.

where

$$t_n = n \Delta t = n \frac{T}{N}.$$

(c) Derive the following implicit scheme:

$$z_{n+1} = \frac{z_n + \Delta t (2t_{n+1} - t_{n+1}^2)}{1 - \Delta t}.$$

(d) Set $\Delta t = 1/10$ and compute, by hand, y_1, y_2, y_3 and z_1, z_2, z_3. Compare these values with the analytical values given by $r(\Delta t), r(2\Delta t)$, and $r(3\Delta t)$.
(e) Write a computer program for the implicit and the explicit schemes.
(f) Test the program for $T = 1$. Evaluate the error for various values of Δt.

◇

Exercise 2.3. Suppose that an isolated island is initially populated by $r_0 = 1{,}000$ rabbits. Assume an exponential growth model and suppose that the growth rate is $a = 50$.

(a) Estimate the number of rabbits at time $t = 1$.
(b) Suppose the initial estimate is uncertain and that repeated counting shows that the actual initial number is between 900 and 1,100. Provide an upper and lower estimate for the number of rabbits at time $t = 1$.
(c) Suppose also that the growth rate is uncertain but that we estimate it to be greater than 40 and less than 60. Give a new upper and lower estimate.

◇

Exercise 2.4. Suppose that an isolated island is initially populated by $r_0 = 1{,}702$ rabbits. Assume a logistic growth model, and suppose that the growth rate $a = 40$ and that the carrying capacity is $R = 8{,}790$.

(a) Estimate the number of rabbits at time $t = 1$.
(b) Suppose the initial estimate is uncertain and that repeated counting shows that the actual initial number is between 1,500 and 1,900. Provide an upper and lower estimate for the number of rabbits at time $t = 1$.
(c) Suppose also that the growth rate is uncertain but that we estimate it to be greater than 30 and less than 50. Give a new upper an lower estimate.
(d) The carrying capacity is also uncertain but is in the interval ranging from 7,000 to 10,000. Give a new upper and lower estimate.

◇

Exercise 2.5. Let us consider the population of the United States[22] between the years 1800 and 1900. The annual growth rate can be estimated to be $a \approx 0.03$.

---

[22] This example is taken from a book by R.B. Banks [5].

Assume that we have an exponential growth model and let $r_0 = 5.3$. Here $t = 0$ corresponds to the year 1800. We want to estimate the population until 1900, i.e., we consider t between 0 and 100. Note that we count the population in millions, i.e., in the year 1800 approximately 5.3 million people lived in the United States.[23]

(a) Explain that an exponential growth model can be written in the form

$$r(t) = 5.3e^{0.03t}$$

where t ranges from 0 to 100.

(b) We want to plot this solution and therefore we want t to go from 1800 to 1900. Show that we can define

$$r(t) = 5.3e^{0.03(t-1800)}$$

for $1800 \leqslant t \leqslant 1900$.

(c) Write a computer program to plot $r(t)$ for $1800 \leqslant t \leqslant 1900$. In the same plot you should include the actual population given in the table below.

Year Population (in millions)
1800 5.3
1810 7.2
1820 9.6
1830 12.9
1840 17.0
1850 23.2
1860 31.4
1870 38.6
1880 50.2
1890 63.0
1900 76.2

(d) Use the model above to predict $r(1980)$. The correct[24] number is about 226.5 million.

(e) Use a logistic model with $r_0 = 5.3$ and $a = 0.03$. Plot the solutions for some values of the carrying capacity. Can you find a value of R such that the logistic model matches the data in the table better than the exponential model did?

---

[23] That is the current population of Denmark and they probably have no idea of what is about to happen. If you do not know much about Denmark, you should find a map and rethink the concept of carrying capacity. Is it reasonable that in the next 100 years the population of Denmark will develop as the U.S. population did from 1800 to 1900?

[24] Unfortunately, it turns out that counting Americans is not much easier than counting rabbits, so the numbers are not at all certain. In fact nobody knows the exact number of Americans at any time. This is, however, not a typical American phenomenon. Nobody knows the exact number of people in any reasonably big country.

Furthermore, use your best value of R to estimate $r(1980)$. Discuss how that relates to the actual figure.

◇

Exercise 2.6. The purpose of this exercise is to study the properties of the implicit scheme for the logistic model. We recall that this scheme can be written in the form

$$\frac{r_{n+1} - r_n}{\Delta t} = a r_{n+1}(1 - \frac{r_{n+1}}{R}).$$

For simplicity, we set $a = R = 1$, and get the scheme

$$r_{n+1} - \Delta t r_{n+1}(1 - r_{n+1}) = r_n, \tag{2.55}$$

so

$$r_{n+1} = \frac{1}{2\Delta t}\left(-1 + \Delta t + \sqrt{(1 - \Delta t)^2 + 4\Delta t r_n}\right). \tag{2.56}$$

(a) Use (2.56) to show that if $r_0 = 1$, then $r_n = 1$, and if $r_0 = 0$, then $r_n = 0$ for all $n \geq 0$.

(b) Define

$$f(r) = r - \Delta t r(1 - r) - r_n$$

for a given value of n, and assume that $0 < r_n < 1$. We want to find r^* such that

$$f(r^*) = 0,$$

and then set

$$r_{n+1} = r^*.$$

Show that

$$f(1) > 0$$

and

$$f(r_n) < 0.$$

Use these two observations to conclude that f is zero for at least one value between r_n and 1.

(c) Show that

$$f(0) < 0$$

and that[25]

$$f(-\infty) > 0.$$

Use these two observations two conclude that f is zero for one value of r between $-\infty$ and 0.

---

[25] We use $f(-\infty)$ as shorthand for the value of $f(r)$ as r tends to minus infinity.

(d) Use the observations in (b) and (c) above to conclude that there is a unique point $r = r^*$ in the interval from r_n to 1 such that $f(r^*) = 0$. Conclude that if $0 \leqslant r_0 \leqslant 1$, then $0 \leqslant r_0 \leqslant r_1 \leqslant \cdots \leqslant 1$.

(e) Suppose $1 < r_n$. Show that

$$f(r_n) > 0$$

and

$$f(1) < 0.$$

Use these observations to argue that f has at least one zero between 1 and r_n.

(f) Show that f has a unique zero between 1 and r_n, and use this to conclude that if $r_0 > 1$, then

$$r_0 \geqslant r_1 \ldots \geqslant 1.$$

◇

Exercise 2.7. In a project below, we will derive further methods for solving initial value problems of the form

$$\begin{aligned} u'(t) &= f(u(t)), \\ u(0) &= u_0, \end{aligned} \tag{2.57}$$

where f is a given function and u_0 is the known initial state. In this exercise, we will derive three schemes based on formulas for numerical integration. Suppose we want to solve (2.57) from $t = 0$ to $t = T$. Let $t_n = n \Delta t$, where, as usual,

$$\Delta t = T/N$$

and $N > 0$ is an integer.

(a) Show that

$$u(t_{n+1}) = u(t_n) + \int_{t_n}^{t_{n+1}} f(u(t))dt. \tag{2.58}$$

(b) Use the trapezoidal rule to motivate the following numerical scheme

$$u_{n+1} = u_n + \frac{\Delta t}{2} \left(f(u_{n+1}) + f(u_n) \right). \tag{2.59}$$

(c) Use the midpoint method to motivate the scheme

$$u_{n+1} = u_n + \Delta t \, f \left(\frac{1}{2}(u_{n+1} + u_n) \right). \tag{2.60}$$

We will derive this scheme below, using another approach. The scheme is often called the Crank–Nicolson scheme, but it really studies a somewhat different problem.

(d) Use Simpson's scheme to motivate

$$u_{n+1} = u_n + \frac{\Delta t}{6}\left(f(u_{n+1}) + 4f\left(\frac{1}{2}(u_{n+1} + u_n)\right) + f(u_n)\right). \quad (2.61)$$

(e) Implement the schemes given by (2.59)–(2.61). Use problem (2.38) to check the accuracy of the schemes.

◇

Exercise 2.8. Differential equations of the form

$$\frac{dy}{dt} = \frac{g(t)}{f(y)} \quad (2.62)$$

are called separable. Such equations can often be solved by direct integration. Suppose that f and y have anti-derivatives given by F and G, i.e.,

$$F(y) = \int f(y)dy \quad (2.63)$$

and

$$G(t) = \int g(t)dt. \quad (2.64)$$

Then the solution of (2.62) satisfies

$$F(y) = G(t) + c, \quad (2.65)$$

where c is a constant that has to be determined using the initial condition.

(a) Consider

$$\begin{aligned} y' &= y, \\ y(0) &= 1. \end{aligned} \quad (2.66)$$

Use the method above to solve this problem. Verify your answer by checking that both conditions in (2.66) hold.

(b) Consider

$$\begin{aligned} y' &= y^2, \\ y(0) &= \alpha. \end{aligned} \quad (2.67)$$

Use the method above to solve (2.67). Verify that the solutions provided in Examples 2.4 and 2.5 are correct.

(c) Consider

$$e^y y' = t,$$
$$y(0) = 1.$$

(2.68)

Use the method above to solve this problem. Verify your answer by checking that both conditions in (2.68) hold.

◇

2.4 Projects

2.4.1 More on Stability

"Two gallons is a great deal of wine, even for two paisanos. Spiritually the jugs may be graduated as this: Just below the shoulder of the first bottle, serious and concentrated conversation. Two inches farther down, sweetly sad memory. Three inches more, thoughts of old and satisfactory loves. An inch, thoughts of bitter loves. Bottom of first jug, general and undirected sadness. Shoulder of the second jug, black, unholy despondency. Two fingers down, a song of death or longing. A thumb, every other song each one knows. The graduations stop here, for the trail splits and there is no certainty. From this point on anything can happen." – John Steinbeck, Tortilla Flat.

In dealing with population models, we have discussed the fact that the parameters involved are based on some sort of measurements and they are therefore impossible to determine exactly. We have to rely on approximations. Often this is fine, when small disturbances in the parameters do not blow up the final results. But, as we all know, real life is not always like that. Imagine two water particles in a river floating very close to each other. We want to predict where these particles will end up and, since they are close initially, we intuitively think that they will remain close for quite a while. And for a smooth river, that is a fair assumption. But suppose we a reach a great waterfall. Obviously, after the waterfall there is absolutely no reason to believe that the two particles will still be close to each other. The problem of computing the position of a particular water-particle after the waterfall based on observations of its position before the waterfall, is completely unstable. In fact, there is no reason to believe that anyone will ever be able to conduct reliable[26] simulations of this phenomenon.

The purpose of this project is to demonstrate that not all differential equations are well behaved. We will do so by studying two very simple equations, one stable

---

[26] There is a difference between qualitatively and quantitatively correct results. We cannot solve the water particle problem correctly quantitatively because it is unstable. But we may very well be able to solve it correctly qualitatively in the sense that the results of our computations may *look* reasonable. This is an important difference.

and one unstable. The issue of stability will also be discussed many times later in
this text.

(a) Consider the logistic model

$$s' = s(1 - s),$$
$$s(0) = x,$$

where $x \geq 0$ is the given initial state. Show, by direct differentiation, that

$$s(t) = \frac{x}{x + e^{-t}(1 - x)}$$

is the analytical solution of this problem.
(b) Make a graph of the solution of this problem for $x = 0, 0.2, 0.4, \ldots, 2.0$ and
for $0 \leq t \leq 5$. Discuss the stability with respect to changes of the initial data
based on the graphs.
(c) Let $S = S(x)$ denote the solution at $t = 1$. Verify that

$$S(x) = \frac{x}{x + e^{-1}(1 - x)},$$

and make a plot of S as a function of x for $0 \leq x \leq 5$.
(d) Show that

$$S'(x) = \frac{1}{e \, (e^{-1}x - x - e^{-1})^2},$$

and plot S' for $0 \leq x \leq 5$.
(e) Let ε be a very small number. Then, by the Taylor series, we have

$$S(x + \varepsilon) \approx S(x) + \varepsilon S'(x).$$

Use this observation to discuss the stability of the solution at time $t = 1$ with
respect to perturbations in the initial data.
(f) Next we consider the initial value problem

$$u' = u(u - 1),$$
$$u(0) = x,$$

where $x \geq 0$ again is the given initial state. Show that the analytical solution of
this problem is given by

$$u(t) = \frac{x}{x + e^t(1 - x)}.$$

(g) Show that

- If $0 < x < 1$, then u goes to zero as t goes to infinity,
- If $x = 1$, then $u = 1$ for all time, and
- If $x > 1$, then u goes to infinity as t approaches

$$\ln(\frac{x}{x-1}).$$

(h) Graph the solution for $x = 0.2, 0.4, \ldots, 1.4$, and discuss the stability of the solution with respect to perturbations in the initial data.

(i) Let $U(x)$ be the solution at time $t = 10$. Observe that

$$U(x) = \frac{x}{x + e^{10}(1-x)}.$$

Compute $U(1)$ and $U(1.0000454)$. Is this initial value problem stable with respect to perturbations in the initial data?

2.4.2 More on Accuracy

In this chapter, we have been concerned with initial value problems of the form

$$u'(t) = f(u(t)),$$ (2.69)
$$u(0) = u_0,$$

where f is a given function and u_0 is the given initial state. In particular, we have studied the exponential growth model where

$$f(u) = au$$

and the logistic model where

$$f(u) = au\left(1 - \frac{u}{R}\right).$$

Here a and R are given parameters. We have introduced two numerical schemes: the *explicit scheme* and the *implicit scheme*. So far we have used these terms, but the two schemes are widely known by other names too: the explicit Euler[27] scheme,

---

[27] Leonhard Paul Euler, 1707–1783, was a pioneering Swiss mathematician and physicist who spent most of his life in Russia and Germany. Euler is one of the greatest scientists of all time and made important contributions to calculus, mechanics, optics, and astronomy. He also introduced much of the modern terminology and notation in mathematics.

the (explicit) Forward Euler scheme, the implicit Euler scheme, and the (implicit) Backward Euler scheme. The two schemes can be summarized by

$$u_{n+1} = u_n + \Delta t f(u_n)$$

and

$$u_{n+1} - \Delta t f(u_{n+1}) = u_n.$$

Here we use the standard notation with $t_n = n\Delta t$, where $\Delta t = T/N$; that is, we want to compute the solution from $t = 0$ to $t = T$ using N time steps. We note that a possibly nonlinear equation has to be solved in order to go from time t_n to time t_{n+1} in the implicit scheme. The numerical experiments reported on page 60 clearly indicates that the accuracy of both schemes is $O(\Delta t)$. The purpose of this project is to study a few schemes that are more accurate.

(a) Suppose u is a sufficiently smooth function. Use the Taylor series to show that

$$u(t + k) = u(t) + ku'(t) + \frac{1}{2}k^2 u''(t) + \frac{1}{6}k^3 u'''(t) + O(k^4).$$

(b) Set $t = t_{n+1/2} = (n + 1/2)\Delta t$ and $k = \Delta t/2$. Show that

$$u(t_{n+1}) = u(t_{n+1/2}) + \frac{\Delta t}{2}u'(t_{n+1/2}) + \frac{1}{2}(\frac{\Delta t}{2})^2 u''(t_{n+1/2})$$
$$+ \frac{1}{6}(\frac{\Delta t}{2})^3 u'''(t_{n+1/2}) + O(\Delta t^4).$$

Set $k = -\Delta t/2$, to show that

$$u(t_n) = u(t_{n+1/2}) - \frac{\Delta t}{2}u'(t_{n+1/2}) + \frac{1}{2}(\frac{\Delta t}{2})^2 u''(t_{n+1/2})$$
$$- \frac{1}{6}(\frac{\Delta t}{2})^3 u'''(t_{n+1/2}) + O(\Delta t^4).$$

(c) Use the two equations in (b) to show that

$$\frac{u(t_{n+1}) - u(t_n)}{\Delta t} = u'(t_{n+1/2}) + O(\Delta t^2).$$

(d) Use the differential equation (2.69) to show that

$$\frac{u(t_{n+1}) - u(t_n)}{\Delta t} = f(u(t_{n+1/2})) + O(\Delta t^2).$$

(e) Use the Taylor series to show that

$$f(u(t_{n+1/2})) = \frac{1}{2}(f(u(t_n)) + f(u(t_{n+1}))) + O(\Delta t^3),$$

and use this to show that

$$\frac{u(t_{n+1}) - u(t_n)}{\Delta t} = \frac{1}{2} \left(f(u(t_n)) + f(u(t_{n+1})) \right) + O(\Delta t^2).$$

(f) Use these observations to derive the Crank–Nicolson scheme

$$u_{n+1} - \frac{\Delta t}{2} f(u_{n+1}) = u_n + \frac{\Delta t}{2} f(u_n). \tag{2.70}$$

(g) Implement the explicit Euler, the implicit Euler and the Crank–Nicolson scheme
for the exponential growth problem

$$u' = u,$$
$$u(0) = 1.$$

Make a table comparing the accuracy of these solutions at $T = 5$. Divide the
errors of the solutions computed by the two Euler schemes by Δt, and the error
of the Crank–Nicolson solution by Δt^2. Use the results to argue that the error
of the Euler schemes is $O(\Delta t)$ and the error of the Crank–Nicolson scheme is
$O(\Delta t^2)$.

(h) Repeat the experiments above for the initial value problem

$$u' = u(1 - u),$$
$$u(0) = 10,$$
$$T = 1.$$

Are the conclusions regarding accuracy the same?

(i) Although the Crank–Nicolson scheme is more accurate than the explicit Euler
scheme, it suffers from the fact that an equation has to be solved at each time
step. The purpose of our next scheme is to maintain the error $O(\Delta t^2)$ but for
an explicit computation, i.e., we want to avoid having to solve an equation at
each time step.

In the derivation of the Crank–Nicolson scheme above, we observed that, in
general,

$$\frac{u(t_{n+1}) - u(t_n)}{\Delta t} = \frac{1}{2} \left(f(u(t_n)) + f(u(t_{n+1})) \right) + O(\Delta t^2).$$

The problem here is that we need to evaluate $f(u(t_{n+1}))$, and this leads to a
possibly[28] nonlinear equation. But note that

---

[28] The equation is nonlinear whenever f is nonlinear, and f is linear if we can write it in the form

$$f(u) = au + b$$

for given constants a and b; otherwise it is nonlinear.

$$u(t_{n+1}) = u(t_n) + \Delta t f(u(t_n)) + O(\Delta t^2).$$

Use this observation to derive the *Heun scheme*

$$u_{n+1} = u_n + \frac{\Delta t}{2} [f(u_n) + f(u_n + \Delta t f(u_n))].$$

(j) Implement the Heun scheme and repeat the experiments in (g) and (h) above. Is the error $O(\Delta t^2)$?

(k) Let us now consider the slightly more general initial value problem

$$u' = f(t, u(t)), \qquad\qquad (2.71)$$
$$u(0) = u_0.$$

We note that, for instance,

$$u'(t) = \sin(t)u^2(t)$$

can be written in this form with $f(t, u) = \sin(t)u^2$, and we note also that this problem cannot be formulated in the form (2.69). Use the Taylor series to derive the explicit Euler scheme,

$$u_{n+1} = u_n + \Delta t f(t_n, u_n)$$

the implicit Euler scheme,

$$u_{n+1} - \Delta t\ f(t_{n+1}, u_{n+1}) = u_n,$$

the Crank–Nicolson scheme

$$u_{n+1} - \frac{\Delta t}{2} f(t_{n+1}, u_{n+1}) = u_n + \frac{\Delta t}{2} f(t_n, u_n),$$

and the Heun scheme for the problem,

$$F_1 = f(t_n, u_n),$$
$$F_2 = f(t_{n+1}, u_n + \Delta t F_1),$$
$$u_{n+1} = u_n + \frac{\Delta t}{2} [F_1 + F_2].$$

(l) Much more accurate schemes can be derived by carrying on in the same spirit as above. Probably the most widespread scheme is the fourth-order Runge–Kutta scheme, which can be formulated as follows:

$$F_1 = f(t_n, u_n),$$

$$F_2 = f(t_{n+1/2}, u_n + \frac{\Delta t}{2} F_1),$$

$$F_3 = f(t_{n+1/2}, u_n + \frac{\Delta t}{2} F_2),$$

$$F_4 = f(t_{n+1}, u_n + \Delta t F_3),$$

$$u_{n+1} = u_n + \frac{\Delta t}{6} [F_1 + 2F_2 + 2F_3 + F_4].$$

Implement this scheme and perform the experiments of (g) and (h) above. Try, using experiments, to figure out the accuracy of this scheme.

(m) Consider the initial value problem

$$u'(t) = \frac{2(1-t)}{\varepsilon^2} u$$

$$u(0) = e^{-1/\varepsilon^2},$$

where ε is given. Show that the solution of this problem is given by

$$u(t) = e^{-\left(\frac{1-t}{\varepsilon}\right)^2}.$$

(n) Set $\varepsilon = 1/4$. We want to compute the solution at time $t = 1.1$. Compare the accuracy of all the schemes above (explicit Euler, implicit Euler, Crank–Nicolson, Heun, and Runge–Kutta) for this problem. Say we want the absolute error to be less than 10^{-5}. Make a ranking of the schemes based on how much CPU time they need to achieve this accuracy.

Chapter 3
Systems of Ordinary Differential Equations

In Chap. 2, we saw that models of the form

$$y'(t) = F(y), \qquad y(0) = y_0, \tag{3.1}$$

can be used to model natural processes. We observed that simple versions of (3.1) can be solved analytically, and we saw that the problem can be solved adequately by using numerical methods. The purpose of the present chapter is to extend our knowledge to the case of systems of ordinary differential equations (ODEs), e.g.,

$$\begin{aligned} y'(t) &= F(y, z), \ y(0) = y_0, \\ z'(t) &= G(y, z), \ z(0) = z_0, \end{aligned} \tag{3.2}$$

where y_0 and z_0 are the given initial states and where F and G are smooth functions. In Sect. 3.1, we derive an interesting model of the form (3.2), which is of relevance for the study of so-called predator–prey systems. Thereafter, we will discuss some analytical aspects and study some numerical methods.

3.1 Rabbits and Foxes; Fish and Sharks

In Sect. 2.1.1 on page 32, we discussed models of population growth for a group of rabbits on an isolated island. We first argued that the growth could be modeled by

$$y' = \alpha y, \qquad y(0) = y_0, \tag{3.3}$$

where $\alpha > 0$ denotes the growth rate and where y_0 is the initial number of rabbits on the island. For relatively small populations we argued that (3.3) is a good model. But as y increases – recall that the solution is $y(t) = y_0 e^{\alpha t}$ – the supply of food will constrain the growth. To model the effect of limited resources, we introduced the concept of carrying capacity β and argued that

$$y' = \alpha y(1 - y/\beta), \qquad y(0) = y_0, \tag{3.4}$$

A. Tveito et al., *Elements of Scientific Computing*, Texts in Computational Science and Engineering 7, DOI 10.1007/978-3-642-11299-7_3,
© Springer-Verlag Berlin Heidelberg 2010

is a better model. We saw, both analytically and numerically, that this model gives predictions that are consistent with our intuition. Data that really support logistic solutions, i.e., solutions of (3.4), can be found in [5].

In this chapter we will introduce a predator–prey model. Imagine that we introduce a group of foxes to the isolated island. The foxes (predators) will eat rabbits (prey) and a decrease in the number of rabbits will result in a decrease in the food resources for the foxes. This will in turn result in a decrease in the number of foxes. Our aim is to model this interaction. A similar situation appears if one studies the growth of fish (prey) and sharks (predators). The model that we will develop is usually referred to as the Lotka–Volterra model.[1]

The background of Volterra's interest in this topic was that fishermen observed a somewhat strange situation immediately after World War I. Fishing was more or less impossible in the upper Adriatic Sea during World War I. After the war, the fishermen expected very rich resources of fish. But contrary to their intuition, they found very little fish. This fact was surprising and Volterra tried to develop a mathematical model that could shed some light on the mystery. He derived a 2×2 system of ODEs of the form (3.2). Volterra's model is now classic in mathematical ecology and it demonstrates the basic features of predator–prey systems.

We will follow Volterra's[2] derivation of a model for the population density of fish and sharks, but keep in mind that exactly the same line of argument is valid for the rabbit–fox system.

Let $F = F(t)$ denote the number of fish located in a specific region of the sea at time t. Similarly, $S = S(t)$ denotes the number of sharks in the same region. Let us first consider the case of $S = 0$, i.e., no sharks are present. Then, the growth of fish can be modeled by

$$F' = \alpha F, \tag{3.5}$$

which predicts exponential growth. Here $\alpha > 0$ is the growth rate. As we observed above, limited resources lead to a logistic model of the form

$$F' = \alpha F(1 - F/\beta), \tag{3.6}$$

where $\beta > 0$ is the carrying capacity of the environment. This equation can be rewritten as

$$\frac{F'}{F} = \alpha \left(1 - F/\beta\right), \tag{3.7}$$

---

[1] Vito Volterra (1860–1940) was an Italian mathematician who worked on functional analysis, integral equations, partial differential equations and mathematical models of biology.

[2] Our presentation is based on the classic book of Richard Haberman [17]. We strongly recommend the reader to read more about population growth in that book or in the book of M. Braun [7], which is also heavily used as background material for the present text.

which states that the *relative growth* of fish decreases as the number of fish increases. When we introduce sharks, it is reasonable for us to assume that the relative growth rate of fish is reduced linearly with respect to S. This gives

$$\frac{F'}{F} = \alpha \left(1 - F/\beta - \gamma S\right), \tag{3.8}$$

where $\gamma > 0$ is a constant. We rewrite this as

$$F' = \alpha \left(1 - F/\beta - \gamma S\right) F. \tag{3.9}$$

Next, we consider the growth or decay of sharks. If there are no fish in the sea, the relative change in sharks can be expressed as

$$\frac{S'}{S} = -\delta, \tag{3.10}$$

where $\delta > 0$ is the decay rate. Since the sharks need fish to survive, $F = 0$ will eventually lead to no sharks as well. On the other hand, the presence of fish will increase the relative change of sharks and we assume that this can be modeled as

$$\frac{S'}{S} = -\delta + \varepsilon F, \tag{3.11}$$

where $\varepsilon > 0$ is a constant.

We can now summarize the 2×2 system as follows,

$$F' = \alpha \left(1 - F/\beta - \gamma S\right) F, \quad F(0) = F_0, \tag{3.12}$$
$$S' = (\varepsilon F - \delta)S, \qquad S(0) = S_0. \tag{3.13}$$

When the parameters α, β, γ and ε are known, by e.g. estimation, and when F_0 and S_0 are given, system (3.12) and (3.13) can predict the number of fish and sharks in the sea.

Volterra studied a version of the system above and found that it could generate periodic[3] solutions. When there was no fishing activity, the number of fish increased. This in turn led to an increase in the number of sharks. But with a large number of sharks, the need for food increased dramatically and thus the number of fish was reduced, which in turn reduced the number of sharks.

---

[3] A periodic solution repeats itself. If $f(t + T) = f(t)$ for some $T \neq 0$ and for all t, then f is periodic.

3.2 A Numerical Method: Unlimited Resources

In order to discuss numerical solutions of system (3.12) and (3.13), we simplify the problem by choosing simple values of the parameters involved. For illustration we choose

$$\alpha = 2, \quad \gamma = \frac{1}{2}, \quad \varepsilon = 1, \quad \text{and} \quad \delta = 1.$$

Let us also start by setting $\beta = \infty$. This means that the number of fish can grow without any limit. Since fish eat plankton, this assumption is valid at least when the number of fish is not too large. To summarize, we have the simplified model

$$F' = (2 - S)F, \quad F(0) = F_0, \tag{3.14}$$
$$S' = (F - 1)S, \quad S(0) = S_0. \tag{3.15}$$

In order to solve this system numerically, we introduce the time step $\Delta t > 0$ and define $t_n = n\Delta t$. Let F_n and S_n denote the approximations of $F(t_n)$ and $S(t_n)$, respectively. Since

$$\frac{F(t_{n+1}) - F(t_n)}{\Delta t} \approx F'(t_n) \quad \text{and} \quad \frac{S(t_{n+1}) - S(t_n)}{\Delta t} \approx S'(t_n),$$

we introduce the numerical scheme

$$\frac{F_{n+1} - F_n}{\Delta t} = (2 - S_n)F_n, \tag{3.16}$$
$$\frac{S_{n+1} - S_n}{\Delta t} = (F_n - 1)S_n. \tag{3.17}$$

The scheme can be rewritten in a computational form,

$$F_{n+1} = F_n + \Delta t(2 - S_n)F_n, \tag{3.18}$$
$$S_{n+1} = S_n + \Delta t(F_n - 1)S_n. \tag{3.19}$$

We observe from this scheme that if F_0 and S_0 are given, then F_1 and S_1 can be computed from (3.18) and (3.19), respectively, using $n = 0$. When F_1 and S_1 have been computed, we set $n = 1$ and compute F_2 and S_2. In this manner, we can compute F_n and S_n for all $n > 0$.

For illustration, we choose $\Delta t = 1/1{,}000$, $F_0 = 1.9$ and $S_0 = 0.1$. This models a situation with a large number of fish and a small number of sharks initially. We have plotted the numerical approximation in Fig. 3.1. Note that, initially, the number of fish increases almost exponentially. This leads to a strong growth in the shark population that subsequently leads to a sharp reduction of the fish, and then of the sharks. We note that this is what Volterra tried to model.

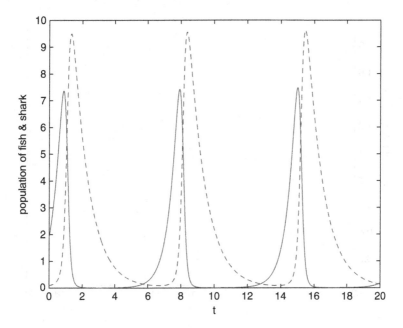

Fig. 3.1 The numerical solution for system (3.14) and (3.15), produced by the explicit scheme (3.18) and (3.19) using $\Delta t = 1/1{,}000$, $F_0 = 1.9$, and $S_0 = 0.1$. The solution for F is represented by the *solid curve*, whereas the solution for S is represented by the *dashed curve*

3.3 A Numerical Method: Limited Resources

In the simple model (3.14) and (3.15) above, we used $\beta = \infty$, which means that we assumed unbounded resources for the fish. If we replace this by $\beta = 2$, which is a fairly strict assumption on the amount of plankton available, we get the system

$$F' = (2 - F - S)F, \quad F(0) = F_0, \tag{3.20}$$
$$S' = (F - 1)S, \qquad S(0) = S_0. \tag{3.21}$$

Following the steps in Sect. 3.2, we define a numerical scheme:

$$F_{n+1} = F_n + \Delta t(2 - F_n - S_n)F_n, \tag{3.22}$$
$$S_{n+1} = S_n + \Delta t(F_n - 1)S_n. \tag{3.23}$$

The solutions can be computed in the same manner as in Sect. 3.2. In Fig. 3.2 we have plotted the numerical solutions using $F_0 = 1.9$, $S_0 = 0.1$ and $\Delta t = 1/1{,}000$. Note that, in the presence of limited resources, the solution quickly converges toward the equilibrium[4] solution represented by $S = F = 1$.

---

[4] Equilibrium solutions stay constant forever. In system (3.20)–(3.21), $F_0 = S_0 = 0$ is an example of an equilibrium solution.

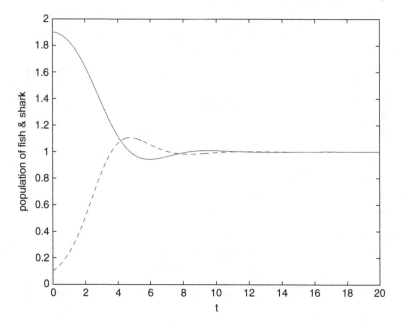

Fig. 3.2 The numerical solution for system (3.20) and (3.21), produced by the explicit scheme (3.22) and (3.23) using $\Delta t = 1/1,000$, $F_0 = 1.9$, and $S_0 = 0.1$. The solution for F is represented by the *solid curve*, whereas the solution for S is represented by the *dashed curve*

3.4 Phase Plane Analysis

3.4.1 A Simplified Model

Let us consider the simplified model

$$
\begin{aligned}
F'(t) &= 1 - S(t), \quad F(0) = F_0, \\
S'(t) &= F(t) - 1, \quad S(0) = S_0.
\end{aligned}
\tag{3.24}
$$

This model represents the very basic properties[5] of the fish–shark interaction. We observe that a large number of sharks ($S > 1$) leads to a decrease in the fish population ($F' < 0$), and that a large number of fish ($F > 1$) leads to an increase in the number of sharks ($S' > 0$).

---

[5] Note that (3.24) is not a realistic model for the fish–shark interaction. It may produce negative values, which are clearly not relevant. We introduce the model here only in order to illustrate properties of ordinary ODEs and their numerical solution.

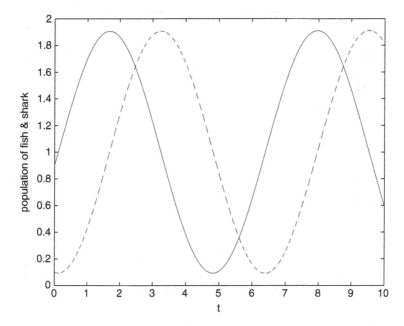

Fig. 3.3 The numerical solution for system (3.24), produced by the explicit scheme (3.25) using $\Delta t = 1/1{,}000$, $F_0 = 0.9$, and $S_0 = 0.1$. The solution for F is represented by the *solid curve*, whereas the solution for S is represented by the *dashed curve*

A numerical scheme for this system, using the notation from Sect. 3.2, reads

$$\begin{aligned}
F_{n+1} &= F_n + \Delta t(1 - S_n), \\
S_{n+1} &= S_n + \Delta t(F_n - 1),
\end{aligned} \tag{3.25}$$

where F_0 and S_0 are the given initial states, see (3.24). In Fig. 3.3 we have used $F_0 = 0.9$, $S_0 = 0.1$, $\Delta t = 1/1{,}000$, and computed the solution from $t = 0$ to $t = 10$. Note also that this simplified system seems to have solutions of a periodic form.

3.4.2 The Phase Plane

In the computation above, we generate (F_n, S_n) for $n = 1, \ldots, 10{,}000$. By plotting these values in an F–S coordinate system, we can study how F and S interact. Consider Fig. 3.4 and observe that the plot depicts an almost perfect circle. In Fig. 3.5 we have done the same calculation, but using $\Delta t = 1/100$. We observe that using a larger Δt results in a less perfect circle.

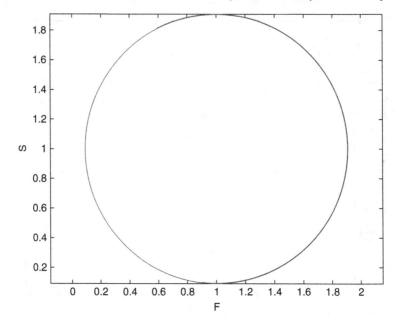

Fig. 3.4 The numerical solution for (3.24) in the F–S coordinate system, produced by the explicit scheme (3.25) using $\Delta t = 1/1{,}000$, $F_0 = 0.9$, and $S_0 = 0.1$

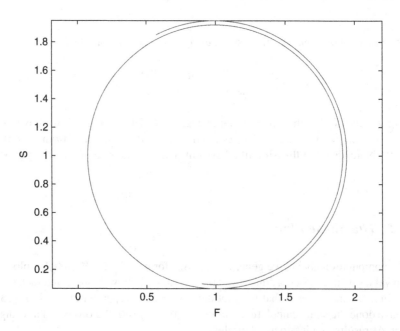

Fig. 3.5 The numerical solution for (3.24) in the F–S coordinate system, produced by the scheme (3.25) using $\Delta t = 1/100$, $F_0 = 0.9$ and $S_0 = 0.1$

Based on these computations we expect that

(a) The analytical solution $(F(t), S(t))$ forms a circle in the F–S coordinate system
(b) A good numerical method generates values (F_n, S_n) that are placed almost exactly on a circle, and they get closer to a circle when Δt is smaller

We will look a bit closer at these two conjectures.

3.4.3 Circles in the Analytical Phase Plane

Our hypothesis, based on the computations above, is that the solution forms a circle that seems to have the point (1,1) as its center. In order to see whether this is correct, we define the function

$$r(t) = (F(t) - 1)^2 + (S(t) - 1)^2. \tag{3.26}$$

In Fig. 3.6 we have plotted an approximation to this function given by

$$r_n = (F_n - 1)^2 + (S_n - 1)^2 \tag{3.27}$$

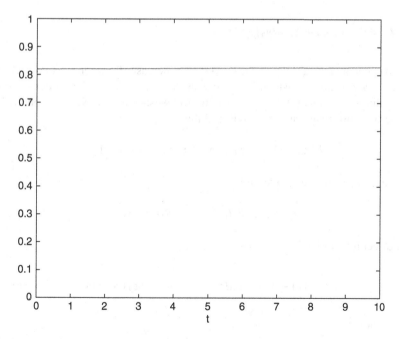

Fig. 3.6 The $r(t)$ function, defined by (3.26), of the numerical solution for (3.24), which is produced by the explicit scheme (3.25) using $\Delta t = 1/1{,}000$, $F_0 = 0.9$, and $S_0 = 0.1$

in the case of $F_0 = 0.9$, $S_0 = 0.1$ and $\Delta t = 1/1{,}000$. We see from the graph that this function is almost a constant. But if it *is* actually a constant in the analytical case, we should be able to see that $r'(t) = 0$ for all t. We differentiate (3.26) with respect to t and get

$$r'(t) = 2(F - 1)F' + 2(S - 1)S'. \tag{3.28}$$

At this point, we have to use the dynamics of F and S given by system (3.24), i.e.,

$$F' = 1 - S \quad \text{and} \quad S' = F - 1. \tag{3.29}$$

By using these expressions for F' and S' in (3.28), we get

$$r'(t) = 2(F - 1)(1 - S) + 2(S - 1)(F - 1) = 0, \tag{3.30}$$

so $r'(t) = 0$ for all t, and thus $r(t)$ is indeed constant in the analytical case.

In general, the solutions of (3.24) form circles in the *state space* (the F–S coordinate system where F and S are both positive) with radius $((F_0 - 1)^2 + (S_0 - 1)^2)^{1/2}$ and centered at (1,1).

3.4.4 Alternative Analysis

We have seen that the graph of $(F(t), S(t))$, as t increases from zero, defines a circle in the F–S coordinate system. This fact was derived in Sect. 3.4.3 by considering the auxiliary function $r(t)$ defined in (3.26). But we can also derive this property by a more straightforward integration. Recall that

$$F'(t) = 1 - S(t) \quad \text{and} \quad S'(t) = F(t) - 1.$$

Consequently, we have the identity

$$(F(t) - 1) F'(t) = (1 - S(t)) S'(t). \tag{3.31}$$

By a direct integration in time from 0 to t, we get

$$\int_0^t (F(\tau) - 1) F'(\tau) d\tau = \int_0^t (1 - S(\tau)) S'(\tau) d\tau, \tag{3.32}$$

which leads to

$$\frac{1}{2} \left[(F(\tau) - 1)^2 \right]_0^t = -\frac{1}{2} \left[(S(\tau) - 1)^2 \right]_0^t, \tag{3.33}$$

so

$$(F(t) - 1)^2 + (S(t) - 1)^2 = (F_0 - 1)^2 + (S_0 - 1)^2 \tag{3.34}$$

for all $t \geq 0$. This is just another way of deriving the fact that

$$r(t) = (F(t) - 1)^2 + (S(t) - 1)^2$$

is constant.

3.4.5 Circles in the Numerical Phase Plane

We still consider the simplified model

$$\begin{aligned}
F'(t) &= 1 - S(t), \quad F(0) = F_0, \\
S'(t) &= F(t) - 1, \quad S(0) = S_0.
\end{aligned} \tag{3.35}$$

We have seen above that for this system the function

$$r(t) = (S(t) - 1)^2 + (F(t) - 1)^2$$

is constant for all $t > 0$. An explicit numerical approximation of (3.35) is given by

$$\begin{aligned}
F_{n+1} &= F_n + \Delta t(1 - S_n), \\
S_{n+1} &= S_n + \Delta t(F_n - 1),
\end{aligned} \tag{3.36}$$

where F_0 and S_0 are given. In Fig. 3.6 above we saw that the discrete function

$$r_n = (F_n - 1)^2 + (S_n - 1)^2$$

was almost constant, i.e., $r_n \approx r_0$ for all $n \geq 0$.

Let us look a bit closer at this. We have $r(t) = r_0$, and thus a perfect numerical scheme should produce $r_n = r_0$ for all $n \geq 0$. Now, of course, the scheme only provides approximations, but this property can give some insight into how accurate the approximation is. This is a general observation: If we have a property of the analytical solution that is computable, this can be used to strengthen, or perhaps weaken, our faith in the numerical solution. More specifically, we can use such a property to compare different numerical approximations.

But let us do some more computations and regard $r_n - r_0$ as a measure of the error in our computations. Suppose we want to solve system (3.35) from $t = 0$ to $t = 10$. We choose $\Delta t = 10/N$, and use $N = 10^k$ for $k = 2, 3, 4, 5$. In Table 3.1 we study $\frac{r_N - r_0}{r_0}$ for these values of Δt.

Table 3.1 The table shows Δt, the number of time steps N, the "relative error" $\frac{r_N - r_0}{r_0}$, and $\frac{r_N - r_0}{r_0 \Delta t}$. Note that the numbers in the last column seem to tend toward a constant

Δt	N	$\dfrac{r_N - r_0}{r_0}$	$\dfrac{r_N - r_0}{r_0 \Delta t}$
10^{-1}	10^2	1.7048	17.0481
10^{-2}	10^3	$1.0517 \cdot 10^{-1}$	10.5165
10^{-3}	10^4	$1.0050 \cdot 10^{-2}$	10.0502
10^{-4}	10^5	$1.0005 \cdot 10^{-3}$	10.0050

We observe that

$$\frac{r_N - r_0}{r_0 \Delta t} \approx 10,$$

and thus $r_N \approx (1 + 10\Delta t)r_0$. Consequently, as Δt goes to zero, the numerical solutions also seem to approach a perfect circle.

Let us consider this issue by analyzing the numerical scheme (3.36). Note that

$$r_{n+1} = (F_{n+1} - 1)^2 + (S_{n+1} - 1)^2, \tag{3.37}$$

and thus, by using the numerical scheme (3.36), we get

$$\begin{aligned}
r_{n+1} &= (F_n - 1 + \Delta t(1 - S_n))^2 + (S_n - 1 + \Delta t(F_n - 1))^2 \\
&= (F_n - 1)^2 + 2\Delta t(F_n - 1)(1 - S_n) + \Delta t^2(1 - S_n)^2 \\
&\quad + (S_n - 1)^2 + 2\Delta t(F_n - 1)(S_n - 1) + \Delta t^2(1 - F_n)^2 \\
&= r_n + \Delta t^2 r_n.
\end{aligned}$$

Hence

$$r_{n+1} = (1 + \Delta t^2)r_n, \tag{3.38}$$

so by induction,

$$r_m = (1 + \Delta t^2)^m r_0. \tag{3.39}$$

Since $\Delta t = 10/N$, we have

$$r_N = \left(1 + \frac{10^2}{N^2}\right)^N r_0. \tag{3.40}$$

Now, in order to estimate $r_N - r_0$, we have to study the asymptotic behavior of $\left(1 + \frac{10^2}{N^2}\right)^N$. Note first that

$$\left(1 + \frac{10^2}{N^2}\right)^N = e^{N \ln(1 + 10^2/N^2)}. \tag{3.41}$$

By Taylor-series expansions we have

$$\ln(1 + x) = x + \mathcal{O}(x^2), \tag{3.42}$$

so

$$\ln\left(1 + \frac{10^2}{N^2}\right) = \frac{10^2}{N^2} + \mathcal{O}\left(10^4/N^4\right), \tag{3.43}$$

and thus

$$N \ln\left(1 + \frac{10^2}{N^2}\right) \approx \frac{10^2}{N} \tag{3.44}$$

for large values of N. By (3.41) we now have

$$\left(1 + \frac{10^2}{N^2}\right)^N \approx e^{\frac{10^2}{N}}. \tag{3.45}$$

Again, by a Taylor expansion,

$$e^x = 1 + x + \mathcal{O}(x^2), \tag{3.46}$$

so for large values of N, it follows that

$$e^{10^2/N} \approx 1 + \frac{10^2}{N}, \tag{3.47}$$

and thus

$$\left(1 + \frac{10^2}{N^2}\right)^N \approx 1 + \frac{10^2}{N}. \tag{3.48}$$

From (3.40), we now find that

$$
\begin{aligned}
r_N - r_0 &= \left((1 + 10^2/N^2)^N - 1\right) r_0 \\
&\approx \left(1 + 10^2/N - 1\right) r_0 \\
&= \frac{10^2}{N} r_0 \\
&= 10\Delta t\, r_0,
\end{aligned}
$$

since $\Delta t = 10/N$. We see that

$$\frac{r_N - r_0}{r_0} \approx 10\Delta t,$$

which is what we expect from Table 3.1 above, where we computed numerically that $\frac{r_N - r_0}{r_0} \approx 10\Delta t$.

We have now demonstrated that the numerical solution also forms a circle as the time step size goes to zero.

3.4.6 More on Numerics

In Project 2.4.2 on page 69, we studied numerical methods for an ODE of the form

$$u'(t) = f(u(t)), \qquad u(0) = u_0. \tag{3.49}$$

By applying Taylor series expansions, we derived the standard explicit (forward Euler) scheme

$$u_{n+1} = u_n + \Delta t\, f(u_n), \tag{3.50}$$

the standard implicit (backward Euler) scheme

$$u_{n+1} - \Delta t\, f(u_{n+1}) = u_n, \tag{3.51}$$

and the Crank–Nicolson scheme

$$u_{n+1} - \frac{\Delta t}{2} f(u_{n+1}) = u_n + \frac{\Delta t}{2} f(u_n). \tag{3.52}$$

We observed, numerically, that the errors for these schemes seem to be $\mathcal{O}(\Delta t)$, $\mathcal{O}(\Delta t)$, and $\mathcal{O}(\Delta t^2)$, respectively. Similar schemes can be derived for the systems considered in the present chapter. We start by considering the simplified system

$$\begin{aligned} F'(t) &= 1 - S(t), \quad F(0) = F_0, \\ S'(t) &= F(t) - 1, \quad S(0) = S_0. \end{aligned} \tag{3.53}$$

We want to use the Crank–Nicolson scheme and see whether we can obtain a higher accuracy than we did with the explicit scheme (3.25) above. The basic form of the Crank–Nicolson scheme for the problem

$$u'(t) = f(u(t))$$

is

$$\frac{u_{n+1} - u_n}{\Delta t} = \frac{1}{2} \left(f(u_{n+1}) + f(u_n) \right), \tag{3.54}$$

from which (3.52) is easily obtained. By applying the form (3.54) to both equations in (3.53), we get

$$\frac{F_{n+1} - F_n}{\Delta t} = \frac{1}{2}[(1 - S_n) + (1 - S_{n+1})],$$

$$\frac{S_{n+1} - S_n}{\Delta t} = \frac{1}{2}[(F_n - 1) + (F_{n+1} - 1)]. \tag{3.55}$$

This system can be rewritten as

$$F_{n+1} + \frac{\Delta t}{2} S_{n+1} = F_n + \Delta t - \frac{\Delta t}{2} S_n,$$

$$-\frac{\Delta t}{2} F_{n+1} + S_{n+1} = S_n - \Delta t + \frac{\Delta t}{2} F_n. \tag{3.56}$$

Note that for each n, this defines a 2×2 system of linear equations. Let us make this clearer by defining the matrix

$$\mathbf{A} = \begin{bmatrix} 1 & \Delta t/2 \\ -\Delta t/2 & 1 \end{bmatrix}, \tag{3.57}$$

and the vector

$$\mathbf{b}_n = \begin{pmatrix} F_n + \Delta t - \frac{\Delta t}{2} S_n \\ S_n - \Delta t + \frac{\Delta t}{2} F_n \end{pmatrix}. \tag{3.58}$$

Then system (3.56) can be written in the form

$$\mathbf{A} \mathbf{x}_{n+1} = \mathbf{b}_n, \tag{3.59}$$

where $\mathbf{x}_{n+1}$ denotes a vector of two components. By solving the linear system (3.59), we find $\mathbf{x}_{n+1}$ and define

$$\begin{pmatrix} F_{n+1} \\ S_{n+1} \end{pmatrix} = \mathbf{x}_{n+1}. \tag{3.60}$$

We note that since

$$\det(\mathbf{A}) = 1 + \Delta t^2/4, \tag{3.61}$$

we have $\det(\mathbf{A}) > 0$ for all values of Δt and the scheme (3.56) is therefore always well defined.

In general, a 2×2 matrix

$$\mathbf{B} = \begin{bmatrix} a & b \\ c & d \end{bmatrix} \tag{3.62}$$

is non-singular if $ad \neq cb$, in which case the inverse is given by

$$\mathbf{B}^{-1} = \frac{1}{ad - bc} \begin{bmatrix} d & -b \\ -c & a \end{bmatrix}. \tag{3.63}$$

When the matrix is given by (3.57), the inverse is

$$\mathbf{A}^{-1} = \frac{1}{1 + \Delta t^2/4} \begin{bmatrix} 1 & -\Delta t/2 \\ \Delta t/2 & 1 \end{bmatrix}. \tag{3.64}$$

Consequently, we have

$$\begin{pmatrix} F_{n+1} \\ S_{n+1} \end{pmatrix} = \frac{1}{1 + \Delta t^2/4} \begin{bmatrix} 1 & -\Delta t/2 \\ \Delta t/2 & 1 \end{bmatrix} \begin{pmatrix} F_n + \Delta t - \frac{\Delta t}{2} S_n \\ S_n - \Delta t + \frac{\Delta t}{2} F_n \end{pmatrix}. \tag{3.65}$$

By multiplying the matrix by the vector and re-arranging terms, we get

$$\begin{aligned} F_{n+1} &= \tfrac{1}{1+\Delta t^2/4} \left[(1 - \Delta t^2/4) F_n + \Delta t \left(\tfrac{\Delta t}{2} + 1 \right) - \Delta t S_n \right], \\ S_{n+1} &= \tfrac{1}{1+\Delta t^2/4} \left[(1 - \Delta t^2/4) S_n + \Delta t \left(\tfrac{\Delta t}{2} - 1 \right) + \Delta t F_n \right]. \end{aligned} \tag{3.66}$$

In Fig. 3.7, we have plotted the state space solution computed by this scheme for $S_0 = 0.1$, $F_0 = 0.9$, and $\Delta t = 1/1{,}000$. Again we note that the numerical solution seems to form a perfect circle. In order to analyze this we define, as above,

$$r_n = (F_n - 1)^2 + (S_n - 1)^2 \tag{3.67}$$

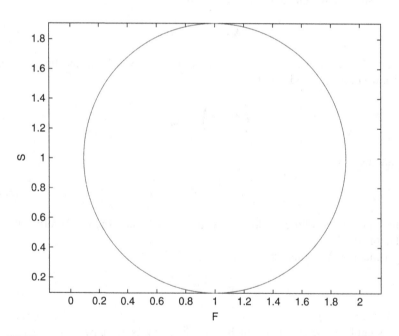

Fig. 3.7 The numerical solution for (3.53) in the F–S coordinate system, produced by the Crank–Nicolson scheme (3.66) using $\Delta t = 1/1{,}000$, $F_0 = 0.9$ and $S_0 = 0.1$

Table 3.2 The table shows Δt, the number of time steps N, and the "error" $\frac{r_N - r_0}{r_0}$

Δt	N	$\dfrac{r_N - r_0}{r_0}$
10^{-1}	10^2	$-2.6682 \cdot 10^{-16}$
10^{-2}	10^3	$-1.59986 \cdot 10^{-17}$
10^{-3}	10^4	$3.97982 \cdot 10^{-17}$
10^{-4}	10^5	$7.06021 \cdot 10^{-15}$

for all $n \geq 0$. In Table 3.2 we display the relative error defined by

$$\frac{r_N - r_0}{r_0}. \tag{3.68}$$

We observe that the errors listed in Table 3.2 are much smaller than those in Table 3.1. Hence, we conclude that the Crank–Nicolson scheme generates better solutions than that of the explicit scheme. In fact, it can be shown that the Crank–Nicolson scheme produces the exact solution in this case. The errors in Table 3.2 are therefore due to round-off errors in the computations.

3.5 Exercises

Exercise 3.1. Consider the system

$$\begin{aligned}
F' &= (2 - S)F, \quad F(0) = F_0, \\
S' &= (F - 1)S, \quad S(0) = S_0,
\end{aligned} \tag{3.69}$$

and the numerical scheme

$$\begin{aligned}
F_{n+1} &= F_n + \Delta t(2 - S_n)F_n, \\
S_{n+1} &= S_n + \Delta t(F_n - 1)S_n.
\end{aligned} \tag{3.70}$$

(a) Write a computer program that implements the scheme (3.70). The program should:

- Accept S_0, F_0, and Δt as input.
- Compute the numerical solution for t ranging from 0 to 10.
- Plot the numerical solution both as a function of t and in the state space (the F–S coordinate system).

(b) Use your program to determine a value of Δt such that if you reduce Δt, you will not see any difference on your screen. In these computations you may use $F_0 = 1.9$, $S_0 = 0.1$.

(c) Choose the value of Δt that you found in (b) and use the program to play around with various values of F_0 and S_0. Are the solutions always periodic? What happens if you choose $F_0 = 1$ and $S_0 = 2$?

$\diamond$

Exercise 3.2. We observed above that the solution of the simple system

$$\begin{aligned} F' &= 1 - S, \quad F(0) = F_0, \\ S' &= F - 1, \quad S(0) = S_0, \end{aligned} \tag{3.71}$$

satisfies

$$r(t) = r(0), \tag{3.72}$$

where

$$r(t) = (F(t) - 1)^2 + (S(t) - 1)^2. \tag{3.73}$$

The purpose of this exercise is to derive a similar property for the system

$$\begin{aligned} F' &= (2 - S)F, \quad F(0) = F_0, \\ S' &= (F - 1)S, \quad S(0) = S_0. \end{aligned} \tag{3.74}$$

(a) Show that

$$\left(1 - \frac{1}{F}\right) F' = \left(\frac{2}{S} - 1\right) S', \tag{3.75}$$

where F and S solve (3.74).

(b) Integrate (3.75) from 0 to t and show that

$$F(t) - \ln(F(t)) - (F_0 - \ln(F_0)) = 2\ln(S(t)) - S(t) - (2\ln(S_0) - S_0). \tag{3.76}$$

(c) Show that

$$\frac{e^F}{F} \frac{e^S}{S^2} = \frac{e_0^F}{F_0} \frac{e_0^S}{S_0^2}. \tag{3.77}$$

(d) Choose $F_0 = 1.9$, $S_0 = 0.1$, and define the constant

$$K_0 = \frac{e_0^F}{F_0} \frac{e_0^S}{S_0^2}. \tag{3.78}$$

Extend your program from Exercise 3.1 to compute

$$K_n = \frac{e_n^F \, e_n^S}{F_n \, S_n^2}$$

and generate a plot of $\frac{K_n - K_0}{K_0}$ for $\Delta t = 1/100$ and $\Delta t = 1/500$. Argue that K_n seems to converge toward a constant, for all n, as Δt becomes smaller.

(e) Let $\Delta t = 10/N$. Use the program you developed in (d) to compute

$$\frac{K_N - K_0}{K_0 \Delta t}$$

for $N = 100, 200, 300, 400, 500$. Argue that

$$\frac{K_N - K_0}{K_0} \approx c \, \Delta t,$$

where c is independent of Δt.

◇

Exercise 3.3. Consider the following system of ODEs:

$$\begin{aligned} u' &= -v^3, \quad u(0) = u_0, \\ v' &= u^3, \quad v(0) = v_0. \end{aligned} \tag{3.79}$$

(a) Define

$$r(t) = u^4(t) + v^4(t), \tag{3.80}$$

and show that

$$r(t) = r(0) \tag{3.81}$$

for all $t \geq 0$.

(b) Derive the following explicit finite difference scheme for system (3.79):

$$\begin{aligned} u_{n+1} &= u_n - \Delta t \, v_n^3, \\ v_{n+1} &= v_n + \Delta t \, u_n^3. \end{aligned} \tag{3.82}$$

(c) Write a program that implements the scheme (3.82). The program should:

- Accept u_0, v_0, N, and T as input.
- Compute the numerical solution defined by (3.82) using $\Delta t = T/N$ for $n = 0, 1, \ldots, N$.
- Plot the numerical solution of u and v both as functions of t and in the state space.

(d) Use your program to check whether (3.81) holds, or holds approximately, for the numerical solutions.

◇

Exercise 3.4. We have seen above that the solution of the system

$$F' = (2 - S)F, \quad F(0) = F_0,$$
$$S' = (F - 1)S, \quad S(0) = S_0, \tag{3.83}$$

is periodic. Consequently, there is a time $T > 0$ such that

$$F(T) = F(0) \quad \text{and} \quad S(T) = S(0). \tag{3.84}$$

(a) Use (3.83) to show that

$$\ln(F(t)) - \ln F_0 = 2t - \int_0^t S(\tau)d\tau, \tag{3.85}$$

and

$$\ln(S(t)) - \ln S_0 = \int_0^t F(\tau)d\tau - t. \tag{3.86}$$

(b) Use (3.85) and (3.86) to show that

$$\frac{1}{T}\int_0^T F(\tau)d\tau = 1 \tag{3.87}$$

and

$$\frac{1}{T}\int_0^T S(\tau)d\tau = 2. \tag{3.88}$$

This means that we are able to compute the average amount of fish and sharks as t goes from 0 to T without actually solving the system (3.83).
(c) Let us introduce fishing to model (3.83). We achieve this by adding a term $-\varepsilon F$ to the first equation and δS to the second equation. Here, ε and δ are positive numbers modeling the relative changes in the fish and shark populations due to fishing. We thus get a new system

$$F' = (2 - S)F - \varepsilon F, \quad F(0) = F_0,$$
$$S' = (F - 1)S - \delta S, \quad S(0) = S_0. \tag{3.89}$$

Show that

$$\ln(F(t)) - \ln F_0 = (2 - \varepsilon)t - \int_0^t S(\tau)d\tau \qquad (3.90)$$

and

$$\ln(S(t)) - \ln S_0 = \int_0^t F(\tau)d\tau - (1 + \delta)t. \qquad (3.91)$$

(d) Assume again that we have a periodic solution such that

$$F(T) = F(0) \quad \text{and} \quad S(T) = S(0) \qquad (3.92)$$

for some $T > 0$. Use (3.90), (3.91), and (3.92) to show that

$$\frac{1}{T} \int_0^T F(\tau)d\tau = 1 + \delta, \qquad (3.93)$$

and

$$\frac{1}{T} \int_0^T S(\tau)d\tau = 2 - \varepsilon. \qquad (3.94)$$

(e) Use (3.93) and (3.94) to conclude that fishing increases the average amount of fish and decreases the average amount of sharks. This helps to explain the effect observed in the upper Adriatic Sea during World War I (see page 76).

$\diamond$

3.6 Project: Analysis of a Simple System

In the text above we observed that we are able to derive quite a number of properties for a certain system of ODEs. The purpose of this project is to teach you to do the same steps on your own. In order to do this, we will study a very simple 2×2 system and try to derive many properties, both numerically and analytically. Throughout this project we will consider the system

$$\begin{aligned}
u'(t) &= -v(t) & u(0) &= u_0, \\
v'(t) &= u(t) & v(0) &= v_0,
\end{aligned} \qquad (3.95)$$

where u_0 and v_0 are given.

(a) Derive, using Taylor series, the following explicit numerical scheme for system (3.95),

$$
\begin{aligned}
u_{n+1} &= u_n - \Delta t v_n, \\
v_{n+1} &= v_n + \Delta t u_n.
\end{aligned}
\tag{3.96}
$$

Write a program that implements this scheme. The program should accept u_0 and v_0 and Δt as input parameters.

(b) Choose $u_0 = 1$ and $v_0 = 0$. Use $\Delta t = 1/100$ and compute the solution from $t = 0$ to $t = 5$. Plot the numerical solution as a function of t and make a graph of it in the u–v space.

(c) Define

$$
r_n = u_n^2 + v_n^2.
\tag{3.97}
$$

Compute r_n for the time interval given above and plot the values. What properties do you think the solution u and v have?

(d) Let N be such that $N\Delta t = 5$. Compute

$$
\frac{r_N - r_0}{r_N \, \Delta t}
\tag{3.98}
$$

for $\Delta t = 1/100, 1/200, 1/300, 1/400$, and $1/500$. Show that there is a constant c such that

$$
\frac{r_N - r_0}{r_N} \approx c \, \Delta t.
\tag{3.99}
$$

(e) You have obtained indications that u and v define a circle in the state space with radius $\left(u_0^2 + v_0^2\right)^{1/2}$ and with center in $(0, 0)$. In order to investigate this from an analytical point of view, we introduce the function

$$
r(t) = u^2(t) + v^2(t).
\tag{3.100}
$$

Use the differential equations in (3.95) to show that

$$
r'(t) = 0
$$

for all t, and conclude that

$$
u^2(t) + v^2(t) = u_0^2 + v_0^2
\tag{3.101}
$$

for all $t \geq 0$.

(f) As we saw in the text above, (3.101) can also be derived directly from system (3.95). Show that

$$u'(t)u(t) = -v'(t)v(t) \tag{3.102}$$

and use direct integration to show that

$$u^2(t) + v^2(t) = u_0^2 + v_0^2. \tag{3.103}$$

(g) Suppose we consider system (3.95) with slightly different initial conditions, i.e., we consider the following system

$$\begin{aligned}
\bar{u}'(t) &= -\bar{v}(t) \quad \bar{u}(0) = \bar{u}_0, \\
\bar{v}'(t) &= \bar{u}(t) \quad \bar{v}(0) = \bar{v}_0.
\end{aligned} \tag{3.104}$$

Define

$$U = u - \bar{u}, \quad V = v - \bar{v} \tag{3.105}$$

and show that

$$\begin{aligned}
U'(t) &= -V(t), \\
V'(t) &= U(t),
\end{aligned} \tag{3.106}$$

and use this to conclude that

$$(u(t) - \bar{u}(t))^2 + (v(t) - \bar{v}(t))^2 = (u_0 - \bar{u}_0)^2 + (v_0 - \bar{v}_0)^2. \tag{3.107}$$

What does this equation tell you about the stability of system (3.95) with respect to perturbations in the initial conditions?

(h) Consider the system

$$\begin{aligned}
u'(t) &= -v(t) \quad u(0) = 1, \\
v'(t) &= u(t) \quad v(0) = 0.
\end{aligned} \tag{3.108}$$

Show that

$$\begin{aligned}
u(t) &= \cos(t), \\
v(t) &= \sin(t)
\end{aligned} \tag{3.109}$$

solves this system.

(i) Use the analytical solution (3.109) to test the accuracy of the scheme considered in (a). More specifically, compute

$$e_{\Delta t} = \frac{|u(5) - u_N|}{|u(5)|} + \frac{|v(5) - v_N|}{|v(5)|}, \tag{3.110}$$

where $N \cdot \Delta t = 5$ for $\Delta t = 1/100, 1/200, 1/300, 1/400,$ and $1/500$. Compute also $\frac{e_{\Delta t}}{\Delta t}$ and show that

$$e_{\Delta t} \approx c \, \Delta t,$$

where c is a constant independent of Δt.

(j) Derive the Crank–Nicolson scheme for system (3.108). Show that the system can be written in the form

$$
\begin{aligned}
u_{n+1} &= \frac{1}{1+\frac{\Delta t^2}{4}} \left(\left(1 - \Delta t^2/4\right) u_n - \Delta t \, v_n \right), \\
v_{n+1} &= \frac{1}{1+\frac{\Delta t^2}{4}} \left(\Delta t \, u_n + \left(1 - \Delta t^2/4\right) v_n \right).
\end{aligned}
\tag{3.111}
$$

(k) Implement scheme (3.111) and redo the computations in (i) using this scheme. Compute $\frac{e_{\Delta t}}{\Delta t^2}$, where $e_{\Delta t}$ is given by (3.110), and conclude that

$$e_{\Delta t} \approx d \, \Delta t^2,$$

where d is independent of Δt.

◇

Chapter 4
Nonlinear Algebraic Equations

Suppose we want to solve the ordinary differential equation (ODE)

$$u' = g(u), \quad u(0) = u_0, \tag{4.1}$$

by an implicit scheme. The standard implicit Euler method for (4.1), cf. Sect. 2.2.4, reads

$$u_{n+1} - \Delta t \, g(u_{n+1}) = u_n, \tag{4.2}$$

where $u_n \approx u(t_n)$ and $t_n = n \Delta t$ for a small time step $\Delta t > 0$. We note that in order to compute u_{n+1} based on u_n, we have to solve (4.2). Let us look a bit closer at this. For simplicity we let c denote u_n and v denote u_{n+1}, so we need to solve

$$v - \Delta t \, g(v) = c, \tag{4.3}$$

where c is given and we want to find v.

Let us first consider the case of

$$g(u) = u, \tag{4.4}$$

which corresponds to the ODE

$$u' = u, \quad u(0) = u_0. \tag{4.5}$$

Then (4.3) can be rewritten as

$$v - \Delta t \, v = c, \tag{4.6}$$

which has the solution

$$v = \frac{1}{1 - \Delta t} c. \tag{4.7}$$

A. Tveito et al., *Elements of Scientific Computing*, Texts in Computational Science and Engineering 7, DOI 10.1007/978-3-642-11299-7_4,
© Springer-Verlag Berlin Heidelberg 2010

Since $u_{n+1} = v$ and $u_n = c$, we get the solution

$$u_{n+1} = \frac{1}{1 - \Delta t} u_n. \tag{4.8}$$

Similarly, for any function g that is in the form

$$g(u) = \alpha + \beta u, \tag{4.9}$$

where α and β are constants, we can solve (4.3) directly and get

$$v = \frac{c + \alpha \Delta t}{1 - \beta \Delta t} \quad \text{or} \quad u_{n+1} = \frac{u_n + \alpha \Delta t}{1 - \beta \Delta t}. \tag{4.10}$$

Moreover, let us also consider the case of

$$u' = u^2, \tag{4.11}$$

that is

$$g(v) = v^2. \tag{4.12}$$

Then (4.3) reads

$$v - \Delta t \, v^2 = c, \tag{4.13}$$

and we have two possible solutions

$$v_+ = \frac{1 + \sqrt{1 - 4\Delta t \, c}}{2\Delta t} \tag{4.14}$$

and

$$v_- = \frac{1 - \sqrt{1 - 4\Delta t \, c}}{2\Delta t}. \tag{4.15}$$

If we expand[1] (4.14) and (4.15) in terms of Δt, we get

$$v_+ = \frac{1}{\Delta t} - c - c^2 \Delta t + \mathcal{O}(\Delta t^2) \tag{4.18}$$

---

[1] What do we mean by "expanding" an expression in terms of something? It means that we derive a series expansion in terms of a certain parameter. Suppose we are interested in the function

$$\sin(e^h - 1) \tag{4.16}$$

for small values of h. We can compute the first few terms in a Taylor series, e.g.,

$$\sin(e^h - 1) \approx h + \frac{1}{2}h^2 + \mathcal{O}(h^3). \tag{4.17}$$

Here, (4.16) is expanded in terms of h.

and

$$v_- = c + c^2 \Delta t + \mathcal{O}(\Delta t^2).$$

(4.19)

Here, (4.18) is an unreasonable solution because $1/\Delta t$ becomes infinite as Δt goes to zero. The solution v_- behaves well, see (4.19). We therefore have, from (4.15), that

$$v = \frac{1 - \sqrt{1 - 4\Delta t\, c}}{2\Delta t}.$$

(4.20)

Based on this, we conclude that the implicit scheme

$$u_{n+1} - \Delta t\, u_{n+1}^2 = u_n$$

(4.21)

can be written in the computational form

$$u_{n+1} = \frac{1 - \sqrt{1 - 4\Delta t\, u_n}}{2\Delta t}.$$

(4.22)

To summarize, we have seen that the equation

$$v - \Delta t\, g(v) = c$$

(4.23)

can easily be solved analytically when

$$g(v) = v$$

(4.24)

or

$$g(v) = v^2.$$

(4.25)

More generally, we can solve (4.23) for linear or quadratic functions g, i.e., functions of the form

$$g(v) = \alpha + \beta v + \gamma v^2.$$

(4.26)

But what happens if

$$g(v) = e^v$$

(4.27)

or

$$g(v) = \sin(v)\,?$$

(4.28)

It turns out that we can solve (4.23) analytically only in very few cases. In other cases, we need to compute the solution by some sort of numerical scheme. The purpose of this chapter is to develop numerical methods for equations that cannot be solved directly by analytical expressions.

Before we start developing methods, let us just consider (4.2) once more. Since

$$u_{n+1} - u_n = \Delta t \, g(u_{n+1}) \qquad (4.29)$$

and Δt is a small number, we know that u_{n+1} is close to u_n, provided that g is bounded. This turns out to be important when we want to construct numerical methods for solving the equation. The reason for this is that we will develop iterative methods. Such methods depend on a good initial guess of the solution. We have just seen that, for finding u_{n+1}, the previous value u_n is a good initial guess.

In the rest of this chapter we study methods for solving equations of the form

$$f(x) = 0, \qquad (4.30)$$

where f is a nonlinear function. We will also study systems of such equations later. Solutions of (4.30) are called zeros, or roots. Motivated by the discussion above, we will assume that we know a value x_0 close to x^*, where

$$f(x^*) = 0. \qquad (4.31)$$

Furthermore, we assume that f has no other zeros in a small region around x^*.

4.1   The Bisection Method

Consider the function

$$f(x) = 2 + x - e^x \qquad (4.32)$$

for x ranging from 0 to 3, see the graph in Fig. 4.1.

We want to compute $x = x^*$ such that

$$f(x^*) = 0. \qquad (4.33)$$

Let

$$x_0 = 0 \qquad (4.34)$$

and

$$x_1 = 3. \qquad (4.35)$$

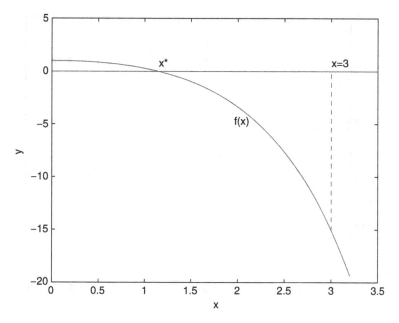

Fig. 4.1 The graph of $f(x) = 2 + x - e^x$

We note that $f(x_0) = f(0) > 0$ and $f(x_1) = f(3) < 0$. Next, we define the mean value

$$x_2 = \frac{1}{2}(x_0 + x_1) = \frac{3}{2}. \tag{4.36}$$

Since we are looking for a point where f is zero, our next step thus depends on the value of $f(x_2)$. Obviously, if $f(x_2)$ is zero, we just put $x^* = x_2$, and our search is completed. But $f(x_2)$ is not zero. In fact,

$$f(x_2) = f\left(\frac{3}{2}\right) = 2 + 3/2 - e^{3/2} < 0, \tag{4.37}$$

see Fig. 4.2.

Since $f(x_0) > 0$ and $f(x_2) < 0$, we know that $x_0 < x^* < x_2$. Next, we define

$$x_3 = \frac{1}{2}(x_0 + x_2) = \frac{3}{4}.$$

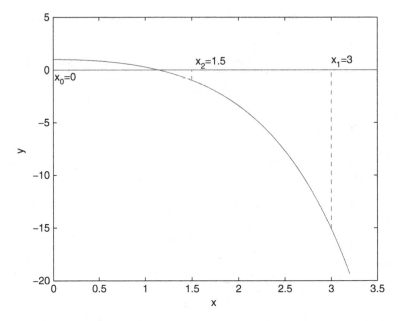

Fig. 4.2 The graph of $f(x) = 2 + x - e^x$ and three values of f: $f(x_0)$, $f(x_1)$, and $f(x_2)$

Since $f(x_3) > 0$, we know that $x_3 < x^* < x_2$, see Fig. 4.3.

Now, we can continue in this way until $|f(x_n)|$ is sufficiently small. For each step in this iterative procedure, we have one point a where $f(a) > 0$ and another point b where $f(b) < 0$. Then, we define $c = \frac{1}{2}(a + b)$. By checking the sign of $f(c)$, we proceed by choosing either a and c or b and c for the next step. In algorithmic form this reads as follows:

Algorithm 4.1

Bisection Method.
Given a, b such that $f(a) \cdot f(b) < 0$ and a tolerance ε.
Define $c = \frac{1}{2}(a + b)$.

while $|f(c)| > \varepsilon$ **do**
  if $f(a) \cdot f(c) < 0$
  then $b = c$
  else $a = c$
  $c := \frac{1}{2}(a + b)$
end

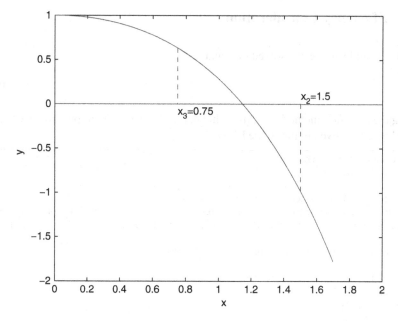

Fig. 4.3 The graph of $f(x) = 2 + x - e^x$ and two values of f: $f(x_2)$ and $f(x_3)$

Table 4.1 Solving the nonlinear equation $f(x) = 2 + x - e^x = 0$ by using Algorithm 4.1; the number of iterations i, c and $f(c)$

i	c	$f(c)$
1	1.500000	−0.981689
2	0.750000	0.633000
4	1.312500	−0.402951
8	1.136719	0.0201933
16	1.146194	$-2.65567 \cdot 10^{-6}$
21	1.146193	$4.14482 \cdot 10^{-7}$

Example 4.1. Consider

$$f(x) = 2 + x - e^x \tag{4.38}$$

and choose $a = 0$, $b = 3$, and $\varepsilon = 10^{-6}$. In Table 4.1, we show the number of iterations i, c and $f(c)$ by Algorithm 4.1 to solve the nonlinear equation

$$f(x) = 0.$$

The number of iterations refers to the number of times we pass through the while-loop of the algorithm.

4.2 Efficiency Consideration

In the example above, we solved the equation

$$2 + x - e^x = 0 \tag{4.39}$$

using the bisection method, see Algorithm 4.1. For the actual computation, we used
the C program given in the frame below.

```c
#include <stdio.h>
#include <math.h>

double f (double x) { return 2.0+x-exp(x); }
/* we define function 'fabs' for calculating absolute values */
inline double fabs (double r) { return ( (r >= 0.0) ? r : -r ); }

int main (int nargs, const char** args)
{
  double epsilon = 1.0e-6;
  double a, b, c, fa, fc;
  a = 0.; b = 3.;
  fa = f(a);
  c = 0.5*(a+b);
  while (fabs(fc=(f(c))) > epsilon) {
    if ((fa*fc) < 0) {
      b = c;
    }
    else {
      a = c;
      fa = fc;
    }
    c = 0.5*(a+b);
  }
  printf("final c=%g, f(c)=%g\n",c,fc);

  return 0;
}
```

The entire computation requires $5.82 \cdot 10^{-6}$ s on a Pentium III 1 GHz processor.
But in actual computations, we often have to solve equations of the form (4.39)
over and over again. Suppose we want to solve a three-dimensional problem with
variables depending on time. If we just guess that we need 10^3 points in all spatial
directions and also 10^3 time steps, then we can end up with solving equations of the
form (4.39) $10^3 \cdot 10^3 \cdot 10^3 \cdot 10^3 = 10^{12}$ times. Now, $10^{12} \times 5.82 \cdot 10^{-6} = 5.82 \cdot 10^6$ s
is about $1,617$ h (67.36 days), so waiting can be quite agonizing. Furthermore, you
may want to use your code in some sort of real-time application where you need your
results very quickly. In such time-dependent applications, you will have better initial
guesses than indicated in the example above, but you will still want to minimize your
CPU efforts. In the next section we will derive a method that converges much faster
than the bisection method.

4.3 Newton's Method

We observed above that we can solve nonlinear algebraic equations of the form

$$f(x) = 0 \tag{4.40}$$

by just searching in a simple and systematic manner. This is fine if you only need to solve the equation once or a few times. But in applications where variants of such equations need to be solved billions of times, we need methods that converge faster than the bisection method.

As discussed above, we search for a value x^* such that

$$f(x^*) = 0, \tag{4.41}$$

and we often know a point x_0 close to x^*. As mentioned above, this is the case for nonlinear equations of the form

$$u_{n+1} - \Delta t \, g(u_{n+1}) = u_n, \tag{4.42}$$

where we know that the unknown u_{n+1} is quite close to u_n, which is given. We can also assume that in a small region around x^*, the function $f = f(x)$ has only one zero and that the derivative is always non-zero in this region. By a Taylor series expansion around the given point $x = x_0$, we have

$$f(x_0 + h) = f(x_0) + hf'(x_0) + \mathcal{O}(h^2). \tag{4.43}$$

So, for small values of h, we have

$$f(x_0 + h) \approx f(x_0) + hf'(x_0). \tag{4.44}$$

Now we want to choose the step h such that $f(x_0 + h) \approx 0$. We can do that by choosing h such that

$$f(x_0) + hf'(x_0) = 0, \tag{4.45}$$

and consequently

$$h = -\frac{f(x_0)}{f'(x_0)}. \tag{4.46}$$

We can now define

$$x_1 = x_0 + h = x - \frac{f(x_0)}{f'(x_0)}. \tag{4.47}$$

Let us try to do this for the example studied above. We have

$$f(x) = 2 + x - e^x$$

and we choose $x_0 = 3$. Since

$$f'(x) = 1 - e^x,$$

so

$$x_1 = x_0 - \frac{f(x_0)}{f'(x_0)} = 3 - \frac{5 - e^3}{1 - e^3} = 2.2096.$$

Note that

$$|f(x_0)| = |f(3)| \approx 15.086$$

and

$$|f(x_1)| = |f(2.2096)| \approx 4.902,$$

so the value of f is significantly reduced. Of course, the steps above can be repeated to obtain the following algorithm.

Algorithm 4.2

Newton's Method.
Given an initial approximation x_0 and a tolerance ε.

$k = 0$
while $|f(x_k)| > \varepsilon$ **do**
$$x_{k+1} = x_k - \frac{f(x_k)}{f'(x_k)}$$
$\quad k = k + 1$
end

In Table 4.2 we give values of the number of iterations k, x_k, and $f(x_k)$. In the computations we have used $x_0 = 3$ and $\varepsilon = 10^{-6}$.

We note that the convergence is much faster than that of the bisection method. This fast convergence does not occur just by accident. In fact, Newton's method usually converges quite rapidly and we will study this closer both by theory and by examples. The theory is presented in Project 4.8.

Table 4.2 Solving the nonlinear equation $f(x) = 2 + x - e^x = 0$ by Algorithm 4.2; the number of iterations k, x_k and $f(x_k)$

k	x_k	$f(x_k)$
1	2.209583	-4.902331
2	1.605246	-1.373837
3	1.259981	-0.265373
4	1.154897	$-1.880020 \cdot 10^{-2}$
5	1.146248	$-1.183617 \cdot 10^{-4}$
6	1.146193	$-4.783945 \cdot 10^{-9}$

Example 4.2. Let

$$f(x) = x^2 - 2.$$

We want to find x^* such that

$$f(x^*) = 0.$$

One of the two solutions of this problem is given by

$$x^* = \sqrt{2},$$

and Newton's method is given by

$$x_{k+1} = x_k - \frac{x_k^2 - 2}{2x_k},$$

or equivalently,

$$x_{k+1} = \frac{x_k^2 + 2}{2x_k},$$

for $k = 0, 1, 2, \ldots$.
   If we choose $x_0 = 1$, we get

$$x_1 = 1.5,$$
$$x_2 = 1.41667,$$
$$x_3 = 1.41422.$$

This should be compared with

$$x^* = \sqrt{2} \approx 1.41421,$$

so in just three iterations we get a very accurate approximation. ∎

An Alternative Derivation

Let us first briefly recall the derivation of Newton's method. We assume that x_0 is a given value close to x^*, where x^* satisfies

$$f(x^*) = 0. \tag{4.48}$$

By a Taylor expansion, we have

$$f(x_0 + h) \approx f(x_0) + hf'(x_0). \tag{4.49}$$

We note that

$$f(x_1) = f(x_0 + h) \approx 0 \tag{4.50}$$

can be obtained by choosing

$$h = -\frac{f(x_0)}{f'(x_0)}. \tag{4.51}$$

This motivates the choice

$$x_1 = x_0 - \frac{f(x_0)}{f'(x_0)}, \tag{4.52}$$

and more generally we have

$$x_{k+1} = x_k - \frac{f(x_k)}{f'(x_k)}. \tag{4.53}$$

The key observation in this derivation is to utilize the Taylor series (4.49).

We can also use the Taylor series in a somewhat different manner. Recall from (1.39) on page 20 that we have the Taylor series of f in the form

$$f(x) = f(x_0) + (x - x_0)f'(x_0) + \mathcal{O}((x - x_0)^2). \tag{4.54}$$

Let us now define the linear model

$$F_0(x) = f(x_0) + (x - x_0)f'(x_0) \tag{4.55}$$

of f around $x = x_0$. In Fig. 4.4 we see that F_0 approximates f around $x = x_0$. What are the properties of $F_0 = F_0(x)$? We first note that

$$F_0(x_0) = f(x_0). \tag{4.56}$$

Second, we observe that

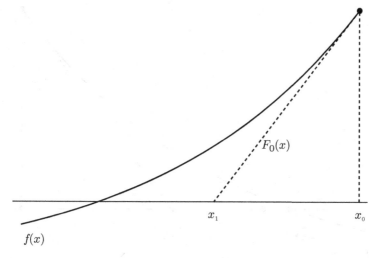

$f(x)$

Fig. 4.4 The figure shows the function f, a linear approximation $F_0(x)$, and how to compute x_1 from x_0

$$F_0'(x) = f'(x_0),$$

so, in particular,

$$F_0'(x_0) = f'(x_0).$$

Hence, both the value and the derivative of f and F_0 coincide at the point $x = x_0$. If we now define x_1 to be such that

$$F_0(x_1) = 0, \tag{4.57}$$

we get

$$f(x_0) + (x_1 - x_0)f'(x_0) = 0, \tag{4.58}$$

and thus

$$x_1 = x_0 - \frac{f(x_0)}{f'(x_0)}, \tag{4.59}$$

which is identical to (4.52). Again, we can repeat this process by defining a linear approximation of f around x_1,

$$F_1(x) = f(x_1) + (x - x_1)f'(x_1). \tag{4.60}$$

By defining x_2 to be such that

$$F_1(x_2) = 0, \tag{4.61}$$

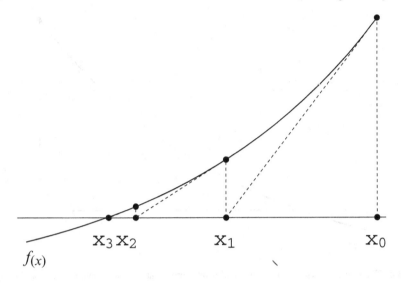

Fig. 4.5 A graphical illustration of Newton's method

we get

$$x_2 = x_1 - \frac{f(x_1)}{f'(x_1)}. \tag{4.62}$$

More generally, we have

$$x_{k+1} = x_k - \frac{f(x_k)}{f'(x_k)}. \tag{4.63}$$

This process is illustrated in Fig. 4.5.

We will see in the next section that this derivation of Newton's method motivates the secant method, for which we do not need an explicit evaluation of $f'(x)$.

4.4 The Secant Method

We noted above that if we have a linear model $F = F(x)$ of $f = f(x)$, we can solve $F(x) = 0$ in order to find the approximation of the solution of $f(x) = 0$. The secant method is based on the same idea, but instead of using the Taylor series, we use interpolation based on a linear function. Suppose we have two values x_0 and x_1 close to x^*, where

$$f(x^*) = 0. \tag{4.64}$$

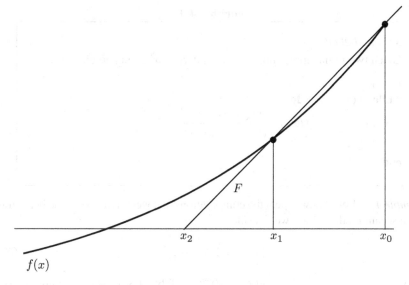

$f(x)$

Fig. 4.6 The figure shows a function $f = f(x)$ and its linear interpolant F between x_0 and x_1

Then we can define the linear function

$$F_0(x) = f(x_1) + \frac{f(x_1) - f(x_0)}{x_1 - x_0}(x - x_1),$$ (4.65)

which satisfies $F_0(x_0) = f(x_0)$ and $F_0(x_1) = f(x_1)$. We have plotted a function f and its linear interpolant F in Fig. 4.6.

Since we have

$$F_0(x) \approx f(x),$$ (4.66)

we can compute an approximation x_2 of x^* by solving

$$F_0(x_2) = 0.$$ (4.67)

The solution of

$$f(x_1) + \frac{f(x_1) - f(x_0)}{x_1 - x_0}(x_2 - x_1) = 0$$

with respect to x_2 is given by

$$x_2 = x_1 - \frac{f(x_1)(x_1 - x_0)}{f(x_1) - f(x_0)}.$$

Generally, we have the following algorithm.

Algorithm 4.3

Secant Method.
Given two initial approximations x_0 and x_1 of x^*, and a tolerance ε.

$k = 1$
while $|f(x_k)| > \varepsilon$ **do**

$$x_{k+1} = x_k - f(x_k) \frac{(x_k - x_{k-1})}{f(x_k) - f(x_{k-1})}$$

$\quad k = k + 1$
end

Example 4.3. Let us now repeat the computations that were used to test the behavior of the secant method. First we consider

$$f(x) = 2 + x - e^x \tag{4.68}$$

and we use $x_0 = 0$, $x_1 = 3$ and $\varepsilon = 10^{-6}$. In Table 4.3 we show the number of iterations k, x_k, and $f(x_k)$ as computed by Algorithm 4.3. Note that convergence is somewhat slower than that of Newton's method (cf. Table 4.2 on page 109), but it is faster than the bisection method (cf. Table 4.1 on page 105). ∎

Example 4.4. Let

$$f(x) = x^2 - 2. \tag{4.69}$$

Our aim is to compute approximations of x^* satisfying

$$f(x^*) = 0, \tag{4.70}$$

Table 4.3 Solving the nonlinear equation $f(x) = 2 + x - e^x = 0$ by Algorithm 4.3: the number of iterations k, x_k, and $f(x_k)$

k	x_k	$f(x_k)$
2	0.186503	0.981475
3	0.358369	0.927375
4	3.304511	−21.930701
5	0.477897	0.865218
6	0.585181	0.789865
7	1.709760	−1.817874
8	0.925808	0.401902
9	1.067746	0.158930
10	1.160589	$-3.122466 \cdot 10^{-2}$
11	1.145344	$1.821544 \cdot 10^{-3}$
12	1.146184	$1.912908 \cdot 10^{-5}$
13	1.146193	$-1.191170 \cdot 10^{-8}$

where one of the zeros is $x^* = \sqrt{2}$. The secant method for this problem reads

$$
\begin{aligned}
x_{k+1} &= x_k - f(x_k) \frac{x_k - x_{k-1}}{f(x_k) - f(x_{k-1})} \\
&= x_k - (x_k^2 - 2) \frac{x_k - x_{k-1}}{x_k^2 - x_{k-1}^2} \\
&= x_k - \frac{x_k^2 - 2}{x_k + x_{k-1}},
\end{aligned}
$$

so

$$
x_{k+1} = \frac{x_k x_{k-1} + 2}{x_k + x_{k-1}}. \tag{4.71}
$$

We choose $x_0 = 1$ and $x_1 = 2$, and get

$$
\begin{aligned}
x_2 &= 1.33333, \\
x_3 &= 1.40000, \\
x_4 &= 1.41463.
\end{aligned}
$$

This should be compared with

$$
x^* = \sqrt{2} \approx 1.41421.
$$

Recall also that three iterations of Newton's method gives 1.41422, so again we note that Newton's method is slightly more accurate than the secant method.

■

4.5 Fixed-Point Iterations

We started this chapter by considering the ODE

$$
u' = g(u), \quad u(0) = u_0. \tag{4.72}
$$

An implicit scheme for this problem reads

$$
u_{n+1} - \Delta t\, g(u_{n+1}) = u_n, \tag{4.73}
$$

where $u_n \approx u(t_n)$ and $t_n = n\Delta t$ for a given time step $\Delta t > 0$. We note that (4.73) is an equation with u_{n+1} as unknown. By setting $v = u_{n+1}$ and $c = u_n$, we obtain the equation

$$
v - \Delta t\, g(v) = c, \tag{4.74}
$$

where c is given and v is unknown. We can rewrite this equation in the form

$$v = h(v), \tag{4.75}$$

where

$$h(v) = c + \Delta t\, g(v).$$

For equations on this form, one can apply a very simple iteration in order to find an approximation of the solution $v = v^*$ satisfying (4.75). The iteration reads

$$v_{k+1} = h(v_k), \tag{4.76}$$

where v_0 is an approximate initial guess.[2]
  A solution v^* of (4.75) satisfies

$$v^* = h(v^*).$$

Since h leaves v^* unchanged, $h(v^*) = v^*$, and the value v^* is referred to as a *fixed-point* of h. Furthermore, since the purpose of iteration (4.76) is to compute v^*, (4.76) is referred to as a *fixed-point iteration*.
  Let us use (4.76) to generate numerical approximations of one specific equation:

$$x = \sin(x/10). \tag{4.77}$$

In Fig. 4.7 we have plotted the graph of $y = x$ and $y = \sin(x/10)$ and we note that $x = 0$ is the only solution of (4.77).

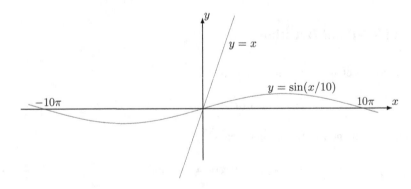

Fig. 4.7 The graph of $y = x$ and $y = \sin(x/10)$

---

[2] It is important to note that (4.76) is an iteration to solve (4.75), i.e. to solve (4.73) at *one* specific time level. Thus, we have an iterative procedure within each time step.

Let us choose $x_0 = 1.0$ and compute approximations to the solution of (4.77) using the iteration

$$x_{k+1} = \sin(x_k/10). \tag{4.78}$$

We get the following results:

$$x_0 = 1.0,$$
$$x_1 = 0.09983,$$
$$x_2 = 0.00998,$$
$$x_3 = 0.00099,$$

which seems to converge nicely toward $x = 0$ being the solution of (4.77). For this particular problem, we can understand this behavior by observing that, for small x,[3]

$$\sin(x/10) \approx x/10, \tag{4.79}$$

and thus, from (4.78),

$$x_{k+1} \approx x_k/10, \tag{4.80}$$

so

$$x_k \approx (1/10)^k, \tag{4.81}$$

which certainly converges toward zero.

4.5.1 Convergence of Fixed-Point Iterations

We have seen that an equation of the form

$$v = h(v) \tag{4.82}$$

can sometimes be solved by an iteration of the form

$$v_{k+1} = h(v_k). \tag{4.83}$$

---

[3] The Taylor series of the sine function gives

$$\sin(y) = y + \mathcal{O}(y^3).$$

Our aim is to analyze when and under what conditions the values $\{v_k\}$ generated by (4.83) converge toward a solution v^* of (4.82). To answer this question we introduce the concept of a *contractive mapping*. Here, we will refer to $h = h(v)$ as a contractive mapping on a closed interval I if

(a) $|h(v) - h(w)| \le \delta |v - w|$ for any $v, w \in I$, where $0 < \delta < 1$, and
(b) $v \in I \Rightarrow h(v) \in I$ for all $v \in I$.

For functions, as we consider in this chapter, we could refer to h as a contractive function. But it turns out that similar properties are useful for much more general mappings and we therefore use the standard term contractive mapping.

But what kind of functions are contractive mappings? To answer this, it is useful to recall the mean value theorem of calculus. It states that if f is a differentiable function defined on an interval $[a, b]$, then there exists a $c \in [a, b]$ such that

$$f(b) - f(a) = f'(c)(b - a). \tag{4.84}$$

It follows from (4.84) that h in (4.75) is a contractive mapping defined on an interval I if

$$\left| h'(\xi) \right| < \delta < 1 \quad \text{for all } \xi \in I, \tag{4.85}$$

and $h(v) \in I$ for all $v \in I$.

We observed above that the equation (see (4.77))

$$x = \sin(x/10) \tag{4.86}$$

could easily be solved by using the fixed-point iteration

$$x_{k+1} = \sin(x_k/10), \tag{4.87}$$

and we note that

$$h(x) = \sin(x/10) \tag{4.88}$$

is indeed contractive on $I = [-1, 1]$ because

$$\left| h'(x) \right| = \left| \frac{1}{10} \cos(x/10) \right| \le \frac{1}{10} \tag{4.89}$$

and also

$$x \in [-1, 1] \quad \Rightarrow \quad \sin(x/10) \in [-1, 1]. \tag{4.90}$$

Let us consider generally the fixed-point iteration

$$v_{k+1} = h(v_k) \tag{4.91}$$

as a method for finding v^* such that

$$v^* = h(v^*). \tag{4.92}$$

Here we assume that h is a contractive mapping, i.e., we assume that for any v, w in a closed interval I we have

$$|h(v) - h(w)| \leq \delta |v - w|, \quad \text{where } 0 < \delta < 1, \tag{4.93}$$
$$v \in I \implies h(v) \in I. \tag{4.94}$$

We define the error

$$e_k = |v_k - v^*| \tag{4.95}$$

and observe that

$$
\begin{aligned}
e_{k+1} &= |v_{k+1} - v^*| \\
&\overset{(4.91),(4.92)}{=} |h(v_k) - h(v^*)| \\
&\overset{(4.93)}{\leq} \delta |v_k - v^*| \\
&\overset{(4.95)}{=} \delta\, e_k.
\end{aligned}
$$

It now follows by induction on k, that

$$e_k \leq \delta^k e_0, \tag{4.96}$$

and since $0 < \delta < 1$, we know that $e_k \to 0$ as $k \to \infty$ and thus we have convergence:

$$\lim_{k \to \infty} v_k = v^*.$$

We have seen that if v^* is the solution of an equation of the form

$$v = h(v),$$

then the iteration

$$v_{n+1} = h(v_n)$$

provides a method of solving the equation numerically, provided that h is a contractive mapping.

4.5.2 Speed of Convergence

If h is a contractive mapping on I and v^* solves

$$v = h(v),\tag{4.97}$$

then we have seen that

$$\frac{e_k}{e_0} \le \delta^k,\tag{4.98}$$

with $e_k = |v_k - v^*|$. Let us now consider the equation

$$x = \sin(x/10).\tag{4.99}$$

We have seen that

$$\delta = \frac{1}{10},\tag{4.100}$$

cf. (4.89). Hence the error is

$$\frac{e_k}{e_0} \le \left(\frac{1}{10}\right)^k,\tag{4.101}$$

which is in agreement with our heuristic arguments on page 117.
    If we want

$$\frac{e_k}{e_0} \le 10^{-6},\tag{4.102}$$

we need $k \ge 6$ iterations. More generally, when

$$\frac{e_k}{e_0} \le \delta^k,\tag{4.103}$$

we have

$$\frac{e_k}{e_0} \le \varepsilon\tag{4.104}$$

if

$$\delta^k \le \varepsilon\tag{4.105}$$

or

$$k \ln(\delta) \le \ln(\varepsilon), \tag{4.106}$$

so we need the number of iterations to satisfy[4]

$$k \ge \frac{\ln(\varepsilon)}{\ln(\delta)}. \tag{4.107}$$

The estimate in (4.107) states that the number of iterations grows as ε is reduced. This is very reasonable: We need to work harder in order to get a more accurate approximation. But it is equally important to note that the number of iterations also increases as $\delta \to 1$. In fact, a smaller δ gives faster convergence. This is also clear from the basic estimate (4.103).

4.5.3 Existence and Uniqueness of a Solution

Given an equation of the form

$$v = h(v), \tag{4.108}$$

it is reasonable to ask the following questions:

(a) Does there exist a value v^* such that

$$v^* = h(v^*)?$$

(b) If so, is v^* unique?
(c) How can we compute v^*?

We now assume that h is a contractive mapping on a closed interval I such that

$$|h(v) - h(w)| \le \delta |v - w|, \quad \text{where } 0 < \delta < 1, \tag{4.109}$$
$$v \in I \implies h(v) \in I \tag{4.110}$$

for all v, w. As noted above, we have a straightforward way of computing approximations of v^*. We know that fixed-point iteration

$$v_{k+1} = h(v_k) \tag{4.111}$$

---

[4] We assume that $0 < \delta \ll 1$ and thus $\ln(\delta) < 0$.

gives good approximations, provided that a solution of (4.111) exists. But we do not know that it exists! We have not discussed existence yet. For a scalar equation of the form (4.108), it is usually sufficient to just graph the functions v and $h(v)$ and observe that they cross each other at one and only one point. This ensures both existence and uniqueness. But we will do this in a much harder way. The reason for this is that later we will study systems of equations of the form (4.108), i.e., equations where v is a vector and h is a vector of functions. In such cases we cannot analyze (a) and (b) above just by graphing. So by presenting a more theoretical argument already in the case of a scalar equation, you will find the argument in the case of systems easier to comprehend. But keep in mind that for any specific equation of the form

$$v = h(v), \tag{4.112}$$

or, more generally on the form,

$$F(v) = 0, \tag{4.113}$$

it is always instructive to graph the involved functions in order to understand what kind of problem you have at hand.

4.5.4 Uniqueness

Let us first see that (4.108) can only have one solution, provided that h is a contractive mapping satisfying (4.109) and (4.110). In order to see this, we simply assume that we have two solutions v^* and w^*, i.e., we assume that

$$v^* = h(v^*) \tag{4.114}$$

and

$$w^* = h(w^*). \tag{4.115}$$

Using (4.109) with $v = v^*$ and $w = w^*$, we have

$$|h(v^*) - h(w^*)| \leq \delta |v^* - w^*| \tag{4.116}$$

where $\delta < 1$. But by (4.114) and (4.115) we have

$$|v^* - w^*| \leq \delta |v^* - w^*|, \tag{4.117}$$

which can only hold when $v^* = w^*$, and consequently we can have only one solution. You should note that this is really highly nontrivial. We have (4.108) and just two very simple conditions (4.109) and (4.110). By using just these conditions, we know that (4.108) can only have *one* solution. This is really quite a remarkable result, but much more remarkable is the fact that it extends to huge systems of equations. We will encounter systems with billions of equations and unknowns, and we can use exactly the same argument to show that only one solution is possible.

4.5.5 Existence

We have seen that if h is a contractive mapping, then the equation

$$v = h(v)$$

can have only one solution. Now we will show an even stronger and more remarkable result: By assuming that (4.109) and (4.110) hold, we will show that (4.108) indeed has a solution. After doing this, we have answered all the questions (a), (b), and (c) on page 121. In order to do so we will need two mathematical results that we will just state here and refer you to read e.g. [6] in order to learn the proof. The first result is trivial and just states that

$$\sum_{k=1}^{n} \alpha^k = \frac{1 - \alpha^{n+1}}{1 - \alpha} \quad \text{and} \quad \sum_{k=1}^{\infty} \alpha^k = \frac{1}{1 - \alpha} \tag{4.118}$$

for $|\alpha| < 1$.

The second result is much more theoretical and is of fundamental importance in mathematics. This result concerns sequences of real numbers, say $\{v_k\}$. Such a collection of numbers is called a *Cauchy sequence* if, for any $\varepsilon > 0$, there exists an integer M such that for all $m, n \geq M$ we have

$$|v_m - v_n| < \varepsilon. \tag{4.119}$$

So for a Cauchy sequence, we can choose a tiny number ε, say 10^{-10}, and be assured that we can find an integer M such that

$$|v_m - v_n| < \varepsilon \tag{4.120}$$

if $m, n \geq M$. It turns out that this property picks up exactly the sequences that are convergent: A sequence $\{v_k\}$ converges if and only if it is a Cauchy sequence.

Our strategy is now to use the simple result (4.118) to show that the sequence generated by

$$v_{k+1} = h(v_k) \tag{4.121}$$

is in fact a Cauchy sequence. When we know this, we know that we have convergence and thus existence of a solution v^*. And since

$$v^* = \lim_{k \to \infty} v_{k+1} \overset{(4.121)}{=} \lim_{k \to \infty} h(v_k) = h(v^*),$$

it follows by the continuity of h that the limit satisfies the proper equation. And by the arguments given above, we know that it is in fact the only solution.

With respect to the existence of a solution, it remains to demonstrate that the sequence $\{v_k\}$ generated by (4.121) is a Cauchy sequence. Since

$$v_{n+1} = h(v_n), \tag{4.122}$$

we have

$$|v_{n+1} - v_n| = |h(v_n) - h(v_{n-1})| \leq \delta \, |v_n - v_{n-1}|. \tag{4.123}$$

By induction, we have

$$|v_{n+1} - v_n| \leq \delta^n \, |v_1 - v_0|. \tag{4.124}$$

In order to see whether $\{v_n\}$ is a Cauchy sequence, we need to bound $|v_m - v_n|$, where we may assume that $m > n$. Note that

$$v_m - v_n = (v_m - v_{m-1}) + (v_{m-1} - v_{m-2}) + \ldots + (v_{n+1} - v_n), \tag{4.125}$$

and thus, by the triangle inequality, we have

$$|v_m - v_n| \leq |v_m - v_{m-1}| + |v_{m-1} - v_{m-2}| + \ldots + |v_{n+1} - v_n|. \tag{4.126}$$

By (4.124), we have

$$|v_m - v_{m-1}| \leq \delta^{m-1} \, |v_1 - v_0|,$$
$$|v_{m-1} - v_{m-2}| \leq \delta^{m-2} \, |v_1 - v_0|,$$
$$\vdots$$
$$|v_{n+1} - v_n| \leq \delta^n \, |v_1 - v_0|.$$

Consequently,

$$|v_m - v_n| \leq |v_m - v_{m-1}| + |v_{m-1} - v_{m-2}| + \ldots + |v_{n+1} - v_n|$$
$$\leq \left(\delta^{m-1} + \delta^{m-2} + \ldots + \delta^n \right) |v_1 - v_0|.$$

Here

$$\delta^{m-1} + \delta^{m-2} + \ldots + \delta^n = \delta^{n-1}\left(\delta + \delta^2 + \ldots + \delta^{m-n}\right)$$

$$\leq \delta^{n-1}\sum_{k=1}^{\infty}\delta^k$$

$$\overset{(4.118)}{=} \delta^{n-1}\frac{1}{1-\delta},$$

so

$$|v_m - v_n| \leq \frac{\delta^{n-1}}{1-\delta}|v_1 - v_0| .$$

This means that for any $\varepsilon > 0$, we can find an integer M such that[5]

$$|v_m - v_n| < \varepsilon,$$

provided that $m, n \geq M$, and consequently $\{v_k\}$ is a Cauchy sequence. A Cauchy sequence is convergent and the limit, as seen above, solves the equation

$$v = h(v). \tag{4.127}$$

We can now conclude that if h is a contractive mapping satisfying (4.109) and (4.110) on page 121, then (4.127) has a unique solution that can be computed by the simple iteration

$$v_{k+1} = h(v_k).$$

4.6   Systems of Nonlinear Equations

We have seen above that equations of the form

$$f(x) = 0 \tag{4.128}$$

---

[5] It is sufficient to choose M such that

$$\frac{\delta^{M-1}}{1-\delta} < \varepsilon,$$

so

$$M \geq \frac{\ln(\varepsilon(1-\delta))}{\ln(\delta)} + 1.$$

can be solved by various methods: bisection, Newton, secant or fixed-point iterations. A good starting point for finding the zeros of $f = f(x)$ on an interval $[a, b]$
is to graph the function on this interval. This is straightforward for scalar equations,
but it is much harder for systems. In this section we will start our study of nonlinear
algebraic equations, a topic we will return to on several occasions. Here, we will
just consider 2×2 systems, but keep in mind that we will encounter systems that
have millions or even billions of unknowns.

4.6.1 A Linear System

What is a system of equations? You have probably seen such equations in your linear
algebra course, where you probably studied linear equations. Let us, for illustration,
start by considering a linear system that arises from the discretization of a linear
2×2 system of ODEs,

$$
\begin{aligned}
u'(t) &= -v(t), \quad u(0) = u_0, \\
v'(t) &= u(t), \quad\ \ v(0) = v_0.
\end{aligned}
\tag{4.129}
$$

Let (u_n, v_n) denote an approximation of $(u(t_n), v(t_n))$, where $t_n = n \Delta t$ for some
$\Delta t > 0$. An implicit Euler scheme for system (4.129) reads

$$
\frac{u_{n+1} - u_n}{\Delta t} = -v_{n+1}, \quad \frac{v_{n+1} - v_n}{\Delta t} = u_{n+1}.
\tag{4.130}
$$

This system can be rewritten in the form

$$
\begin{aligned}
u_{n+1} + \Delta t\, v_{n+1} &= u_n, \\
-\Delta t\, u_{n+1} + v_{n+1} &= v_n.
\end{aligned}
\tag{4.131}
$$

By introducing the matrix

$$
\mathbf{A} = \begin{pmatrix} 1 & \Delta t \\ -\Delta t & 1 \end{pmatrix},
\tag{4.132}
$$

we can write (4.131) in the form

$$
\mathbf{A} \mathbf{w}_{n+1} = \mathbf{w}_n,
\tag{4.133}
$$

where $\mathbf{w}_n$ is a vector

$$
\mathbf{w}_n = \begin{pmatrix} u_n \\ v_n \end{pmatrix}.
\tag{4.134}
$$

We observe that in order to compute $\mathbf{w}_{n+1} = (u_{n+1}, v_{n+1})^T$ based on $\mathbf{w}_n = (u_n, v_n)$, we have to solve the linear system (4.133). Since

$$\det(\mathbf{A}) = 1 + \Delta t^2 > 0, \tag{4.135}$$

we know that (4.133) has a unique solution given by

$$\mathbf{w}_{n+1} = \mathbf{A}^{-1}\mathbf{w}_n. \tag{4.136}$$

Here

$$\mathbf{A}^{-1} = \frac{1}{1 + \Delta t^2}\begin{pmatrix} 1 & -\Delta t \\ \Delta t & 1 \end{pmatrix}, \tag{4.137}$$

see (3.63) on page 85. From (4.136) and (4.137) we have

$$\begin{pmatrix} u_{n+1} \\ v_{n+1} \end{pmatrix} = \frac{1}{1+\Delta t^2}\begin{pmatrix} 1 & -\Delta t \\ \Delta t & 1 \end{pmatrix}\begin{pmatrix} u_n \\ v_n \end{pmatrix} = \frac{1}{1+\Delta t^2}\begin{pmatrix} u_n - \Delta t\, v_n \\ \Delta t\, u_n + v_n \end{pmatrix}, \tag{4.138}$$

or

$$u_{n+1} = \frac{1}{1 + \Delta t^2}(u_n - \Delta t\, v_n),$$

$$v_{n+1} = \frac{1}{1 + \Delta t^2}(v_n + \Delta t\, u_n). \tag{4.139}$$

By choosing $u_0 = 1$ and $v_0 = 0$, we have the analytical solutions

$$u(t) = \cos(t), \quad v(t) = \sin(t). \tag{4.140}$$

In Fig. 4.8 we have plotted (u, v) and (u_n, v_n) for $0 \le t \le 2\pi$, $\Delta t = \pi/500$. We observe that the scheme provides good approximations.

4.6.2 A Nonlinear System

We saw that the fully implicit discretization of the linear system of ODEs (4.129) leads to a linear algebraic system of equations, see (4.133). Let us now consider the following nonlinear system of ODEs:

$$\begin{aligned} u' &= -v^3, & u(0) &= u_0, \\ v' &= u^3, & v(0) &= v_0. \end{aligned} \tag{4.141}$$

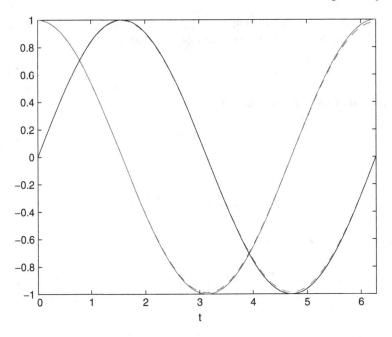

Fig. 4.8 The analytical solution ($u = \cos(t), v = \sin(t)$) and the numerical solution (u_n, v_n), in *dashed lines*, produced by the implicit Euler scheme (4.130)

An implicit Euler scheme for this system reads

$$\frac{u_{n+1} - u_n}{\Delta t} = -v_{n+1}^3, \quad \frac{v_{n+1} - v_n}{\Delta t} = u_{n+1}^3, \tag{4.142}$$

which can be rewritten in the form

$$\begin{aligned} u_{n+1} + \Delta t \, v_{n+1}^3 - u_n &= 0, \\ v_{n+1} - \Delta t \, u_{n+1}^3 - v_n &= 0. \end{aligned} \tag{4.143}$$

Observe that in order to compute (u_{n+1}, v_{n+1}) based on (u_n, v_n), we need to solve a nonlinear system of equations. To clarify matters, we study (4.143) for a fixed value of n. Let $\alpha = u_n$ and $\beta = v_n$. We want to compute $x = u_{n+1}$ and $y = v_{n+1}$. Let

$$\begin{aligned} f(x, y) &= x + \Delta t \, y^3 - \alpha, \\ g(x, y) &= y - \Delta t \, x^3 - \beta. \end{aligned} \tag{4.144}$$

Then system (4.143) can be written on the generic form

$$\begin{aligned} f(x, y) &= 0, \\ g(x, y) &= 0. \end{aligned} \tag{4.145}$$

We will derive Newton's method for this system along the lines used in the scalar case. But before we do that, we observe that the first equation of (4.145) implies that

$$x = \alpha - \Delta t \, y^3 \tag{4.146}$$

and thus the second equation reads

$$y - \Delta t \, (\alpha - \Delta t \, y^3)^3 - \beta = 0. \tag{4.147}$$

Here we note that (4.147) is a scalar equation. We can therefore solve (4.147) by the techniques derived above. The resulting approximation $y \approx y^*$ can then be used in (4.146) to compute an approximation $x \approx x^*$.

For the present simple 2×2 system, this is a possible approach. But keep in mind that we are preparing for really big systems. And for such systems, this substitution technique is not always workable and leads to horrible complications and an extremely high likelihood of errors. Therefore, we should solve (4.145) as a system.

4.6.3 Newton's Method

Our aim is now to derive Newton's method for solving the system

$$\begin{aligned} f(x, y) &= 0, \\ g(x, y) &= 0. \end{aligned} \tag{4.148}$$

But let us start by briefly recalling the scalar derivation, where we consider the scalar equation

$$p(x) = 0, \tag{4.149}$$

and we assume that x_0 is an approximation of x^* that solves (4.149). By a Taylor series expansion, we have

$$p(x_0 + h) = p(x_0) + hp'(x_0) + \mathcal{O}(h^2). \tag{4.150}$$

Since we want h to be such that $p(x_0 + h) \approx 0$, we can define h to be the solution of the linear equation

$$p(x_0) + hp'(x_0) = 0, \tag{4.151}$$

and thus

$$h = -p(x_0)/p'(x_0). \tag{4.152}$$

Consequently, it is natural to define

$$x_1 = x_0 - \frac{p(x_0)}{p'(x_0)},$$
(4.153)

and generally we have

$$x_{k+1} = x_k - \frac{p(x_k)}{p'(x_k)}.$$
(4.154)

Let us now turn our attention to system (4.148). The key point in deriving Newton's method for this system is the multi-dimensional version of the Taylor series. Consult e.g. Apostol [4] to see that for a smooth function $F(x, y)$, we have[6]

$$F(x + \Delta x, y + \Delta y) = F(x, y) + \Delta x \frac{\partial F}{\partial x}(x, y) + \Delta y \frac{\partial F}{\partial y}(x, y)$$
$$+ \mathcal{O}(\Delta x^2, \Delta x \Delta y, \Delta y^2).$$
(4.155)

Now, we can proceed as in the scalar case. Suppose that (x_0, y_0) are approximations of (x^*, y^*) that solve the system

$$\begin{aligned} f(x, y) &= 0, \\ g(x, y) &= 0. \end{aligned}$$
(4.156)

Using (4.155), we get

$$f(x_0 + \Delta x, y_0 + \Delta y) = f(x_0, y_0) + \Delta x \frac{\partial f}{\partial x}(x_0, y_0) + \Delta y \frac{\partial f}{\partial y}(x_0, y_0)$$
$$+ \mathcal{O}(\Delta x^2, \Delta x \Delta y, \Delta y^2),$$
(4.157)

and

$$g(x_0 + \Delta x, y_0 + \Delta y) = g(x_0, y_0) + \Delta x \frac{\partial g}{\partial x}(x_0, y_0) + \Delta y \frac{\partial g}{\partial y}(x_0, y_0)$$
$$+ \mathcal{O}(\Delta x^2, \Delta x \Delta y, \Delta y^2).$$
(4.158)

Since we want Δx and Δy to be such that

$$\begin{aligned} f(x_0 + \Delta x, y_0 + \Delta y) &\approx 0, \\ g(x_0 + \Delta x, y_0 + \Delta y) &\approx 0, \end{aligned}$$
(4.159)

it is reasonable to define Δx and Δy to be the solution of the linear system

---

[6] Note that $\mathcal{O}(\varepsilon, \delta)$ is shorthand for $\mathcal{O}(\varepsilon) + \mathcal{O}(\delta)$.

$$f(x_0, y_0) + \Delta x \frac{\partial f}{\partial x}(x_0, y_0) + \Delta y \frac{\partial f}{\partial y}(x_0, y_0) = 0,$$

$$g(x_0, y_0) + \Delta x \frac{\partial g}{\partial x}(x_0, y_0) + \Delta y \frac{\partial g}{\partial y}(x_0, y_0) = 0. \tag{4.160}$$

Do you understand that (4.160) is a linear system of equations with the unknowns Δx and Δy? To do that, keep in mind that x_0 and y_0 are known. It means that $f(x_0, y_0)$ is just a known number. Similarly, every term in (4.160) evaluated at the point (x_0, y_0) is known. So the only unknowns are Δx and Δy. To simplify the notation we let $f_0 = f(x_0, y_0)$, and so on. Then, system (4.160) can be rewritten in the form[7]

$$\begin{pmatrix} \frac{\partial f_0}{\partial x} & \frac{\partial f_0}{\partial y} \\ \frac{\partial g_0}{\partial x} & \frac{\partial g_0}{\partial y} \end{pmatrix} \begin{pmatrix} \Delta x \\ \Delta y \end{pmatrix} = - \begin{pmatrix} f_0 \\ g_0 \end{pmatrix}. \tag{4.161}$$

Let us assume that the matrix

$$\mathbf{A} = \begin{pmatrix} \frac{\partial f_0}{\partial x} & \frac{\partial f_0}{\partial y} \\ \frac{\partial g_0}{\partial x} & \frac{\partial g_0}{\partial y} \end{pmatrix} \tag{4.162}$$

is nonsingular. Then

$$\begin{pmatrix} \Delta x \\ \Delta y \end{pmatrix} = - \begin{pmatrix} \frac{\partial f_0}{\partial x} & \frac{\partial f_0}{\partial y} \\ \frac{\partial g_0}{\partial x} & \frac{\partial g_0}{\partial y} \end{pmatrix}^{-1} \begin{pmatrix} f_0 \\ g_0 \end{pmatrix}, \tag{4.163}$$

and we can define

$$\begin{pmatrix} x_1 \\ y_1 \end{pmatrix} = \begin{pmatrix} x_0 \\ y_0 \end{pmatrix} + \begin{pmatrix} \Delta x \\ \Delta y \end{pmatrix} = \begin{pmatrix} x_0 \\ y_0 \end{pmatrix} - \begin{pmatrix} \frac{\partial f_0}{\partial x} & \frac{\partial f_0}{\partial y} \\ \frac{\partial g_0}{\partial x} & \frac{\partial g_0}{\partial y} \end{pmatrix}^{-1} \begin{pmatrix} f_0 \\ g_0 \end{pmatrix}. \tag{4.164}$$

As in the scalar case, this argument can be repeated and we obtain Newton's method for the 2×2 system (4.156),

$$\begin{pmatrix} x_{k+1} \\ y_{k+1} \end{pmatrix} = \begin{pmatrix} x_k \\ y_k \end{pmatrix} - \begin{pmatrix} \frac{\partial f_k}{\partial x} & \frac{\partial f_k}{\partial y} \\ \frac{\partial g_k}{\partial x} & \frac{\partial g_k}{\partial y} \end{pmatrix}^{-1} \begin{pmatrix} f_k \\ g_k \end{pmatrix}. \tag{4.165}$$

Here,

$$f_k = f(x_k, y_k), \qquad \frac{\partial f_k}{\partial x} = \frac{\partial f}{\partial x}(x_k, y_k),$$

and so on.

----

[7] Note that $\dfrac{\partial f_0}{\partial x} = \dfrac{\partial f}{\partial x}(x_0, y_0)$, and so on.

4.6.4 A Nonlinear Example

Before we return to the nonlinear system (4.143), we will consider an example of a nonlinear algebraic system just to test method (4.165). The system

$$e^x - e^y = 0,$$
$$\ln(1 + x + y) = 0, \qquad (4.166)$$

can be solved analytically. It follows from the first equation that $x = y$ and thus the second equation is reduced to

$$\ln(1 + 2x) = 0.$$

Note that $\ln(z) = 0 \Rightarrow z = 1$, thus we have

$$1 + 2x = 1,$$

which gives $x = 0$ and so $y = 0$. Since we have the exact solution, we can use this example to check the properties of Newton's method applied to this system. By defining

$$f(x, y) = e^x - e^y$$

and

$$g(x, y) = \ln(1 + x + y),$$

we can specify Newton's method (4.165) as

$$\begin{pmatrix} x_{k+1} \\ y_{k+1} \end{pmatrix} = \begin{pmatrix} x_k \\ y_k \end{pmatrix} - \begin{pmatrix} e^{x_k} & -e^{y_k} \\ \frac{1}{1+x_k+y_k} & \frac{1}{1+x_k+y_k} \end{pmatrix}^{-1} \begin{pmatrix} e^{x_k} - e^{y_k} \\ \ln(1 + x_k + y_k) \end{pmatrix} \qquad (4.167)$$

Recall that, in general, if $ad - bc \neq 0$,

$$\begin{pmatrix} a & b \\ c & d \end{pmatrix}^{-1} = \frac{1}{ad - bc} \begin{pmatrix} d & -b \\ -c & a \end{pmatrix}, \qquad (4.168)$$

so the inverse needed in (4.167) is readily computed. By choosing $x_0 = y_0 = \frac{1}{2}$, we get the following results:

k	x_k	y_k
0	0.5	0.5
1	−0.193147	−0.193147
2	−0.043329	−0.043329
3	−0.001934	−0.001934
4	$−3.75 \cdot 10^{-6}$	$−3.75 \cdot 10^{-6}$
5	$−1.40 \cdot 10^{-11}$	$−1.40 \cdot 10^{-11}$

We observe that, as in the scalar case, Newton's method gives very rapid convergence toward the solution $x = y = 0$.

4.6.5 The Nonlinear System Revisited

In Sect. 4.6.2 on page 127 we introduced the system

$$u' = -v^3, \quad u(0) = u_0,$$
$$v' = u^3, \quad v(0) = v_0, \tag{4.169}$$

and the implicit Euler scheme

$$\frac{u_{n+1} - u_n}{\Delta t} = -v_{n+1}^3, \quad \frac{v_{n+1} - v_n}{\Delta t} = u_{n+1}^3, \tag{4.170}$$

where we have used standard notation. The nonlinear system (4.170) can be rewritten on the form

$$u_{n+1} + \Delta t\, v_{n+1}^3 - u_n = 0,$$
$$v_{n+1} - \Delta t\, u_{n+1}^3 - v_n = 0. \tag{4.171}$$

We note that for a given time level $t_n = n\Delta t$, the (4.171) define a 2×2 system of nonlinear algebraic equations. By setting $\alpha = u_n$, $\beta = v_n$, $x = u_{n+1}$, and $y = v_{n+1}$, we can write the system in the form

$$f(x, y) = 0,$$
$$g(x, y) = 0, \tag{4.172}$$

where

$$f(x, y) = x + \Delta t\, y^3 - \alpha,$$
$$g(x, y) = y - \Delta t\, x^3 - \beta. \tag{4.173}$$

We can now solve system (4.172) using Newton's method. We set $x_0 = \alpha$, $y_0 = \beta$ and iterate as

$$\begin{pmatrix} x_{k+1} \\ y_{k+1} \end{pmatrix} = \begin{pmatrix} x_k \\ y_k \end{pmatrix} - \begin{pmatrix} \frac{\partial f_k}{\partial x} & \frac{\partial f_k}{\partial y} \\ \frac{\partial g_k}{\partial x} & \frac{\partial g_k}{\partial y} \end{pmatrix}^{-1} \begin{pmatrix} f_k \\ g_k \end{pmatrix}, \tag{4.174}$$

where

$$f_k = f(x_k, y_k), \quad g_k = g(x_k, y_k),$$
$$\frac{\partial f_k}{\partial x} = \frac{\partial f}{\partial x}(x_k, y_k) = 1, \quad \frac{\partial f_k}{\partial y} = \frac{\partial f}{\partial y}(x_k, y_k) = 3\Delta t \, y_k^2, \tag{4.175}$$
$$\frac{\partial g_k}{\partial x} = \frac{\partial g}{\partial x}(x_k, y_k) = -3\Delta t \, x_k^2, \quad \frac{\partial g_k}{\partial y} = \frac{\partial g}{\partial y}(x_k, y_k) = 1.$$

Note that the matrix

$$\mathbf{A} = \begin{pmatrix} \frac{\partial f_k}{\partial x} & \frac{\partial f_k}{\partial y} \\ \frac{\partial g_k}{\partial x} & \frac{\partial g_k}{\partial y} \end{pmatrix} = \begin{pmatrix} 1 & 3\Delta t \, y_k^2 \\ -3\Delta t \, x_k^2 & 1 \end{pmatrix} \tag{4.176}$$

has a determinant given by

$$\det(\mathbf{A}) = 1 + 9\Delta t^2 \, x_k^2 \, y_k^2 > 0, \tag{4.177}$$

and so $\mathbf{A}^{-1}$ is well defined and is given by

$$\mathbf{A}^{-1} = \frac{1}{1 + 9\Delta t^2 \, x_k^2 \, y_k^2} \begin{pmatrix} 1 & -3\Delta t \, y_k^2 \\ 3\Delta t \, x_k^2 & 1 \end{pmatrix}. \tag{4.178}$$

In order to test the method, we consider system (4.169) with $u_0 = 1$ and $v_0 = 0$. At each time level, we iterate using (4.174), until

$$|f(x_k, y_k)| + |g(x_k, y_k)| < \varepsilon = 10^{-6}. \tag{4.179}$$

We choose $\Delta t = 1/100$ and compute the numerical solution from $t = 0$ to $t = 1$. The numerical approximations, as functions of t, are plotted in Figs. 4.9 and 4.10 in the (u, v)-coordinate system. In Fig. 4.11 we have plotted the number of Newton iterations needed to reach the stopping criterion (4.179) at each time level. Observe that we need no more than two iterations at all time levels.

4.7 Exercises

Exercise 4.1. Consider the function

$$f(x) = e^x - 1. \tag{4.180}$$

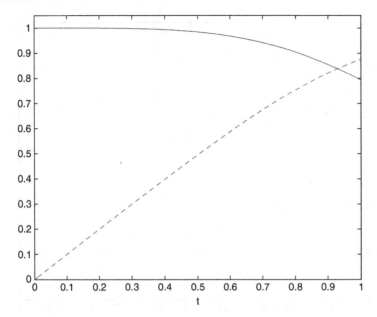

Fig. 4.9 The numerical approximations u_n (*solid line*) and v_n (*dashed line*) of the solutions of (4.169) produced by the implicit Euler scheme (4.170) using $u_0 = 1$, $v_0 = 0$ and $\Delta t = 1/100$

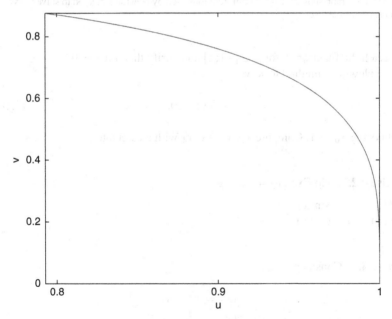

Fig. 4.10 The numerical approximations of the solutions of (4.169) in the (u, v)-coordinate system, resulting from the implicit Euler scheme (4.170) using $u_0 = 1$, $v_0 = 0$ and $\Delta t = 1/100$

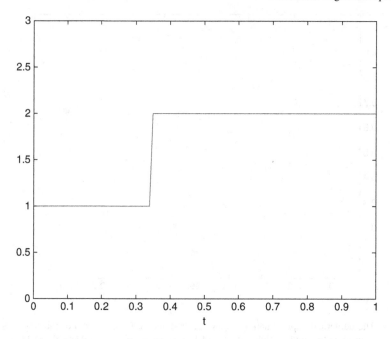

Fig. 4.11 The graph shows the number of iterations used by Newton's method to solve the system (4.171) at each time level

(a) Sketch the function f for $x \in [-1, 1]$ and verify that $f(0) = 0$.
(b) Use Newton's method to solve

$$f(x) = 0, \qquad (4.181)$$

choosing $x_0 = 1$. Compute x_1, x_2, x_3, x_4 with a calculator.

◇

Exercise 4.2. Redo Exercise 4.1 using

(a) $f(x) = x - \sin(x)$.
(b) $f(x) = 1 - \cos(x)$.

◇

Exercise 4.3. Consider

$$f(x) = x^2 - 4,$$
$$g(x) = x^2.$$

(a) Set $x_0 = 3$ and compute $x_1, \ldots, x_4$ in Newton's method to solve

$$f(x) = 0. \qquad (4.182)$$

(b) Set $x_0 = 1$ and compute $x_1, \ldots, x_4$ in Newton's method to solve

$$g(x) = 0. \tag{4.183}$$

(c) You have observed that Newton's method converges much faster for (4.182) than for (4.183). Use the graphical interpretation of Newton's method, see Fig. 4.5 on page 112, to try to understand the convergence behavior for these two problems.

(d) Consider

$$h(x) = x^6.$$

Set $x_0 = 1$ and compute $x_1, \ldots, x_4$ in Newton's method for solving

$$h(x) = 0.$$

Discuss the convergence speed along the lines suggested in (c).

◇

Exercise 4.4. Consider the system

$$e^y - x = 1,$$
$$x^2 - y = 0.$$

(a) Show that $x = y = 0$ solves the system.
(b) Set $x_0 = y_0 = 1/2$ and compute (x_1, y_1) and (x_2, y_2) in Newton's method.

◇

Exercise 4.5. Consider the ODE

$$u' = e^{-u}, \quad u(0) = 0. \tag{4.184}$$

(a) Derive and implement an explicit Euler scheme for (4.184).
(b) Derive an implicit Euler scheme for (4.184). Use Newton's method to solve the nonlinear equation arising at each time level. Implement your scheme on a computer.
(c) Set $\Delta t = 1/100$ and compute numerical solutions from $t = 0$ to $t = 1$, using the programs developed in (a) and (b).
(d) Show that

$$u(t) = \ln(1 + t)$$

solves (4.184). Use this to check the accuracy of the schemes discussed above.
(e) Consider the scheme

$$\frac{u_{n+1} - u_n}{\Delta t} = \frac{1}{2} \left(e^{-u_n} + e^{-u_{n+1}} \right).$$

Derive Newton's method for solving the nonlinear equation arising at each time level. Implement the scheme and discuss its accuracy and compare the results with those obtained by the implicit and explicit Euler schemes.

(f) Redo (e) using the scheme

$$\frac{u_{n+1} - u_n}{\Delta t} = e^{-\frac{1}{2}(u_{n+1}+u_n)}.$$

◇

Exercise 4.6. Suppose you have a number c and you want to find its reciprocal $1/c$. Can you compute that number only by applying multiplication? In order to do this we define

$$f(x) = c - \frac{1}{x}$$

and seek a solution of

$$f(x) = 0. \tag{4.185}$$

(a) Show that Newton's method for (4.185) can be written in the form

$$x_{k+1} = (2 - c\, x_k)x_k. \tag{4.186}$$

(b) Set $c = 4$, $x_0 = 0.2$ and compute $x_1, \ldots, x_4$ by (4.186). Comment the accuracy of the scheme.

◇

4.8 Project: Convergence of Newton's Method

There are basically three things you need to know about the convergence of Newton's method:

(a) Usually, the method converges[8] very fast, typically in just a few iterations.
(b) Sometimes the convergence is slow.
(c) From time to time, the method does not converge at all.

It is possible to derive sharp results on how fast and in which cases Newton's method converges, see e.g. Stoer and Bulirsch [27]. These results, however, tend to

---

[8] This is not completely precise. We should write something like "the sequence of approximations generated by Newton's method usually converges very fast". The method does not converge, however; it is the numbers generated by the method that, hopefully, converge. But that is really a whole lot of words, so we will sacrifice precision for clarity and readability.

be hard to use in practice. For scalar equations, the results are fairly easy to use, but for such equations, we do not really need the help of theoretical results. The reason is because just by graphing the function, we are able to choose a very good initial guess that usually leads to fast convergence. For systems of equations with many unknowns, the situation is much more complex. But for such systems, it is very hard to check all the conditions of a precise convergence theorem. The most common computational practice is just to use the method and see what happens. In preparation for such a practice, we will try to explain (a), (b) and (c) above. Our explanation is certainly not a full theory, however; we will rely on experiments and elements of analysis.

We will first look at some examples where the convergence is fast. In fact, we will show that the convergence is often quadratic, which means that there exists a constant c such that

$$|e_{k+1}| \le c\, e_k^2, \tag{4.187}$$

where e_k is the error of the kth iterate of Newton's method. Suppose, just for illustration, that $c = 1$ and $e_0 = 1/10$. Then

$$|e_1| \le (1/10)^2 = 1/10^2,$$
$$|e_2| \le \left(1/10^2\right)^2 = 1/10^4,$$
$$|e_3| \le \left(1/10^4\right)^2 = 1/10^8,$$

so we see that (4.187) gives fast convergence.

(a) Consider

$$f(x) = x^2 - 4. \tag{4.188}$$

Use $x_0 = 3$ and compute x_1, x_2, x_3, and x_4 in Newton's method for the equation

$$f(x) = 0. \tag{4.189}$$

(b) Define

$$e_k = x_k - x^*, \tag{4.190}$$

where $x^* = 2$ is a root of f, and consequently $e_k = x_k - 2$. Use the numbers you computed in (a) to compute the ratio

$$c_k = \frac{|e_{k+1}|}{e_k^2} \tag{4.191}$$

for $k = 0, 1, 2, 3$. Use the results to argue that

$$|e_{k+1}| \approx c\, e_k^2$$

for a fixed value of c.

(c) Make a sketch of the functions e^x and $\cos(x)$ for $-1/2 \le x \le 1/2$. Use this sketch to argue that

$$g(x) = e^x - \cos(x) \tag{4.192}$$

has only one zero in this interval. Verify that

$$g(0) = 0. \tag{4.193}$$

(d) Compute $x_1, \ldots, x_4$ in Newton's method with $x_0 = 1/4$. Let $e_k = x_k - x^* = x_k$ and compute c_k as defined by (4.191). Conclude, again, that the convergence seems to be quadratic.

Let us now try to understand the convergence behavior of these cases in a bit more mathematical way. We consider

$$f(x) = x^2 - 4, \tag{4.194}$$

for which we saw, computationally, in (b) that

$$|e_{k+1}| \approx c\, e_k^2. \tag{4.195}$$

Our aim is now to derive (4.195) analytically.

(e) Show that Newton's method for (4.194) can be written in the form

$$x_{k+1} = \frac{x_k^2 + 4}{2x_k}. \tag{4.196}$$

(f) Define

$$h(x) = \frac{x^2 + 4}{2x} \tag{4.197}$$

and show that

$$h'(x) = \frac{x^2 - 4}{2x^2}. \tag{4.198}$$

(g) Show that

$$h(x) \ge 2 \tag{4.199}$$

for any $x \ge 2$.

(h) Suppose $x_0 \geq 2$. Show that

$$x_k \geq 2 \tag{4.200}$$

for $k \geq 0$, where x_k is generated by (4.196).

(i) Recall that

$$e_k = x_k - 2. \tag{4.201}$$

Use (4.200) to argue that

$$|e_k| = x_k - 2, \tag{4.202}$$

since $x_0 \geq 2$.

(j) Use (4.196) to show that

$$|e_{k+1}| = \frac{e_k^2}{2x_k}. \tag{4.203}$$

(k) Use (4.200) to conclude that

$$|e_{k+1}| \leq \frac{1}{4} e_k^2. \tag{4.204}$$

How does this result compare with your computational results in (b)?

We will now derive a result stating that the convergence of Newton's method, under certain assumptions, is quadratic. This result shows that the quadratic convergence encountered in numerous exercises above are not just a coincidence; we have quadratic convergence for a whole class of problems. On the other hand, our derivation will also give us some hints about when the method can get into trouble.

Before we start the analysis, let us just recall that for a smooth function f, we have

$$f(x) = f(a) + (x - a)f'(a) + \frac{1}{2}(x - a)^2 f''(\xi), \tag{4.205}$$

where ξ is a number in the interval bounded by x and a, see (1.30) on page 7. We consider the problem of solving the equation

$$f(x) = 0 \tag{4.206}$$

using Newton's method,

$$x_{k+1} = x_k - \frac{f(x_k)}{f'(x_k)}, \tag{4.207}$$

where x_0 is given. Recall that the error is defined by

$$e_k = x_k - x^*, \tag{4.208}$$

where x^* is a root of f. Our aim is now to show that, under proper conditions,

$$|e_{k+1}| \le c\, e_k^2. \tag{4.209}$$

Let us assume that the $\{x_k\}$ generated by (4.207) are all located in a bounded interval I. We assume also that f is a smooth function defined on I, and that

$$|f'(x)| \ge \alpha > 0, \tag{4.210}$$

and

$$|f''(x)| \le \beta < \infty, \tag{4.211}$$

for all $x \in I$.

(l) Use (4.207) and (4.208) to show that

$$e_{k+1} = \frac{e_k f'(x_k) - f(x_k)}{f'(x_k)}. \tag{4.212}$$

(m) Use the Taylor series (4.205) to show that

$$f(x^*) = f(x_k) - e_k f'(x_k) + \frac{1}{2} e_k^2 f''(\xi), \tag{4.213}$$

where ξ is a number in the interval bounded by x_k and x^*.

(n) Show that

$$e_k f'(x_k) - f(x_k) = \frac{1}{2} e_k^2 f''(\xi). \tag{4.214}$$

(o) Combine (4.212) and (4.214) to show that

$$e_{k+1} = \frac{f''(\xi)}{2 f'(x_k)} e_k^2. \tag{4.215}$$

(p) Use (4.210) and (4.211) to conclude that

$$|e_{k+1}| \le \frac{\beta}{2\alpha} e_k^2. \tag{4.216}$$

The estimate (4.216) states that when the iterates stay in a region where f' is non-zero and f'' is bounded– (4.210) and (4.211) – then the convergence is quadratic. We have now concluded the discussion of quadratic convergence; we claimed initially that the convergence is usually quadratic, and you can now understand roughly when this is the case.

Next, we will discuss slow convergence. We see from (4.215) that $f'(x^*) = 0$ is a problem. This was already discussed in Exercise 4.3 above. But we also note from (4.215) that if $f''(x^*)$ is zero, we can expect faster convergence. In order to study these two effects, we define two functions

$$f(x) = \cosh(x) - 1, \tag{4.217}$$
$$g(x) = \sinh(x), \tag{4.218}$$

and we want to use Newton's method to solve

$$f(x) = 0 \tag{4.219}$$

and

$$g(x) = 0. \tag{4.220}$$

Since

$$\cosh(0) = 1 \tag{4.221}$$

and

$$\sinh(0) = 0, \tag{4.222}$$

we have $x^* = 0$ for both (4.219) and (4.220). Let us also recall that

$$\frac{d}{dx} \cosh(x) = \sinh(x) \tag{4.223}$$

and

$$\frac{d}{dx} \sinh(x) = \cosh(x). \tag{4.224}$$

(q) Sketch the functions $f(x)$ and $g(x)$ for $x \in I = [-1, 1]$.

(r) Set $x_0 = 1$ and use the graphical illustration of Newton's method, see page 112, to explain that you expect fast convergence for (4.218) and slow convergence for (4.217).

(s) Verify that

$$f'(0) = 0, \quad f''(0) = 1, \tag{4.225}$$

and

$$g'(0) = 1, \quad g''(0) = 0. \tag{4.226}$$

Use these two facts and (4.215) to explain once more that you expect slow and fast convergence for (4.217) and (4.218) respectively.

(t) Compute x_1, x_2, x_3, x_4 in Newton's method for (4.217) and (4.218) using $x_0 = 1$. For (4.217) you should also compute e_{k+1}/e_k, and for (4.218) you should compute[9] e_{k+1}/e_k^3.

We have now seen one example where the convergence becomes slow and also one example of very fast convergence. In both cases we can explain these effects by either graphing or analytically using (4.215). Finally, we shall discuss divergence. In fact, several things can go terribly wrong with Newton's method:

(I) We can generate a sequence $\{x_k\}$ where $|x_k| \to \infty$ as k increases.
(II) We can obtain convergence toward an unwanted solution.
(III) We can generate an infinite loop, reiterating the same values over and over again without converging toward a solution.

The purpose of the rest of this project is to illustrate these possibilities.

(u) To illustrate (I) above, we consider

$$f(x) = 1 - x^2. \tag{4.227}$$

Use graphical analysis to see that if $x_0 \approx 0$, then $|x_1|$ is very large. Furthermore, if $x_0 = 0$, then x_1 becomes infinite.

(v) To illustrate (II), we consider

$$f(x) = (x + 1)(x - 1)(x - 2). \tag{4.228}$$

Set $x_0 = 0$. Since, both $x = -1$ and $x = 1$ are zeros of f, we would expect Newton's method to converge toward one of these values. This is however, not the case. Compute x_1, graph f, and explain what happens.

(w) To illustrate (III), we consider

$$f(x) = x - x^3, \tag{4.229}$$

which has zeros in $-1, 0$ and 1. Show that Newton's method for solving (4.229) can be written in the form

---

[9] If $|e_{k+1}| \le c|e_k|^3$ for a constant c not depending on k, we have cubic convergence. If $|e_{k+1}| \le c|e_k|$, with $c < 1$, we have linear convergence.

$$x_{k+1} = \frac{-2x_k^3}{1 - 3x_k^2}.$$ (4.230)

Set $x_0 = 1/\sqrt{5}$. Show that $x_1 = -1/\sqrt{5}$ and that $x_2 = x_0 = 1/\sqrt{5}$. Conclude that

$$x_k = (-1)^k \frac{1}{\sqrt{5}}$$

and thus we do not have convergence. Graph $f = f(x)$ and try to understand what you have just observed using graphical analysis.

Chapter 5
The Method of Least Squares

Suppose you have a set of discrete data (t_i, y_i), $i = 1, \ldots, n$. How can you determine a function $p(t)$ that approximates the data set in the sense that

$$p(t_i) \approx y_i, \quad \text{for } i = 1, \ldots, n?$$

In Fig. 5.1, we have plotted some data points (t_i, y_i) and a linear approximation $p = p(t)$. In Fig. 5.2, we use a quadratic function to model the same set of data points.

In these examples, the data are given by discrete points (t_i, y_i) for $i = 1, \ldots, n$. The problem is to determine a function representing these data in a reasonable way. A similar problem arises if the data are given by a continuous function on an interval. In Fig. 5.3, we have graphed a function $y = y(t)$ and a linear approximation $p = p(t)$. Similarly, we can approximate a function using a quadratic function; see Fig. 5.4. Of course, higher-order polynomials can be also used to model more complex functions.

The problem that we want to solve in this chapter is how to compute approximations of either discrete data or data represented by a function. In the discrete case, we will use a realistic data set representing changes in the mean temperature of the globe. We will also apply our techniques to compute coefficients used in population models. But why do we need such approximations? One reason is that data tend to be noisy and we just want the "big picture". What is essentially going on? Another reason is to find one specific value at a point where we have no data. We can use an approximated function to fill the gap between discrete data points. A third reason is that we may be able to detect a certain trend that can be valid also outside the range of the measurements. Of course, such estimates of the trend must be used with great care.

The basic idea we want to pursue here is that of computing the best approximation in the sense of *least squares*. This term will become clearer in the following text. Basically, we try to compute a simple function (constant, linear, etc.) that minimizes the distance between the approximation and the data. Since the minimization is based on using the squares of the distance, this approach is referred to as the method of least squares.

A. Tveito et al., *Elements of Scientific Computing*, Texts in Computational Science and Engineering 7, DOI 10.1007/978-3-642-11299-7_5,
© Springer-Verlag Berlin Heidelberg 2010

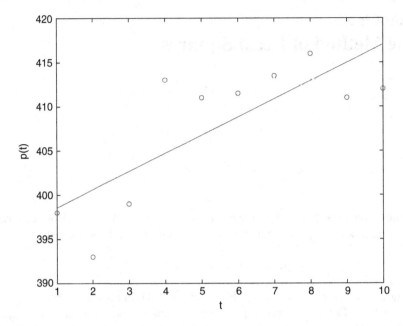

Fig. 5.1 A set of discrete data marked by *small circles* and a linear function $p = p(t)$ represented by the *solid line*

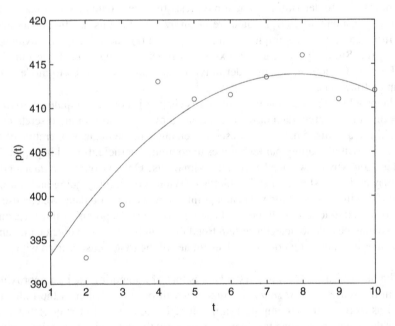

Fig. 5.2 A set of discrete data marked by *small circles* and a quadratic function $p = p(t)$ represented by the *solid curve*

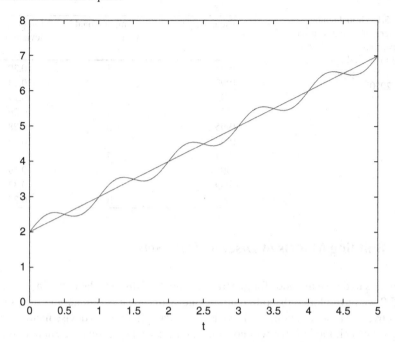

Fig. 5.3 An oscillating function $y = y(t)$ and a linear approximation $p = p(t)$

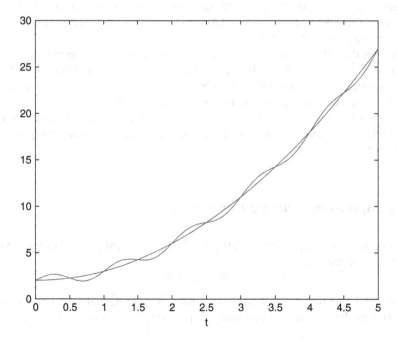

Fig. 5.4 An oscillating function $y = y(t)$ and a quadratic approximation $p = p(t)$

Table 5.1 Global annual mean temperature deviation (relative to the 1961–1990 mean) measured in degrees centigrade for years 1991–2000

Calendar year	Computational year t_i	Temperature deviation y_i
1991	1	0.29
1992	2	0.14
1993	3	0.19
1994	4	0.26
1995	5	0.28
1996	6	0.22
1997	7	0.43
1998	8	0.59
1999	9	0.33
2000	10	0.29

5.1 Building Models of Discrete Data Sets

Our aim is to derive methods for generating models of discrete data sets. In Figs. 5.1 and 5.2 we have plotted discrete data together with linear and quadratic functions, respectively. These functions are models of the data. In order to devise methods for computing such models, we will consider a data set representing the change in the mean temperature for the globe from 1856 to 2000. The data set gives the history of the deviation of the annual mean temperature from the 1961 to 1990 mean.[1] Let us first concentrate on modeling the annual mean temperature between 1991 and 2000. The data are given in Table 5.1.

We note that all the y_i-values in Table 5.1 are positive. This means that the global mean temperature is consistently higher in the 1991–2000 period than that of the reference time period 1961–1990. This observation has generated political concerns about consequences of a global increase in the temperature.

In Fig. 5.5, we have plotted the data points (t_i, y_i), $i = 1, \ldots, 10$ based on the data in Table 5.1. Our objective is now to compute constant, linear and quadratic models of these data.

5.1.1 Approximation by a Constant

Suppose we want to represent the data given in Table 5.1 and graphed in Fig. 5.5 by a constant function. Let

$$p = p(t) = \alpha$$

denote a constant function that takes the value α for all t. Our aim is to find α such that $p(t)$ models the data as accurately as possible. That sounds fair enough. But

---

[1] See http://cdiac.ornl.gov/ftp/trends/temp/jonescru/global.dat.

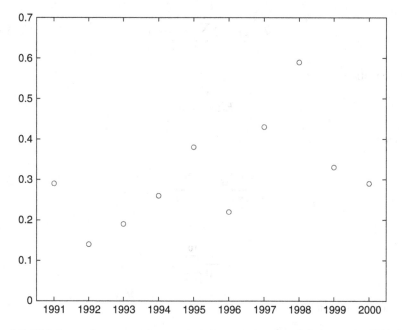

Fig. 5.5 Global annual mean temperature deviation measurements (relative to the 1961–1990 mean) for the period between 1991 and 2000

what do we really mean by "as accurately as possible"? Is that uniquely determined? No, it is not, and we shall discuss some ways of estimating α. An intuitive way of estimating α is to compute the average:

$$\alpha = \frac{1}{10} \sum_{i=1}^{10} y_i = 0.312. \tag{5.1}$$

But is the average a fair approximation? To answer this, we consider the function

$$F(\alpha) = \sum_{i=1}^{10} (\alpha - y_i)^2. \tag{5.2}$$

Note that F measures the total deviation from α to the whole data set $(t_i, y_i)_{i=1}^{10}$. All these deviations, each in its square, are added up and the sum is a measure of quality. One way of determining $\alpha = \alpha^*$ is to minimize F as a function of α and let α^* denote the value that minimizes $F(\alpha)$. We observe that F is a second-order polynomial in α. Therefore, we can minimize F by computing the zero of F', i.e., finding α^* such that $F'(\alpha^*) = 0$. We have

$$F'(\alpha) = 2 \sum_{i=1}^{10} (\alpha - y_i), \qquad (5.3)$$

so the equation

$$F'(\alpha) = 0$$

leads to

$$2 \sum_{i=1}^{10} \alpha = 2 \sum_{i=1}^{10} y_i, \qquad (5.4)$$

or

$$\alpha = \frac{1}{10} \sum_{i=1}^{10} y_i. \qquad (5.5)$$

You should also note that

$$F''(\alpha) = 2 \sum_{i=1}^{10} 1 = 20 > 0, \qquad (5.6)$$

therefore, α given by (5.5) is a unique minimum of F and we have

$$\alpha = \frac{1}{10} \sum_{i=1}^{10} y_i = 0.312, \qquad (5.7)$$

which is exactly the average given by (5.1). In Fig. 5.6, we have plotted $F = F(\alpha)$ for α ranging from 0.1 to 0.6. We have also marked $\alpha = 0.312$, which is the value at which F attains its minimum.

So this explains why the average is a fair value; it minimizes the square of the deviations from the data. This also explains why we call this way of computing $p(t) = \alpha$ the method of least squares. But this is not the only way of defining a constant approximation. We could also define

$$G(\alpha) = \sum_{i=1}^{10} (\alpha - y_i)^4, \qquad (5.8)$$

and then try to minimize this function. In this case, we have

$$G'(\alpha) = 4 \sum_{i=1}^{10} (\alpha - y_i)^3 \qquad (5.9)$$

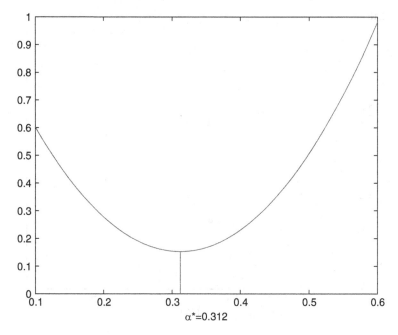

$\alpha^* = 0.312$

Fig. 5.6 A graph of $F = F(\alpha)$ given by (5.2)

and also

$$G''(\alpha) = 12 \sum_{i=1}^{10} (\alpha - y_i)^2. \tag{5.10}$$

Since $G''(\alpha) > 0$,[2] we have a unique minimum and we compute this by solving

$$G'(\alpha^*) = 0. \tag{5.11}$$

Now, (5.11) is a third-order equation and such equations are in general difficult to solve analytically. Instead, we use Newton's method. Consider Algorithm 4.2 on page 108. By setting $\varepsilon = 10^{-8}$, $\alpha_0 = 0.312$ and using G and G' as defined by (5.8) and (5.9), respectively, we can run Newton's method to find

$$\alpha^* \approx 0.345$$

in three iterations. In Fig. 5.7, we have plotted G as function of α for $\alpha \in [0.1, 0.6]$ and we have marked the minimizing value $\alpha^* \approx 0.345$.

---

[2] This is not completely accurate. If $G(\alpha) = e^\alpha$, then $G''(\alpha) > 0$, but we do not have a minimum. A more precise statement is as follows: If G is a smooth function with the property that $G''(\alpha) > 0$ for all relevant α and $G'(\alpha^*) = 0$, then α^* is a global minimum of G.

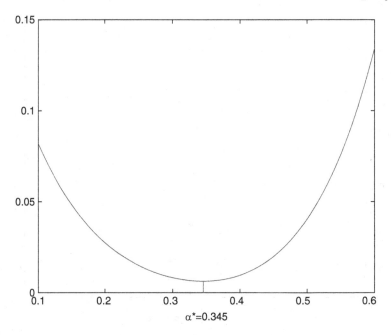

Fig. 5.7 A graph of $G = G(\alpha)$ given by (5.8)

Note that there arise two different constant approximations from minimizing F given by (5.2) and G given by (5.8). It is not correct to say one approximation is better than the other. They are both the best approximation we can get according to two different criteria. There exists many other criteria as well, each leading to slightly different constant approximations.

In Fig. 5.8 we have plotted all the data in Table 5.1 together with the constant functions

$$p(t) = 0.312 \quad \text{and} \quad q(t) = 0.345$$

obtained by minimizing $F(\alpha)$ (see (5.2)) and $G(\alpha)$ (see (5.8)), respectively.

We notice that minimizing the quadratic function F leads to a problem that can easily be solved. The alternative, G, results in a nonlinear problem that needs to be solved and we had to appeal to Newton's method. This observation is fairly general and we will therefore rely on minimizing quadratic deviations in the rest of this chapter. That is, we will focus on the method of least squares.

5.1.2 Approximation by a Linear Function

Next, we will model the data in Table 5.1 using a linear function. More precisely, we will try to determine two constants α and β such that

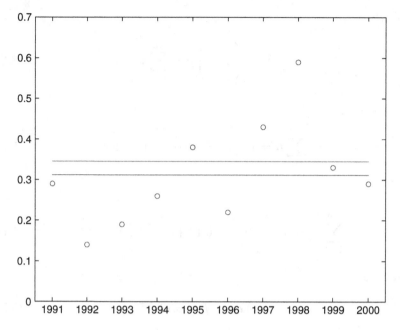

Fig. 5.8 Two constant approximations of the global annual mean temperature deviation measurements (relative to the 1961–1990 mean); from year 1991 to 2000

$$p(t) = \alpha + \beta t \tag{5.12}$$

fits the data as well as possible in the sense of least squares. Let us define

$$F(\alpha, \beta) = \sum_{i=1}^{10} (\alpha + \beta t_i - y_i)^2. \tag{5.13}$$

We want to minimize F with respect to α and β. A necessary condition for a minimum is

$$\frac{\partial F}{\partial \alpha} = \frac{\partial F}{\partial \beta} = 0. \tag{5.14}$$

Since

$$\frac{\partial F}{\partial \alpha} = 2 \sum_{i=1}^{10} (\alpha + \beta t_i - y_i), \tag{5.15}$$

the condition

$$\frac{\partial F}{\partial \alpha} = 0 \tag{5.16}$$

leads to

$$10\alpha + \left(\sum_{i=1}^{10} t_i\right)\beta = \sum_{i=1}^{10} y_i. \tag{5.17}$$

Here

$$\sum_{i=1}^{10} t_i = 1 + 2 + 3 + \cdots + 10 = 55 \tag{5.18}$$

and

$$\sum_{i=1}^{10} y_i = 0.29 + 0.14 + 0.19 + \cdots + 0.29 = 3.12, \tag{5.19}$$

so we have

$$10\alpha + 55\beta = 3.12. \tag{5.20}$$

Next, since

$$\frac{\partial F}{\partial \beta} = 2 \sum_{i=1}^{10} (\alpha + \beta t_i - y_i)t_i, \tag{5.21}$$

the condition

$$\frac{\partial F}{\partial \beta} = 0 \tag{5.22}$$

gives

$$\left(\sum_{i=1}^{10} t_i\right)\alpha + \left(\sum_{i=1}^{10} t_i^2\right)\beta = \sum_{i=1}^{10} y_i t_i. \tag{5.23}$$

Here,

$$\sum_{i=1}^{10} t_i^2 = 1 + 2^2 + 3^2 + \cdots + 10^2 = 385 \tag{5.24}$$

and

$$\sum_{i=1}^{10} t_i y_i = 1 \cdot 0.29 + 2 \cdot 0.14 + 3 \cdot 0.19 + \cdots + 10 \cdot 0.29 = 20, \tag{5.25}$$

so we arrive at the equation

$$55\alpha + 385\beta = 20. \tag{5.26}$$

We now have a 2×2 system of linear equations that determines α and β:

$$\begin{pmatrix} 10 & 55 \\ 55 & 385 \end{pmatrix} \begin{pmatrix} \alpha \\ \beta \end{pmatrix} = \begin{pmatrix} 3.12 \\ 20 \end{pmatrix}. \tag{5.27}$$

By the formula (3.63) on page 89, we have

$$\begin{pmatrix} 10 & 55 \\ 55 & 385 \end{pmatrix}^{-1} = \frac{1}{825} \begin{pmatrix} 385 & -55 \\ -55 & 10 \end{pmatrix}, \tag{5.28}$$

so

$$\begin{pmatrix} \alpha \\ \beta \end{pmatrix} = \frac{1}{825} \begin{pmatrix} 385 & -55 \\ -55 & 10 \end{pmatrix} \begin{pmatrix} 3.12 \\ 20 \end{pmatrix} \approx \begin{pmatrix} 0.123 \\ 0.034 \end{pmatrix}. \tag{5.29}$$

Hence we have the linear model

$$p(t) = 0.123 + 0.034t. \tag{5.30}$$

In Fig. 5.9 we have plotted all the data from 1991 to 2000 together with the constant approximation

$$p_0(t) = 0.312 \tag{5.31}$$

and the linear approximation

$$p_1(t) = 0.123 + 0.034t. \tag{5.32}$$

5.1.3 Approximation by a Quadratic Function

We have now seen approximations by constants and linear functions, and we proceed by using a quadratic function. It is tempting to assume that we can simply go on forever and increase the degree of the approximating polynomial as high as we want. In practice, that is not a good idea. When more accuracy is needed, it is common to glue together pieces of polynomials of rather low degree. It turns out that this gives computations that are better suited for computers. However, quadratic polynomials are just fine and we now consider approximations of the form

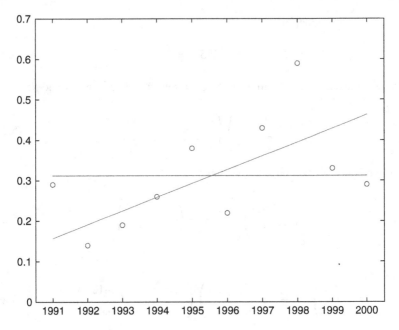

Fig. 5.9 Constant and linear approximations of global annual mean temperature deviation measurements (relative to the 1961–1990 mean); from 1991 to 2000

$$p(t) = \alpha + \beta t + \gamma t^2. \tag{5.33}$$

That is, we want to determine constants α, β and γ, such that p fits the data in Table 5.1 as accurately as possible. This will eventually lead to a 3×3 system of linear equations. In order to see that, we define

$$F(\alpha, \beta, \gamma) = \sum_{i=1}^{10} (\alpha + \beta t_i + \gamma t_i^2 - y_i)^2. \tag{5.34}$$

Again, a necessary condition for a minimum of F is

$$\frac{\partial F}{\partial \alpha} = \frac{\partial F}{\partial \beta} = \frac{\partial F}{\partial \gamma} = 0. \tag{5.35}$$

Here,

$$\frac{\partial F}{\partial \alpha} = 2 \sum_{i=1}^{10} \left(\alpha + \beta t_i + \gamma t_i^2 - y_i \right) = 0 \tag{5.36}$$

leads to the equation

$$10\alpha + \left(\sum_{i=1}^{10} t_i\right)\beta + \left(\sum_{i=1}^{10} t_i^2\right)\gamma = \sum_{i=1}^{10} y_i. \tag{5.37}$$

Furthermore,

$$\frac{\partial F}{\partial \beta} = 2\sum_{i=1}^{10}\left(\alpha + \beta t_i + \gamma t_i^2 - y_i\right) t_i = 0 \tag{5.38}$$

leads to

$$\left(\sum_{i=1}^{10} t_i\right)\alpha + \left(\sum_{i=1}^{10} t_i^2\right)\beta + \left(\sum_{i=1}^{10} t_i^3\right)\gamma = \sum_{i=1}^{10} y_i t_i, \tag{5.39}$$

and finally

$$\frac{\partial F}{\partial \gamma} = 2\sum_{i=1}^{10}\left(\alpha + \beta t_i + \gamma t_i^2 - y_i\right) t_i^2 = 0 \tag{5.40}$$

leads to

$$\left(\sum_{i=1}^{10} t_i^2\right)\alpha + \left(\sum_{i=1}^{10} t_i^3\right)\beta + \left(\sum_{i=1}^{10} t_i^4\right)\gamma = \sum_{i=1}^{10} y_i t_i^2. \tag{5.41}$$

We notice that

$$\sum_{i=1}^{10} t_i = 55, \quad \sum_{i=1}^{10} t_i^2 = 385, \quad \sum_{i=1}^{10} t_i^3 = 3025,$$

$$\sum_{i=1}^{10} t_i^4 = 25330, \quad \sum_{i=1}^{10} y_i = 3.12, \quad \sum_{i=1}^{10} t_i y_i = 20,$$

$$\sum_{i=1}^{10} t_i^2 y_i = 138.7.$$

From (5.37), (5.39), (5.41), and the calculations above, we obtain the following linear system

$$\begin{pmatrix} 10 & 55 & 385 \\ 55 & 385 & 3025 \\ 385 & 3025 & 25330 \end{pmatrix} \begin{pmatrix} \alpha \\ \beta \\ \gamma \end{pmatrix} = \begin{pmatrix} 3.12 \\ 20 \\ 138.7 \end{pmatrix}. \tag{5.42}$$

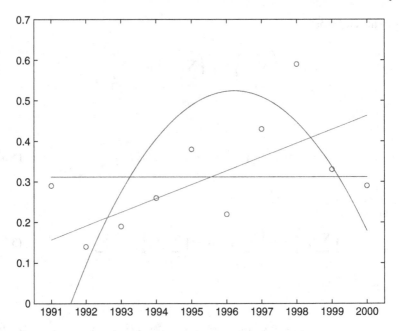

Fig. 5.10 Constant, linear and quadratic approximations of global annual mean temperature deviation measurements (relative to the 1961–1990 mean), year 1991 to 2000

Using, e.g., Matlab, we find that the solution of this linear system is given by

$$\begin{aligned}
\alpha &\approx -0.4078, \\
\beta &\approx 0.2997, \\
\gamma &\approx -0.0241.
\end{aligned} \tag{5.43}$$

So far, we have not discussed how to solve linear systems. Let us, for the time being, just accept that such a system can be solved under proper conditions on the matrix of the system. We will return to this topic in later chapters. In Fig. 5.10, we have plotted the data given in Table 5.1 together with the constant approximation

$$p_0(t) = 0.312, \tag{5.44}$$

the linear approximation

$$p_1(t) = 0.123 + 0.034t, \tag{5.45}$$

and the quadratic approximation

$$p_2(t) = -0.4078 + 0.2997t - 0.0241t^2. \tag{5.46}$$

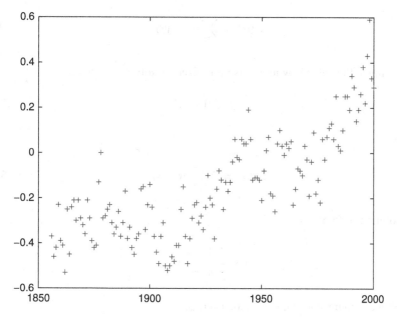

Fig. 5.11 Global annual mean temperature deviation measurements (relative to the 1961–1990 mean); from 1856 to 2000

5.1.4 Large Data Sets

Above, we derived methods for generating models of discrete data using either constant, linear, or quadratic functions. We used the method of least squares to determine the coefficients of the approximating functions. All the calculations were done for the data given in Table 5.1. Here we will first extend the methods to the general case of a data set containing n points,

$$(t_i, y_i) \quad i = 1, \ldots, n, \tag{5.47}$$

and then apply our new formulas to the large data set presented in Fig. 5.11.

Approximation by a Constant

We assume that the data (5.47) are given and that we want to determine a constant α such that

$$p_0(t) = \alpha \tag{5.48}$$

is the best constant approximation of these data in the sense of least squares. To this end, we define

$$F(\alpha) = \sum_{i=1}^{n} (\alpha - y_i)^2. \tag{5.49}$$

We compute the value α where F is minimized by solving the equation

$$F'(\alpha) = 0. \tag{5.50}$$

Since

$$F'(\alpha) = 2 \sum_{i=1}^{n} (\alpha - y_i), \tag{5.51}$$

the solution of (5.50) is

$$\alpha = \frac{1}{n} \sum_{i=1}^{n} y_i, \tag{5.52}$$

which we recognize as the arithmetic average.

Approximation by a Linear Function

Next, we seek a linear function

$$p_1(t) = \alpha + \beta t \tag{5.53}$$

approximating the data (5.47) in the sense of least squares. We define

$$F(\alpha) = \sum_{i=1}^{n} (\alpha + \beta t_i - y_i)^2. \tag{5.54}$$

and determine the constants α and β by requiring that

$$\frac{\partial F}{\partial \alpha} = \frac{\partial F}{\partial \beta} = 0. \tag{5.55}$$

Since

$$\frac{\partial F}{\partial \alpha} = 2 \sum_{i=1}^{n} (\alpha + \beta t_i - y_i) \tag{5.56}$$

and

$$\frac{\partial F}{\partial \beta} = 2 \sum_{i=1}^{n} (\alpha + \beta t_i - y_i) t_i, \tag{5.57}$$

we can derive from (5.55) the following 2×2 system of linear equations determining α and β:

$$\begin{pmatrix} n & \sum\limits_{i=1}^{n} t_i \\ \sum\limits_{i=1}^{n} t_i & \sum\limits_{i=1}^{n} t_i^2 \end{pmatrix} \begin{pmatrix} \alpha \\ \beta \end{pmatrix} = \begin{pmatrix} \sum\limits_{i=1}^{n} y_i \\ \sum\limits_{i=1}^{n} t_i y_i \end{pmatrix}. \tag{5.58}$$

Approximation by a Quadratic Function

Finally, we will seek a quadratic approximation on the form

$$p_2(t) = \alpha + \beta t + \gamma t^2 \tag{5.59}$$

of the data given in (5.47). Using

$$F(\alpha, \beta, \gamma) = \sum_{i=1}^{n} \left(\alpha + \beta t_i + \gamma t_i^2 - y_i \right)^2, \tag{5.60}$$

we appeal to the principle of least squares and require that α, β, and γ are such that

$$\frac{\partial F}{\partial \alpha} = \frac{\partial F}{\partial \beta} = \frac{\partial F}{\partial \gamma} = 0. \tag{5.61}$$

Since

$$\frac{\partial F}{\partial \alpha} = 2 \sum_{i=1}^{n} (\alpha + \beta t_i + \gamma t_i^2 - y_i) \tag{5.62}$$

$$\frac{\partial F}{\partial \beta} = 2 \sum_{i=1}^{n} (\alpha + \beta t_i + \gamma t_i^2 - y_i) t_i, \tag{5.63}$$

$$\frac{\partial F}{\partial \gamma} = 2 \sum_{i=1}^{n} (\alpha + \beta t_i + \gamma t_i^2 - y_i) t_i^2, \tag{5.64}$$

(5.61) leads to the following 3×3 linear system of equations that determines α, β, and γ:

$$\begin{pmatrix} n & \sum\limits_{i=1}^{n} t_i & \sum\limits_{i=1}^{n} t_i^2 \\ \sum\limits_{i=1}^{n} t_i & \sum\limits_{i=1}^{n} t_i^2 & \sum\limits_{i=1}^{n} t_i^3 \\ \sum\limits_{i=1}^{n} t_i^2 & \sum\limits_{i=1}^{n} t_i^3 & \sum\limits_{i=1}^{n} t_i^4 \end{pmatrix} \begin{pmatrix} \alpha \\ \beta \\ \gamma \end{pmatrix} = \begin{pmatrix} \sum\limits_{i=1}^{n} y_i \\ \sum\limits_{i=1}^{n} y_i t_i \\ \sum\limits_{i=1}^{n} y_i t_i^2 \end{pmatrix}. \tag{5.65}$$

For a given set of data (t_i, y_i), the linear system (5.65) can be solved using, e.g., Matlab.

Application to the Temperature Data

We will now use the methods derived above to model the data describing the change in the global temperature from 1856 to 2000. [3]

Based on the data, we need to compute the entries in (5.52), (5.58) and (5.65):

$$n = 145, \qquad \sum_{i=1}^{145} t_i = 10585, \qquad \sum_{i=1}^{145} t_i^2 = 1026745,$$

$$\sum_{i=1}^{145} t_i^3 = 1.12042 \cdot 10^8, \ \sum_{i=1}^{145} t_i^4 = 1.30415 \cdot 10^{10}, \ \sum_{i=1}^{145} y_i = -21.82, \quad (5.66)$$

$$\sum_{i=1}^{145} t_i y_i = -502.43, \qquad \sum_{i=1}^{145} t_i^2 y_i = 19649.8,$$

where we have used $t_i = i$, i.e., $t_1 = 1$ corresponds to the year of 1856, $t_2 = 2$ corresponds to the year of 1857, and so forth.

From (5.48) and (5.52), we now have

$$p_0(t) \approx -0.1505. \tag{5.67}$$

The coefficients α and β of the linear model (5.53) are obtained by solving the linear system (5.58), i.e.,

$$\begin{pmatrix} 145 & 10585 \\ 10585 & 1026745 \end{pmatrix} \begin{pmatrix} \alpha \\ \beta \end{pmatrix} = \begin{pmatrix} -21.82 \\ -502.43 \end{pmatrix}. \tag{5.68}$$

Consequently,

$$\alpha \approx -0.4638 \quad \text{and} \quad \beta \approx 0.0043,$$

so the linear model is given by

$$p_1(t) \approx -0.4638 + 0.0043\, t. \tag{5.69}$$

Similarly, the coefficients α, β and γ of the quadratic model (5.59) are obtained by solving the linear system (5.65), i.e.,

---

[3] See http://cdiac.ornl.gov/ftp/trends/temp/jonescru/global.dat.

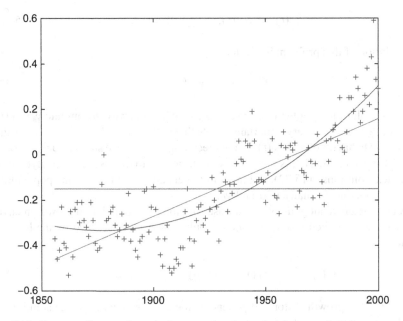

Fig. 5.12 Constant, linear and quadratic approximations of global annual mean temperature deviation measurements (relative to the 1961–1990 mean); from 1856 to 2000

$$
\begin{pmatrix}
145 & 10585 & 1026745 \cdot 10^6 \\
10585 & 1026745 \cdot 10^6 & 1.12042 \cdot 10^8 \\
1026745 \cdot 10^6 & 1.12042 \cdot 10^8 & 1.30415 \cdot 10^{10}
\end{pmatrix}
\begin{pmatrix}
\alpha \\ \beta \\ \gamma
\end{pmatrix}
=
\begin{pmatrix}
-21.82 \\ -502.43 \\ 19649.8
\end{pmatrix}. \quad (5.70)
$$

The solution of this system is given by

$$
\alpha \approx -0.3136, \quad \beta \approx -1.8404 \cdot 10^{-3}, \quad \text{and} \quad \gamma \approx 4.2005 \cdot 10^{-5},
$$

so the quadratic model is given by

$$
p_2(t) \approx -0.3136 - 1.8404 \cdot 10^{-3}\, t + 4.2005 \cdot 10^{-5}\, t^2. \quad (5.71)
$$

In Fig. 5.12 we have plotted the data points, the constant approximation $p_0(t)$, the linear approximation $p_1(t)$ and the quadratic approximation $p_2(t)$.

5.2 Application to Population Models

In Chaps. 2 and 3 above, we studied various population models. For instance, we suggested that the growth of populations in environments with unlimited resources can be modeled by the ordinary differential equation

$$y'(t) = \alpha y(t), \qquad y(0) = y_0. \tag{5.72}$$

The solution of this problem is given by

$$y(t) = y_0 e^{\alpha t}. \tag{5.73}$$

From this solution we note that once α and y_0 are known, we can compute $y = y(t)$ for any $t \geq 0$. But it is very important to note that without α and y_0 we only know the qualitative behavior of $y = y(t)$. In the preceding sections we simply guessed some values for α and y_0. That is fine when our purpose is to discuss either analytical or numerical solutions of (5.72). But in order to get actual numbers for the population, we need concrete estimates of α and y_0.

A completely analogous situation arises in models of population growth in environments with limited resources. Such populations can be modeled by the logistic model

$$y'(t) = \alpha y(t) (1 - y(t)/\beta), \qquad y(0) = y_0, \tag{5.74}$$

where α is the growth factor and β denotes the carrying capacity. The solution of (5.74) is given by

$$y(t) = \frac{y_0 \beta}{y_0 + e^{-\alpha t}(\beta - y_0)}, \tag{5.75}$$

cf. Sect. 2.1.4. Again we note that y_0, α and β have to be estimated in order to know how the population evolves according to the logistic model.

The purpose of this section is to explain how to estimate α in (5.72) and α and β in (5.74). We will illustrate the methods using data for the development of the world population over the last 50 years.[4]

5.2.1 Exponential Growth of the World Population?

Does the number of people living on the Earth follow an exponential growth law? This question has been discussed for centuries. A famous contribution to the discussion was given by Malthus,[5] who was concerned that mankind's unlimited growth could eventually threaten life on the Earth.

Here, we will try to analyze the population growth over the last 50 years. In Table 5.2 we have listed the total world population from 1950 to 1955, measured

---

[4] The data can be found at www.census.gov.

[5] See http://desip.igc.org/malthus for information about Thomas R. Malthus (1766–1834).

Table 5.2 The total world
population from 1950 to 1955

Year	Population (billions)
1950	2.555
1951	2.593
1952	2.635
1953	2.680
1954	2.728
1955	2.780

in billions of people. First, we will assume exponential growth and determine the growth rate α. Thereafter, we will consider a logistic model.

Let us now assume that the population $p = p(t)$ is governed by

$$p'(t) = \alpha p(t), \qquad p(0) = p_0. \tag{5.76}$$

Let us set $t = 0$ at 1950 and measure time in years such that $t = 1$ corresponds to 1951, $t = 2$ corresponds to 1952, and so on. From Table 5.2, we have that

$$p_0 = 2.555.$$

Now, the parameter α needs to be determined using the data in Table 5.2. From (5.76), we have

$$\frac{p'(t)}{p(t)} = \alpha. \tag{5.77}$$

Since only p is available, we have to approximate $p'(t)$ using the standard formula[6]

$$p'(t) \approx \frac{p(t + \Delta t) - p(t)}{\Delta t}. \tag{5.78}$$

By choosing $\Delta t = 1$, we estimate α to be

$$\alpha_n = \frac{p(n + 1) - p(n)}{p(n)} \tag{5.79}$$

for $n = 0, 1, 2, 3, 4$ corresponding to the years from 1950 to 1954. Since these numbers are small, we multiply them by 100 and compute

$$b_n = 100 \frac{p(n + 1) - p(n)}{p(n)}. \tag{5.80}$$

The results are given in Table 5.3 below.

---

[6] Another approach to this problem is discussed in Project 1.

Table 5.3 The calculated values of b_n using (5.80) based on the numbers of the world's population from 1950 to 1955

Year	n	$p(n)$	$b_n = 100 \dfrac{p(n+1) - p(n)}{p(n)}$
1950	0	2.555	1.49
1951	1	2.593	1.62
1952	2	2.635	1.71
1953	3	2.680	1.79
1954	4	2.728	1.91
1955	5	2.780	

Based on the results in Table 5.3, we want to compute a constant approximation to b. We use the arithmetic average, see (5.1)–(5.5), and define

$$b = \frac{1}{5} \sum_{n=0}^{4} b_n = \frac{1}{5}(1.49 + 1.62 + 1.71 + 1.79 + 1.91) = 1.704.$$

Since

$$b_n = 100 \frac{p(n+1) - p(n)}{p(n)} = 100\,\alpha_n, \tag{5.81}$$

we define

$$\alpha = \frac{1}{100}\, b = 0.01704. \tag{5.82}$$

Recall that the solution of (5.76) is given by

$$p(t) = p_0 e^{\alpha t},$$

so we have

$$p(t) = 2.555\, e^{0.01704 t}. \tag{5.83}$$

Since the year 2000 corresponds to $t = 50$, we have

$$p(50) = 2.555\, e^{0.01704 \times 50} \approx 5.990. \tag{5.84}$$

This should be compared with the actual population in the year of 2000, which is 6.080 billion. So the relative error is

$$\frac{6.080 - 5.990}{6.080} \cdot 100\% = 1.48\%. \tag{5.85}$$

Based on the population of 1950–1955, we computed the number of people living on the Earth in the year 2000 with an error of about 1.5%. That is quite remarkable!

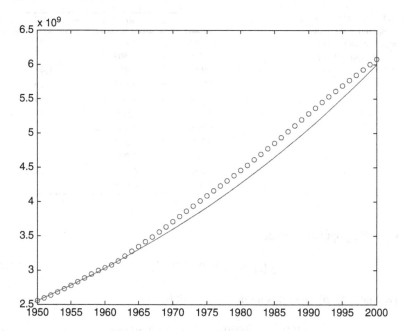

Fig. 5.13 The figure shows the graph of an exponential population model $p = p(t)$ together with the actual measurements marked by the open marks

In Fig. 5.13, we have plotted the estimated population $p(t)$, given by (5.83), together with the actual population from 1955 to 2000. Again we note that the model gives extremely good predictions.

What about the future? How many people will there be on the planet 100 years from now? This is of course very hard to tell and it is a research topic on its own. Recent estimates have suggested an upper bound of about 8–9 billion people. But let us just pursue our exponential approach a bit further and see what we get. In Table 5.4 below we analyze data in the time period 1990–2000.

Now, the average of the b_n values is

$$b = \frac{1}{10} \sum_{n=0}^{9} b_n = 1.42,$$

so

$$\alpha = \frac{b}{100} = 0.0142.$$

Using this growth factor and starting in the year of 2000, we have the model

$$p'(t) = 0.0142\,p(t), \qquad p(0) = 6.080. \qquad (5.86)$$

Table 5.4 The calculated b_n values associated with an exponential model for the world population between 1990 and 2000

Year	n	$p(n)$	$b_n = 100 \, \frac{p(n+1)-p(n)}{p(n)}$
1990	0	5.284	1.57
1991	1	5.367	1.55
1992	2	5.450	1.49
1993	3	5.531	1.46
1994	4	5.611	1.43
1995	5	5.691	1.37
1996	6	5.769	1.35
1997	7	5.847	1.33
1998	8	5.925	1.32
1999	9	6.003	1.28
2000	10	6.080	

This gives the population model

$$p(t) = 6.080 \, e^{0.0142t}. \tag{5.87}$$

Observe that, according to this model, there will be

$$p(100) = 6.080 \, e^{1.42} \approx 25.2 \text{ billion} \tag{5.88}$$

people living on the Earth in the year 2100.

5.2.2 Logistic Growth of the World Population

As we discussed earlier in this chapter, exponential growth is usually only a good model for a limited period of time. When the population starts increasing, resources will eventually limit the growth. Looking at Table 5.4, you may notice that b_n seems to decrease as n increases. This means that the constant approximation of b_n is not fully justified. This may be due to limitations on the carrying capacity of the Earth. Such limitations are incorporated in logistic models of the form

$$p'(t) = \alpha \, p(t) \, (1 - p(t)/\beta), \tag{5.89}$$

where β denotes the so-called carrying capacity. Let us now discuss[7] how we can estimate both α and β in (5.89) in order to obtain a more realistic model of the population growth in the future.

It follows from (5.89) that

$$\frac{p'(t)}{p(t)} = \alpha \, (1 - p(t)/\beta). \tag{5.90}$$

---

[7] Another method is discussed in Project 1.

Let $\gamma = -\alpha/\beta$. Then (5.90) reads

$$\frac{p'(t)}{p(t)} = \alpha + \gamma\, p(t). \tag{5.91}$$

Hence, we can determine constants α and γ by fitting a linear function to the observations of

$$\frac{p'(t)}{p(t)}. \tag{5.92}$$

As above, we let

$$b_n = 100\,\frac{p(n+1) - p(n)}{p(n)}. \tag{5.93}$$

The values of the data set (n, b_n) for $n = 0, 1, \ldots, n$ are given in Table 5.4. We now want to determine two constants A and B, such that the data are modeled as accurately as possible by a linear function

$$b_n \approx A + Bp(n), \tag{5.94}$$

in the sense of least squares. It is important to note that we model the data (n, b_n) as a function of p, not as a function of time. We see this from (5.91), where it is clear that if we can make a linear model of p'/p of the form (5.94), we have a logistic model.

By using the standard formulas for approximation by a linear function, see page 162, we find that A and B are determined by the following 2×2 linear system:

$$\begin{pmatrix} 10 & \sum\limits_{n=0}^{9} p(n) \\ \sum\limits_{n=0}^{9} p(n) & \sum\limits_{n=0}^{9} p^2(n) \end{pmatrix} \begin{pmatrix} A \\ B \end{pmatrix} = \begin{pmatrix} \sum\limits_{n=0}^{9} b_n \\ \sum\limits_{n=0}^{9} p(n)b_n \end{pmatrix}. \tag{5.95}$$

Here

$$\sum_{n=0}^{9} p(n) \approx 56.5, \qquad \sum_{n=0}^{9} p^2(n) \approx 319.5,$$

$$\sum_{n=0}^{9} b_n \approx 14.1, \qquad \sum_{n=0}^{9} p(n)b_n \approx 79.6,$$

so we get the linear system

$$\begin{pmatrix} 10 & 56.5 \\ 56.5 & 319.5 \end{pmatrix} \begin{pmatrix} A \\ B \end{pmatrix} = \begin{pmatrix} 14.1 \\ 79.6 \end{pmatrix}, \tag{5.96}$$

which has the solution

$$\begin{aligned} A &= 2.7455, \\ B &= -0.2364. \end{aligned} \tag{5.97}$$

Consequently, the data indicate that

$$100 \, \frac{p'(t)}{p(t)} \approx 2.7455 - 0.2364 \, p(t), \tag{5.98}$$

or

$$p'(t) \approx 0.027 \, p(t) - 0.00236 \, p^2(t). \tag{5.99}$$

So, the logistic model takes the form

$$p'(t) \approx 0.027 \, p(t) \, (1 - p(t)/11.44). \tag{5.100}$$

Here, we note that, according to this model, 11.44 billion seems to be the carrying capacity of the Earth. We also note that the analytical solution of (5.100) with the initial condition

$$p(0) \approx 6.08,$$

where $t = 0$ corresponds to the year 2000, is given by

$$p(t) \approx \frac{69.5}{6.08 + 5.36 \, e^{-0.027t}}, \tag{5.101}$$

see (5.75). Note that the model (5.101) predicts that there are

$$p(100) \approx 10.79 \text{ billion}$$

people on the Earth in the year of 2100. This should be compared with 25.2 billion, which was predicted by the purely exponential model. In Fig. 5.14 we graph the analytical solution of the exponential model (5.87) and the logistic model (5.101), respectively.

Let us conclude this section with a warning. Do not put your money on these estimates. They are just meant to briefly sketch how such computations can be done. There are serious and continuous research efforts to find out how the population evolves in various parts of the world. In some of the fastest-growing regions, serious

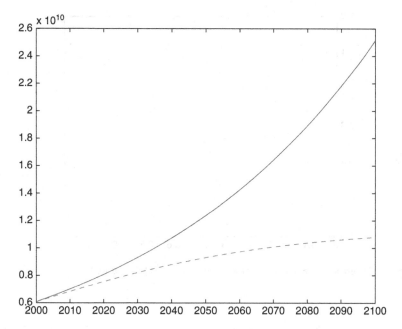

Fig. 5.14 Models of the population growth on the Earth based on an exponential model (*solid curve*) and a logistic model (*dashed curve*)

efforts at birth control have limited the population growth. This may be the reason for the rather low estimates on future growth issued recently. Note also that a different way of estimating the growth rate α and the carrying capacity β leads to much lower estimates, see Project 5.5.

5.3 Least Squares Approximations of Functions

We have seen how to make models of discrete data sets and how to apply these techniques to generate coefficients in models of population growth. Now we will face a different but related problem: Suppose we have data given by a function $y = y(t)$. How can we make a linear model of such a function? Suppose we are interested in the function

$$y(t) = \ln\left(\frac{1}{10}\sin(t) + e^t\right). \tag{5.102}$$

This function is plotted in Fig. 5.15 and we observe that y is almost linear. It is not completely trivial to read this from (5.102). But based on the graph, it seems appropriate to approximate y by a linear function. Just by guessing

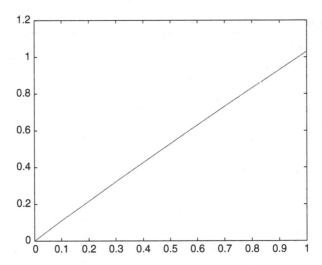

Fig. 5.15 The graph of the function $y(t) = \ln\left(\frac{1}{10}\sin(t) + e^t\right)$ for $t \in [0, 1]$

$$p(t) = t, \tag{5.103}$$

we see from Fig. 5.16 that p is a good approximation of y on the interval $[0, 1]$. But is it the best possible linear approximation? In Fig. 5.17 we have plotted y and p for $t \in [0, 10]$ and now we see that the approximation is really good. If we consider y a bit closer, we see that as t increases,

$$\frac{1}{10}\sin(t) + e^t \approx e^t.$$

For instance, at $t = 10$, we have

$$\frac{1}{10}\sin(10) + e^{10} \approx 22026.41$$

and

$$e^{10} \approx 22026.46.$$

So for large t we have

$$y(t) = \ln\left(\frac{1}{10}\sin(t) + e^t\right) \approx \ln(e^t) = t.$$

This explains the good approximation.

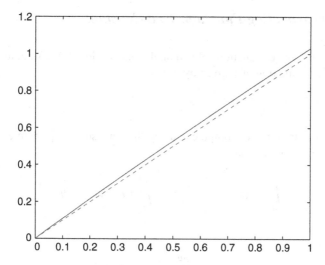

Fig. 5.16 The function $y(t) = \ln\left(\frac{1}{10}\sin(t) + e^t\right)$ (*solid curve*) and a linear approximation (*dashed line*) on the interval $t \in [0, 1]$

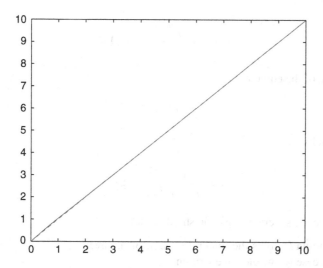

Fig. 5.17 The function $y(t) = \ln\left(\frac{1}{10}\sin(t) + e^t\right)$ (*solid curve*) and a linear approximation (*dashed line*) on the interval $t \in [0, 10]$

But this is, of course, just pure luck and the example was constructed to obtain exactly this effect. We obviously need a more systematic way of computing approximations of functions. And that is the purpose of this section: to provide methods for computing constant, linear and quadratic approximations of functions.

5.3.1 *Approximating Functions by a Constant*

Let $y = y(t)$ be a given function defined on the interval $[a, b]$. We want to compute a constant approximation of y given by

$$p(t) = \alpha \tag{5.104}$$

for $t \in [a, b]$. Our aim is to compute α using the least squares principle. Thus we want to minimize the integral

$$\int_a^b (p(t) - y(t))^2 \, dt = \int_a^b (\alpha - y(t))^2 \, dt. \tag{5.105}$$

Define

$$F(\alpha) = \int_a^b (\alpha - y(t))^2 \, dt \tag{5.106}$$

and compute

$$F'(\alpha) = 2 \int_a^b (\alpha - y(t)) \, dt. \tag{5.107}$$

The solution of the equation

$$F'(\alpha) = 0 \tag{5.108}$$

is thus given by

$$\alpha = \frac{1}{b - a} \int_a^b y(t) dt. \tag{5.109}$$

Here there are several things you should note:

(a) The formula for α is the integral version of the average of y on $[a, b]$. In the discrete case we would have written

$$\alpha = \frac{1}{n} \sum_{i=1}^n y_i, \tag{5.110}$$

If y_i in (5.110) is $y(t_i)$, where $t_i = a + i \Delta t$ and $\Delta t = \frac{b-a}{n}$, then

$$\frac{1}{n} \sum_{i=1}^n y_i = \frac{1}{b - a} \Delta t \sum_{i=1}^n y(t_i) \approx \frac{1}{b - a} \int_a^b y(t) \, dt.$$

So it is fair to conclude that (5.109) is a natural continuous version of the discrete average given in (5.110).

(b) Did you notice that we have used

$$\frac{d}{d\alpha} \int_a^b (\alpha - y(t))^2 dt = \int_a^b \frac{\partial}{\partial \alpha}(\alpha - y(t))^2 dt ?$$

Is that a legal operation? We discuss this in Exercise 5.5.

(c) Is α given by (5.109) really a minimum? Yes it is, since

$$F''(\alpha) = 2(b - a) > 0.$$

Example 5.1. Consider

$$y(t) = \sin(t)$$

defined on $0 \le t \le \pi/2$. A constant approximation of y is given by

$$p(t) = \alpha \overset{(5.109)}{=} \frac{2}{\pi} \int_0^{\pi/2} \sin(t)\, dt = \frac{-2}{\pi} [\cos(t)]_0^{\pi/2}$$

$$= \frac{-2}{\pi}(0 - 1) = \frac{2}{\pi}.$$

∎

Example 5.2. Consider

$$y(t) = e^t - e^{-t}$$

defined on $0 \le t \le 1$. A constant approximation of y is given by

$$p(t) = \alpha \overset{(5.109)}{=} \int_0^1 e^t - e^{-t}\, dt = [e^t + e^{-t}]_0^1 = e + e^{-1} - (1 + 1)$$

$$= e + e^{-1} - 2 \approx 1.086.$$

∎

Example 5.3. Consider

$$y(t) = t^2 + \frac{1}{10}\cos(t)$$

defined on $0 \le t \le 1$. A constant approximation of y is given by

$$p(t) = \alpha \overset{(5.109)}{=} \int_0^1 \left(t^2 + \frac{1}{10}\cos(t) \right) dt = \left[\frac{1}{3}t^3 + \frac{1}{10}\sin(t) \right]_0^1$$

$$= \frac{1}{3} + \frac{1}{10}\sin(1) \approx 0.417.$$

∎

Below, we will also compute linear and quadratic approximations of the functions in examples 5.1–5.3. We will graph the functions and their approximations there (see pages 184–185).

5.3.2 Approximation Using Linear Functions

Again we consider a function $y = y(t)$ defined on an interval $[a, b]$. Our aim is to compute a linear approximation

$$p(t) = \alpha + \beta t \tag{5.111}$$

of y using the principle of least squares.

Define

$$F(\alpha, \beta) = \int_a^b (\alpha + \beta t - y(t))^2 dt. \tag{5.112}$$

A minimum of F is obtained by finding α and β such that

$$\frac{\partial F}{\partial \alpha} = \frac{\partial F}{\partial \beta} = 0. \tag{5.113}$$

Since

$$\frac{\partial F}{\partial \alpha} = 2 \int_a^b (\alpha + \beta t - y(t)) \, dt, \tag{5.114}$$

and

$$\frac{\partial F}{\partial \beta} = 2 \int_a^b (\alpha + \beta t - y(t)) t \, dt, \tag{5.115}$$

the requirements given by (5.113) now read

$$(b-a)\alpha + \frac{1}{2}(b^2 - a^2)\beta = \int_a^b y(t)\,dt,$$

$$\frac{1}{2}(b^2 - a^2)\alpha + \frac{1}{3}(b^3 - a^3)\beta = \int_a^b t\,y(t)\,dt. \tag{5.116}$$

The linear system (5.116) determines α and β. Let us now apply the method to the same functions as we considered for constant approximations.

Example 5.4. Consider

$$y(t) = \sin(t) \tag{5.117}$$

defined on $0 \le t \le \pi/2$. Since

$$\int_0^{\pi/2} \sin(t)\,dt = 1$$

and

$$\int_0^{\pi/2} t\,\sin(t)\,dt = 1,$$

the linear system (5.116) now reads

$$\begin{pmatrix} \pi/2 & \pi^2/8 \\ \pi^2/8 & \pi^3/24 \end{pmatrix} \begin{pmatrix} \alpha \\ \beta \end{pmatrix} = \begin{pmatrix} 1 \\ 1 \end{pmatrix} \tag{5.118}$$

and the solution is given by

$$\begin{pmatrix} \alpha \\ \beta \end{pmatrix} = \frac{1}{\pi^2}\begin{pmatrix} 8\pi - 24 \\ \dfrac{96}{\pi} - 24 \end{pmatrix} \approx \begin{pmatrix} 0.115 \\ 0.664 \end{pmatrix}, \tag{5.119}$$

so

$$p(t) \approx 0.115 + 0.664\,t.$$

■

Example 5.5. Consider

$$y(t) = e^t - e^{-t} \tag{5.120}$$

defined on $0 \leq t \leq 1$. The linear system (5.116) now reads

$$\begin{pmatrix} 1 & \frac{1}{2} \\ \frac{1}{2} & \frac{1}{3} \end{pmatrix} \begin{pmatrix} \alpha \\ \beta \end{pmatrix} = \begin{pmatrix} e + e^{-1} - 2 \\ 2e^{-1} \end{pmatrix} \tag{5.121}$$

and the solution is

$$\alpha = 4e - 8e^{-1} - 8 \approx -0.070,$$
$$\beta = 18e^{-1} - 6e + 12 \approx 2.312.$$

Hence, the linear least squares approximation of (5.120) is given by

$$p(t) = -0.070 + 2.312\,t. \tag{5.122}$$

∎

Example 5.6. Consider

$$y(t) = t^2 + \frac{1}{10}\cos(t) \tag{5.123}$$

defined on $0 \leq t \leq 1$. The linear system (5.116) then reads

$$\begin{pmatrix} 1 & \frac{1}{2} \\ \frac{1}{2} & \frac{1}{3} \end{pmatrix} \begin{pmatrix} \alpha \\ \beta \end{pmatrix} = \begin{pmatrix} \frac{1}{3} + \frac{1}{10}\sin(1) \\ \frac{3}{20} + \frac{1}{10}\cos(1) + \frac{1}{10}\sin(1) \end{pmatrix}$$

and thus

$$\alpha \approx -0.059,$$
$$\beta \approx 0.953,$$

and we conclude that the linear least squares approximation of (5.123) is given by

$$p(t) \approx -0.059 + 0.953\,t. \tag{5.124}$$

∎

5.3.3 Approximation Using Quadratic Functions

We seek a quadratic function

$$p(t) = \alpha + \beta + \gamma\,t^2 \tag{5.125}$$

that approximates a given function $y = y(t)$, $a \le t \le b$, in the sense of least squares. Let

$$F(\alpha, \beta, \gamma) = \int_a^b (\alpha + \beta t + \gamma t^2 - y(t))^2 dt \qquad (5.126)$$

and define α, β, and γ to be the solution of the three equations:

$$\frac{\partial F}{\partial \alpha} = \frac{\partial F}{\partial \beta} = \frac{\partial F}{\partial \gamma} = 0. \qquad (5.127)$$

Here,

$$\frac{\partial F}{\partial \alpha} = 2 \int_a^b (\alpha + \beta t + \gamma t^2 - y(t)) \, dt,$$

$$\frac{\partial F}{\partial \beta} = 2 \int_a^b (\alpha + \beta t + \gamma t^2 - y(t)) t \, dt,$$

and

$$\frac{\partial F}{\partial \gamma} = 2 \int_a^b (\alpha + \beta t + \gamma t^2 - y(t)) t^2 \, dt.$$

From (5.127), it follows that

$$(b-a)\alpha + \frac{1}{2}(b^2 - a^2)\beta + \frac{1}{3}(b^3 - a^3)\gamma = \int_a^b y(t) \, dt,$$

$$\frac{1}{2}(b^2 - a^2)\alpha + \frac{1}{3}(b^3 - a^3)\beta + \frac{1}{4}(b^4 - a^4)\gamma = \int_a^b t \, y(t) \, dt, \qquad (5.128)$$

$$\frac{1}{3}(b^3 - a^3)\alpha + \frac{1}{4}(b^4 - a^4)\beta + \frac{1}{5}(b^5 - a^5)\gamma = \int_a^b t^2 \, y(t) \, dt.$$

The linear system (5.128) determines the coefficients α, β, and γ in the quadratic least squares approximation (5.125) of $y(t)$.

Example 5.7. For the function

$$y(t) = \sin(t), \quad 0 \le t \le \pi/2, \qquad (5.129)$$

the linear system (5.128) reads

$$\begin{pmatrix} \pi/2 & \pi^2/8 & \pi^3/24 \\ \pi^2/8 & \pi^3/24 & \pi^4/64 \\ \pi^3/24 & \pi^4/64 & \pi^5/160 \end{pmatrix} \begin{pmatrix} \alpha \\ \beta \\ \gamma \end{pmatrix} = \begin{pmatrix} 1 \\ 1 \\ \pi - 2 \end{pmatrix} \qquad (5.130)$$

and the solution is given by

$$\alpha \approx -0.024, \quad \beta \approx 1.196, \quad \gamma \approx -0.338,$$

which gives the quadratic approximation

$$p(t) = -0.024 + 1.196\,t - 0.338\,t^2. \tag{5.131}$$

∎

Example 5.8. Let

$$y(t) = e^t - e^{-t} \tag{5.132}$$

be defined on $0 \leq t \leq 1$. Then (5.128) takes the form

$$\begin{pmatrix} 1 & 1/2 & 1/3 \\ 1/2 & 1/3 & 1/4 \\ 1/3 & 1/4 & 1/5 \end{pmatrix} \begin{pmatrix} \alpha \\ \beta \\ \gamma \end{pmatrix} = \begin{pmatrix} e + e^{-1} - 2 \\ 2e^{-1} \\ 5e^{-1} + e - 4 \end{pmatrix} \tag{5.133}$$

and the coefficients are given by

$$\alpha \approx 0.019, \quad \beta \approx 1.782, \quad \gamma \approx 0.531.$$

The quadratic least squares approximation of (5.132) is therefore given by

$$p(t) = 0.019 + 1.782\,t + 0.531\,t^2. \tag{5.134}$$

∎

Example 5.9. Let us consider

$$y(t) = t^2 + \frac{1}{10}\cos(t) \tag{5.135}$$

for $0 \leq t \leq 1$. The linear system (5.128) takes the form

$$\begin{pmatrix} 1 & 1/2 & 1/3 \\ 1/2 & 1/3 & 1/4 \\ 1/3 & 1/4 & 1/5 \end{pmatrix} \begin{pmatrix} \alpha \\ \beta \\ \gamma \end{pmatrix} = \begin{pmatrix} \frac{1}{3} + \frac{1}{10}\sin(1) \\ \frac{3}{20} + \frac{1}{10}\cos(1) + \frac{1}{10}\sin(1) \\ \frac{1}{5} + \frac{1}{5}\cos(1) - \frac{1}{10}\sin(1) \end{pmatrix}$$

and the solution is given by

$$\alpha \approx 0.100, \quad \beta \approx -0.004, \quad \gamma \approx 0.957.$$

The quadratic least squares approximation of (5.135) is therefore given by

$$p(t) = 0.100 - 0.004\,t + 0.957\,t^2.$$

■

5.3.4 Summary of the Examples

In the examples above we computed constant, linear and quadratic least squares approximations of three given functions. Here we will give a summary and plot these functions together with their approximations.

Function 1

Let

$$y(t) = \sin(t)$$

be defined on $0 \le t \le \pi/2$. Then the constant, linear and quadratic least squares approximations are given by

$$p_0(t) \approx 0.637,$$
$$p_1(t) \approx 0.115 + 0.664\,t,$$
$$p_2(t) \approx -0.024 + 1.196\,t - 0.338\,t^2,$$

respectively. All the functions are plotted in Fig. 5.18.

Function 2

Let

$$y(t) = e^t - e^{-t}$$

be defined on $0 \le t \le 1$. Then the constant, linear, and quadratic least squares approximations are given by

$$p_0(t) \approx 1.086,$$
$$p_1(t) \approx -0.070 + 2.312\,t,$$
$$p_2(t) \approx 0.019 + 1.782\,t + 0.531\,t^2,$$

respectively, see Fig. 5.19.

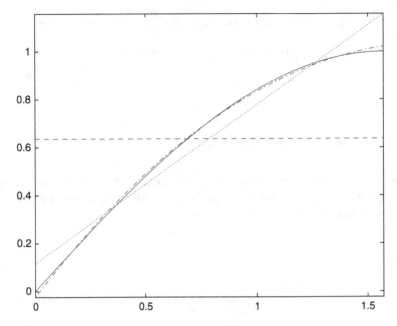

Fig. 5.18 The function $y(t) = \sin(t)$ (*solid curve*) and its least squares approximations: constant (*dashed line*), linear (*dotted line*) and quadratic (*dash-dotted curve*)

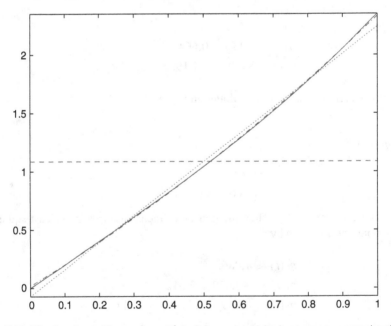

Fig. 5.19 The function $y(t) = e^t - e^{-t}$ (*solid curve*) and its least squares approximations: constant (*dashed line*), linear (*dotted line*) and quadratic (*dash-dotted curve*)

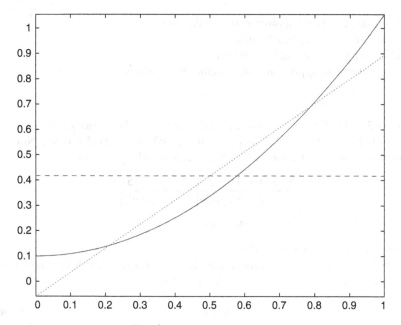

Fig. 5.20 The function $y(t) = t^2 + \frac{1}{10}\cos(t)$ (*solid curve*) and its least squares approximations: constant (*dashed line*), linear (*dotted line*) and quadratic (*dash-dotted curve*)

Function 3

Let

$$y(t) = t^2 + \frac{1}{10}\cos(t)$$

be defined on $0 \le t \le 1$. Then

$$p_0(t) \approx 0.417,$$
$$p_1(t) \approx -0.059 + 0.953\,t,$$
$$\text{and} \quad p_2(t) \approx 0.100 - 0.004\,t + 0.957\,t^2$$

denote the constant, linear and quadratic least squares approximations; see Fig. 5.20. In this case, it is hard to distinguish between p_2 and y. This is not surprising, since y is dominated by t^2.

5.4 Exercises

Exercise 5.1. Consider the data set

t_i	5	6	7
y_i	8	4	0

(a) Compute a constant approximation to the data.
(b) Compute a linear approximation.
(c) Compute a quadratic approximation.
(d) Graph the three approximations together with the data points.

◇

Exercise 5.2. On the Internet, you can find information about prices of used cars. In the table below, we have listed the prices of used BMWs of the 3-series in Norway. We give the average price for a car that is 1, 2, 3, 4 and 5 years old.

Years t_i	1	2	3	4	5
Price y_i	55.2	44.9	37.9	35.3	30.1

The prices are given in thousands of euros.

(a) Build a linear model of the data.
(b) Graph the data and the linear model. Instead of a standard linear model, we assume that the price can be modeled as follows,

$$p(t) = \alpha e^{\beta t}, \tag{5.136}$$

where we expect β to be negative. By taking the logarithm of both side of (5.136), we get

$$\ln(p(t)) = \ln(\alpha) + \beta t. \tag{5.137}$$

This shows that we can compute α and β in (5.136) by constructing a linear model of the data given by $(t_i, \ln(y_i))$.
(c) Compute the entries of the following table:

Years t_i	1	2	3	4	5
$z_i = \ln(y_i)$					

(d) Compute a linear model

$$q(t) = \gamma + \beta t$$

of the data in (c).
(e) Since $\ln(\alpha) = \gamma$, compute $\alpha = e^{\gamma}$ and define

$$p(t) = \alpha e^{\beta t}.$$

(f) Plot the price-time data, the linear model and the exponential model. Which model seems to fit the data better?

◇

Table 5.5 The CPU time (in seconds) needed for solving (5.138) using an explicit Euler scheme on a 600MHz Pentium III processor

n	CPU time	y_n
100,000	0.05	9.9181
200,000	0.09	10.549
300,000	0.13	10.919
400,000	0.18	11.183

Exercise 5.3. When you are using programs to construct simulations, you often need an estimate of the CPU time needed for running the simulation, since certain parameters have changed. Sometimes, this may be a very hard problem to solve, because it is not clear how the CPU effort depends on the parameters. In this exercise, we will consider a very simple program in order to illustrate how timing estimates can be generated.

Suppose we want a numerical solution of the problem

$$y'(t) = e^{y(t)}, \qquad y(0) = 0 \tag{5.138}$$

for $0 \le t \le 1$. We try to solve (5.138) using a standard explicit Euler scheme,

$$y_{k+1} = y_k + \Delta t \, e^{y_k}, \qquad k = 0, 1, \dots, n - 1, \tag{5.139}$$

with $y_0 = 0$. Here, $\Delta t = 1/n$, where $n > 0$ is an integer. In Table 5.5 we have listed the CPU time that is needed by a simple C program to compute y_n at time $t = 1$ on a 600 MHz Pentium III processor.

From scheme (5.139), it is reasonable to assume that the CPU time, $c(n)$, can be appropriately modeled using a linear function, i.e.,

$$c(n) = \alpha + \beta n. \tag{5.140}$$

(a) Use the data in the table to determine α and β by the method of least squares.
(b) Estimate the CPU time needed in the cases of $n = 10^6$ and $n = 10^7$.

$\diamond$

Exercise 5.4. Really big problems in scientific computing are solved by coupling several computers together in a network. These computers have to exchange information during the computation. The overall computing time thus depends on how fast each computer does its job and how fast they are able to communicate with each other. In this exercise, we will discuss a model of the communication speed. Let us assume that we want to exchange vectors of real numbers (stored in double precision) between a computer A and a computer B. The vector y has n entries.

Let $T = T_n$ be the time needed to send the vector y from A to B. Actually, we send y first from A to B and then back to A. The cost of the "one-way" communication can be found by dividing the "round-trip" time by a factor 2. In this way we

Table 5.6 The one-way communication time (in seconds) for sending a vector containing n double-precision numbers on a fast Ethernet network

n (Length of the vector y)	T_n (one-way communication time)
65536	0.048755
131072	0.097074
262144	0.194003
524288	0.386721
1048576	0.771487

only need to do time measurement on one computer, say A. In the table below we have listed values of n and the communication time T_n.

(a) Construct a linear least squares model of the form

$$T(n) = \alpha + \beta n.$$

(b) Try to explain why $T(0) = \alpha \neq 0$.
(c) In computer terms, α is referred to as the *latency* of the communication network, and $1/\beta$ can be regarded as the practical *bandwidth*, which should be somewhat lower than the peak-performance bandwidth. Normally, bandwidth is measured in gigabits (10^9) per second. Use the fact that one double-precision number has 64 bits and report the bandwidth of the network that has been used for producing the measurements given in Table 5.6.

$\diamond$

Exercise 5.5. In the text above we used the fact that

$$\frac{d}{d\alpha} \int_a^b (\alpha - p(t))^2 dt = \int_a^b \frac{\partial}{\partial \alpha}(\alpha - p(t))^2 dt. \qquad (5.141)$$

We will indicate a proof of this. But let us first note that for sufficiently smooth functions $\{g_\varepsilon(t)\}$, we have that

$$\lim_{\varepsilon \to 0} \int_a^b g_\varepsilon(t)\, dt = \int_a^b \lim_{\varepsilon \to 0} g_\varepsilon(t)\, dt, \qquad (5.142)$$

see e.g. Royden [25]. Next we consider the problem of computing

$$\frac{d}{d\alpha} \int_a^b F(\alpha, t) dt \qquad (5.143)$$

for a smooth function F. By definition, we have

$$\frac{d}{d\alpha} \int_a^b F(\alpha, t) dt = \lim_{\varepsilon \to 0} \frac{1}{\varepsilon} \left[\int_a^b F(\alpha + \varepsilon, t)\, dt - \int_a^b F(\alpha, t)\, dt \right]. \qquad (5.144)$$

(a) Use (5.142) and (5.144) to show that

$$\frac{d}{d\alpha}\int_a^b F(\alpha,t)dt = \int_a^b \frac{\partial}{\partial\alpha}F(\alpha,t)\,dt. \qquad (5.145)$$

(b) Strictly speaking, the derivation above is only valid when (5.142) holds, and it holds for sufficiently smooth functions and we do not want to be precise on that issue. But the relation (5.141) may also be derived by a direct calculation. Define

$$L(\alpha) = \frac{d}{d\alpha}\int_a^b (\alpha - p(t))^2 dt \qquad (5.146)$$

and

$$R(\alpha) = \int_a^b \frac{\partial}{\partial\alpha}(\alpha - p(t))^2 dt. \qquad (5.147)$$

Show, by direct computations, that

$$L(\alpha) = 2\alpha(b-a) - 2\int_a^b p(t)\,dt$$

and

$$R(\alpha) = 2\alpha(b-a) - 2\int_a^b p(t)\,dt.$$

Conclude that (5.141) thus holds.

◇

Exercise 5.6. Compute a constant approximation, using the least squares method, for the following functions:

(a) $y(t) = 1 + \frac{1}{100}\sin(t), \quad 0 \le t \le \pi.$
(b) $y(t) = e^t, \quad 0 \le t \le 1/e.$
(c) $y(t) = \sqrt{t}, \quad 0 \le t \le \pi.$

◇

Exercise 5.7. Compute a linear least squares approximation for the following functions:

(a) $y(t) = e^t, \quad 0 \le t \le 1.$
(b) $y(t) = e^{-t}, \quad 0 \le t \le 1.$
(c) $y(t) = 1/t, \quad 1 \le t \le 2.$

◇

Exercise 5.8. Compute quadratic least squares approximations of the functions in Exercise 5.7.   ◇

Exercise 5.9. Consider

$$y(t) = e^t$$

for $-1 \leq t \leq 1$.

(a) Make a quadratic least squares approximation of y.
(b) Compute three terms of the Taylor series expansion of y around $t = 0$.
(c) Graph $y(t)$ together with the approximations generated in (a) and (b) and discuss the properties of the approximations.

◇

5.5 Project: Computing Coefficients

In Sect. 5.2 we discussed how to determine the parameters in the exponential model

$$y'(t) = \alpha \, y(t) \tag{5.148}$$

and in the logistic model

$$y'(t) = \alpha \, y(t) \, (1 - y(t)/\beta) . \tag{5.149}$$

The problem is to compute α in (5.148) and α, β in (5.149) based on observations of only y (not $y'(t)$) for some values of $t = t_n$. In the analysis of this problem, we basically introduced the following two ideas:

(a) Approximate $y'(t_n)$ by

$$\frac{y(t_{n+1}) - y(t_n)}{\Delta t} . \tag{5.150}$$

Using this approximation in (5.148), we get

$$\alpha \approx \frac{y(t_{n+1}) - y(t_n)}{y(t_n) \, \Delta t} \tag{5.151}$$

and then we can use measurements of y to get an estimate of α.

(b) In (5.149) we can use (5.150) to get

$$\frac{y(t_{n+1}) - y(t_n)}{y(t_n) \, \Delta t} \approx \alpha - \frac{\alpha}{\beta} \, y(t_n). \tag{5.152}$$

Since we have observations of y, we can compute the left-hand side of (5.152) and use the results to build a linear model of the right-hand side of the equation, cf. Sect. 5.2.2.

The purpose of this project is to introduce a third idea. Using this idea, we eliminate the use of (5.150) and thus avoid the inaccuracies introduced by numerical differentiation. We do, however, introduce errors due to numerical integration. The essence of the third idea is to observe that

$$\frac{d}{dt} \ln y(t) = \frac{y'(t)}{y(t)}. \tag{5.153}$$

In this project we will assume that we have data given by $(t_n, y(t_n))$.

(a) Use (5.153) to show that

$$\int_{t_n}^{t_{n+1}} \frac{y'(t)}{y(t)} \, dt = \ln y(t_{n+1}) - \ln y(t_n). \tag{5.154}$$

(b) Consider (5.148) and show that

$$\alpha = \frac{\ln y(t_{n+1}) - \ln y(t_n)}{t_{n+1} - t_n}. \tag{5.155}$$

(c) Define

$$\alpha_n = \frac{\ln y(t_{n+1}) - \ln y(t_n)}{t_{n+1} - t_n} \tag{5.156}$$

and compute α_0, α_1, α_3, and α_4 using the data in Table 5.2 on page 167. Note that $t_0 = 0$ corresponds to 1950, $t_1 = 1$ corresponds to 1951, and so on.
(d) Verify that

$$\alpha = \frac{1}{5} \sum_{n=0}^{4} \alpha_n \approx 0.01688 \tag{5.157}$$

and compare your result with the result given in (5.82) on page 168.
(e) Define

$$p(t) = 2.555 \, e^{0.01688 \, t} \tag{5.158}$$

and graph both p given by (5.158) and p given by (5.83) for t ranging from 0 to 50.

(f) Set $\gamma = -\alpha/\beta$ and show that (5.149) can be written in the form

$$\frac{y'(t)}{y(t)} = \alpha + \gamma \, y(t). \tag{5.159}$$

(g) Use (5.153) and (5.159) to show that

$$\ln(y(t_{n+1})) - \ln(y(t_n)) = \alpha(t_{n+1} - t_n) + \gamma \int_{t_n}^{t_{n+1}} y(t)\, dt. \tag{5.160}$$

(h) Use (5.160) to show that

$$\frac{\ln\left(y(t_{n+1})/y(t_n)\right)}{t_{n+1} - t_n} \approx \alpha + \frac{\gamma}{2}\left(y(t_{n+1}) + y(t_n)\right). \tag{5.161}$$

(i) We want to reconsider the problem of computing a model of the future growth of the world population. We therefore want to use the data in Table 5.4 on page 170 in order to determine the coefficients α and γ in (5.161). In (5.161) we note that

$$t_{n+1} - t_n = 1.$$

In addition, we define

$$c_n = \ln\left(y(t_{n+1})/y(t_n)\right) \tag{5.162}$$

and

$$d_n = \frac{1}{2}\left(y(t_{n+1}) + y(t_n)\right). \tag{5.163}$$

From the data in Table 5.3, c_n and d_n are readily computed. Make a table consisting of columns for the year, n, $y(t_n)$, c_n, and d_n, in the same style as in Table 5.3.

(j) Using the definitions (5.162) and (5.163) together with (5.161), we have

$$c_n \approx \alpha + \gamma \, d_n, \qquad n = 0, 1, \ldots, 9, \tag{5.164}$$

where c_n, d_n are given and where α and γ are to be determined. We want to compute α and β using the method of least squares. In order to do this, we define

$$F(\alpha, \gamma) = \sum_{n=0}^{9} (\alpha + \gamma \, d_n - c_n)^2. \tag{5.165}$$

Explain why we want to compute the values of α and γ that minimize F.

(k) Show that the requirement

$$\frac{\partial F}{\partial \alpha} = \frac{\partial F}{\partial \gamma} = 0$$

leads to the linear equations

$$10\alpha + \left(\sum_{n=0}^{9} d_n\right)\gamma = \sum_{n=0}^{9} c_n$$

$$\left(\sum_{n=0}^{9} d_n\right)\alpha + \left(\sum_{n=0}^{9} d_n^2\right)\gamma = \sum_{n=0}^{9} c_n d_n.$$

(5.166)

(l) Solve the linear system (5.166) with respect to α and γ. Verify that

$$\alpha \approx 0.0368 \quad \text{and} \quad \gamma \approx -0.0040.$$

(5.167)

(m) Use the fact that $\gamma = -\alpha/\beta$ to show that

$$\beta \approx 9.2.$$

(5.168)

(n) Show that the solution of the logistic model with an initial data $y_0 = 6.08$, which corresponds to the population on the Earth in the year 2000, is given by

$$y(t) = \frac{55.94}{6.08 + 3.12\, e^{-0.0368\, t}}.$$

(5.169)

(o) Plot $y(t)$ given by (5.169) and $p(t)$ given by (5.101).

Chapter 6
About Scientific Software

When a science problem is solved with the aid of numerical computations, the solution procedure involves several steps:

1. Understanding the problem and formulating a mathematical model
2. Using numerical methods to solve the mathematical problems
3. Implementing the numerical methods in a computer program
4. Verifying that the results from the program are mathematically correct
5. Applying the program to the scientific problem and interpreting the results

Normally, this is a repetitive cycle: interpretation of the results often leads to adjustments in the mathematical model, numerical methods, and the computer program.

The cycle listed above typically has a "theory" part and 'a 'practice" part. Most books emphasize theory, i.e., deriving and analyzing mathematical models and numerical methods. The practice part, consisting of translating the models and methods to running code, producing numbers, and verifying the results, is equally important and requires skills that must be developed systematically. The present chapter provides the first steps toward gaining the necessary skills.

Scientific Software Requirements

Software performing scientific computations must be

1. Mathematically correct
2. Efficient (speed, memory usage)
3. Easy to maintain and extend

If there is an error in the program, the calculations will most likely be wrong and the results will become useless. Many types of numerical computations demand days or weeks of computing time and the combined memory of a large collection of computers. Efficiency with respect to speed and memory usage is thus of utmost importance. Unfortunately, many efficiency improvements also easily introduce errors in the code. The complexity of scientific software has reached the limit

A. Tveito et al., *Elements of Scientific Computing*, Texts in Computational Science and Engineering 7, DOI 10.1007/978-3-642-11299-7_6,

where maintenance and future extensions have become very difficult, and the modification of complicated codes easily leads to a significant danger of introducing errors. Careful design of scientific software systems is therefore necessary.

Software Development Skills

The scientific software developer needs several skills to meet the demands of the previous paragraph. These skills include

1. Understanding the mathematical problem to be solved
2. Understanding the numerical methods to be used
3. Designing appropriate algorithms and data structures
4. Selecting the most suitable programming language and tools
5. Using libraries
6. Verifying the correctness of the results

The first two points are critical to the last point. Ideally, the software developer should also have an understanding of the physical problem being solved, but as long as the mathematical model for the physical problem is specified completely, the software development (in terms of programming and verification) is decoupled from the original problem. In fact, such a decoupling encourages the production of software that can be applied to a range of different physical problem areas. Points 3–5 are closely tied and have often been ignored in the literature. One reason may be that up until the 1990s almost all scientific software developers used Fortran as the programming tool. The tendency now is to use a collection of tools to solve a given problem, i.e., the developer needs to select the right tool for each subtask. This requires a knowledge of a range of tools.

Scientific software development, and especially the testing phase, is known to be very time consuming and the number one reason why budgets are so frequently exceeded in scientific computing projects. Students also tend to spend much time on going from the mathematics to a working code. The rule of thumb is therefore to avoid developing numerical software if possible, i.e., one should reuse existing software to as large extent as possible. For some problem areas, there are software packages providing all functionality you need in the solution process, and there is no need for software development. Most scientific problems, however, demand some kind of programming, fortunately in the form of calling up available functionality in various libraries. How to do this efficiently again requires knowledge of different programming tools.

This Chapter

Fortunately, there are many techniques to speed up the development of numerical software and increase the reliability of an implementation. The present chapter gives a first introduction to writing scientific software, with emphasis on two aspects:

– The reliable translation of mathematical algorithms into working code
– A glimpse of some relevant programming languages and tools

We work with very simple numerical sample problems such that the type of problem and its solution methods are hopefully well understood. This enables us to concentrate on software issues. However, many of the important software issues listed above are beyond the scope of this chapter, because they do not become evident before one tries to implement sets of complicated algorithms.

6.1 Algorithms Expressed as Pseudo Code

Let us first address the issue of *creating a program without errors*. We suggest approaching the challenge in two steps:

1. Express the numerical problem to be solved and the methods to be used in a *complete algorithm*.
2. Translate the algorithm into a computer code using a specific programming language.

Jumping directly from a compact description of the numerical problem to computer programming is often an error-prone procedure, even if you are very experienced with both the numerics of the problem and software development. The smaller you can make the gap between an algorithm, expressed in mathematical terms on a piece of paper, and a computer program, the easier it will be to develop the program and check it for correctness. A direct consequence of this strategy is that we end most of the discussion of numerical methods in this book with detailed, complete algorithms. In most cases it will be straightforward to translate the algorithm into a program, provided you are experienced with programming, which is taken as a prerequisite when working with this book.

We emphasize the adjective *complete* in our phrase *complete algorithms*; this means that all numerical details are covered in the algorithm such that the implementation only involves a translation and no numerical or mathematical issues.

The art of formulating algorithms sufficiently close to computer programs is best illustrated through examples. Two simple problem areas are considered here: numerical integration by the trapezoidal rule and solution of an ordinary differential equation (ODE) by the Heun scheme.

6.1.1 Basic Ingredients of Pseudo Codes

The mapping of a physical problem into computer code has some characteristics that are independent of the problem being solved. First, the physical problem must be expressed as a mathematical model, i.e., a set of mathematical problems to be solved. For each mathematical problem we need to select an appropriate numerical solution method. Each method is expressed as an algorithm and implemented as a piece of code, normally a function. The code is then a collection of functions implementing the various algorithms (solution steps) of the whole problem. The communication between the functions depends on what type of data structures we use. Therefore, creating numerical software is an interplay between algorithms and data structures. Typical data structures are scalar variables, arrays, and functions.

A complete algorithm is normally expressed as *pseudo code*, which is a mixture of mathematical formulas and instructions known from computer languages. The purpose is to get an algorithm that easily translates into a computer program. Some important instructions used in pseudo code (and computer programs) are

1. Assignments; the notation $s \leftarrow s + 2$ means that we assign the value of the expression $s + 2$ to s, i.e., s is over-written by a new value,[1]
2. For loops[2]; loops controlled by a counter running between two values with a certain step length,
3. While loops; loops controlled by a boolean condition,
4. Functions[3]; subprograms taking a set of variables as input (arguments) and returning a set of variables,
5. Arrays; sequences of numbers, such as $u_1, u_2, \ldots, u_{10}$,

As an example, consider the sum

$$\sum_{i=1}^{n-1} f(a + ih),$$

where f is a function of a scalar variable, and a and h are constants. We want to express this sum in a way that translates more or less directly to statements in a computer program. Normally, a sum is computed by a *for loop*, using a variable (here s) that accumulates the individual contributions $f(a + ih)$ to the sum:

```
s = 0
for i = 1,...,n − 1
    s ← s + f(a + ih)
end for
```

Note that the parameters a and h must be pre-computed for this algorithm to work. After the loop, s equals $\sum_{i=1}^{n-1} f(a + ih)$.

Instead of a for loop, we can use a *while loop* to implement the sum:

```
s = 0
i = 1
```

---

[1] Most programming languages would just use a notation such as s=s+2, but this notation is mathematically incorrect, so we prefer to use the left arrow $\leftarrow$ in our mathematically-oriented algorithmic notation.

[2] Also called do loops (named after the corresponding Fortran construction).

[3] Also called subprogram, subroutine, or procedure.

```
while i ≤ n − 1
    s ← s + f(a + ih)
    i ← i + 1
end while
```

The forthcoming sections provide more examples.

6.1.2 Integration

Suppose we want to integrate a function $f(x)$ from $x = a$ to $x = b$ with the aid of the composite trapezoidal RULE (derived in Sect. 1.3). Mathematically, we can express this method for computing an integral as follows:

$$\int_a^b f(x)dx \approx \frac{h}{2}f(a) + \frac{h}{2}f(b) + \sum_{i=1}^{n-1} hf(a + ih), \quad h = \frac{b-a}{n}. \tag{6.1}$$

This is a compact formulation that contains all the necessary information for calculating the integral on a computer. There are some steps from (6.1) to running code, but you have hopefully already tried to implement methods such as (6.1) while working with exercises in the introductory chapter. Now it is time to revise such implementational work and adopt good habits.

Before thinking of an implementation, we should express a compact formula such as (6.1) as an algorithm, i.e., as a set of steps that naturally translate into similar steps in a computer code. The complete algorithm corresponding to the mathematical formulation (6.1) can be expressed as follows:

```
h = b−a
    n
s = 0
for i = 1, . . . , n − 1
    s ← s + hf(a + ih)
end for
s ← s + h/2 f(a) + h/2 f(b)
```

The nature of this algorithm is that we provide a, b, n, and $f(x)$ as *input* and get the answer s as *output*. Such input–output algorithms are often conveniently expressed in a way that resembles the concept of functions or subprograms in programming languages.

For example, we could write the algorithms in a notation such as

```
trapezoidal (a, b, f, n)
    ... do something ... store final result in s
    return s
```

This notation is hopefully self-explanatory for anyone having some experience with computer programming; we explicitly indicate that a, b, f, and n are input data and s the output, i.e., the result of the algorithm.

The complete algorithm for the trapezoidal rule, using the proposed notation, appears separately as Algorithm 6.1.

Algorithm 6.1

Trapezoidal Integration.

trapezoidal (a, b, f, n)
$\quad h = \frac{b-a}{n}$
$\quad s = 0$
$\quad$ for $i = 1, \ldots, n - 1$
$\quad\quad s \leftarrow s + hf(a + ih)$
$\quad$ end for
$\quad s \leftarrow s + \frac{h}{2} f(a) + \frac{h}{2} f(b)$
$\quad$ return s

The notation used in Algorithm 6.1 is a typical example of mathematical pseudo code.

Having the algorithm available as mathematical pseudo code makes the step toward a working computer code quite small. Some readers might think that the step is so small that we could equally well have written the lines directly in a specific computer language. This argument is relevant in the present example, where there is hardly any complexity of the algorithm implied by (6.1). In more complicated problems, however, the mathematical pseudo code is still quite close to the mathematical exposition of the numerical methods, whereas a computer code will contain many more details specific to the chosen language. Experience shows that the two-step procedure of first deriving a correct mathematical pseudo code and then using it as a reference when translating, and later checking, the computer implementation simplifies software development considerably.

Another attractive feature of the mathematical pseudo code is that we turn numerical solution procedures into a form that is ready for implementation. However, the choice of programming language, programming style, data types, and computing instructions (which differ highly between individuals and organizations) remains open.

6.1.3 Optimization of Algorithms and Implementations

There is a strong tradition in scientific computing of carefully examining algorithms with the aim of reducing the number of arithmetic operations and the need for storing computed information. This is because advanced problems in scientific computing require optimization of speed and memory usage.

Saving Arithmetic Operations

Examining the algorithm above, we see that the factor h appears in all terms. This is perhaps even more obvious from the mathematical formula (6.1). We can avoid n multiplications by h through a factorization of (6.1). We can also be more careful with the factor $1/2$; division is often slower than multiplication (perhaps by a factor of 4, depending on the hardware). A computationally more efficient version of (6.1) therefore reads

$$\int_a^b f(x)dx \approx h \left\{ 0.5(f(a) + f(b)) + \sum_{i=1}^{n-1} f(a + ih) \right\} , \quad h = \frac{b-a}{n}. \quad (6.2)$$

To really see why (6.2) is more efficient than (6.1) we can count the number of operations: additions, subtractions, multiplications, divisions, and $f(x)$ function calls. We find that (6.1) has $2n$ multiplications, $2n$ additions, three divisions, and $n + 1$ function calls. The formula in (6.2) implies $n + 1$ multiplications, $2n$ additions, one division, and $n + 1$ function calls. If $f(x)$ is a complicated function requiring many arithmetic operations, such as $\sin x$ or e^x, evaluation of f will dominate the work of the algorithm. In that case there are minor differences between (6.1) and (6.2); the total work is well approximated by $n + 1$ calls to the function f.

It is common to introduce *floating-point operation* (FLOP) as a synonym for addition, subtraction, multiplication, or division. However, this can be misleading on some hardware where division is significantly slower than the three other operations. Counting one multiplication/division and one addition/subtraction as a single floating-point operation is also common, because modern CPUs often have the possibility of running pieces of a compound expression in parallel.

We can avoid $n - 1$ multiplications ih in (6.2) by incrementing the function argument by h in each pass in the for loop used to compute the sum. The mathematical pseudo code incorporating this trick in (6.2) is expressed in Algorithm 6.2. The total work is now one multiplication, $2n$ additions, one division, and $n + 1$ function calls.

Algorithm 6.2

Optimized Trapezoidal Integration.

```
trapezoidal (a, b, f, n)
    h = b−a
        n
    s = 0
    x = a
    for i = 1, ..., n − 1
        x ← x + h
        s ← s + f(x)
    end for
    s ← s + 0.5(f(a) + f(b))
    s ← hs
    return s
```

When performing a single numerical integration of a function $f(x)$ with a "reasonable" n, the computations will, on today's machines, be so fast and use so little memory that this latter optimization has no practical advantage; you get the answer immediately anyway. In fact, rewriting from (6.1) to (6.2) and then to Algorithm 6.2 just increases the probability of introducing errors. Nevertheless, if this integration is used inside a huge scientific computing code and is called billions of times, even a small speed improvement may be significant.

Example 6.1. The payoff of the optimization depends heavily on the function $f(x)$. If $f(x)$ is complicated, say, an evaluation of f requires 100 multiplications (which is relevant when f contains fundamental functions such as sin and log), saving a multiplication by h can only improve the speed by 1%. To be specific, we tried to integrate

$$f_1(x) = e^{-x^2} \log(1 + x \sin x)$$

and

$$f_2(x) = 1 + x$$

from $a = 0$ to $b = 2$ with $n = 1,000$, repeated 10,000 times. In Fortran 77 implementations of Algorithms 6.1 and 6.2 there were no differences between the two versions of the algorithms when integrating $f_1(x)$. Even with $f_2(x)$, Algorithm 6.2 was only slightly faster than Algorithms 6.1. On the other hand, the total execution time increased by a factor of 10 when we switched from the simple f_2 function to the more complicated f_1 function.

∎

Rely on Compiler Optimization Technology

Why was the benefit of our hand optimizations so small in the previous example? Administering the loop $i = 1, \ldots, n - 1$ and calling a function are expensive operations in computer programs, so saving a couple of multiplications inside a loop can drown in other tasks. Nevertheless, a much more important explanation why the benefit was so small has to do with compiler optimization technology. We turned on the maximum optimization of the GNU Fortran 77 compiler when compiling the codes corresponding to Algorithms 6.1 and 6.2. The Fortran compiler has 50 years of experience with optimizing loops involving mathematical expressions. Loops of the very simple type encountered in the present example can probably be analyzed to the fullest extent by most compilers. This means that the compiler will detect when we perform unnecessary multiplications by h. The compiler will probably also see that the function evaluation $f(a + ih)$ can be optimized, as we did by hand. These assertions are supported by compiling the codes without optimization; the CPU times[4] of Algorithms 6.1 and 6.2 differed by a factor of almost 2, when integrating f_2. Negligible differences arose when integrating f_1, since the function evaluations dominate in this case.

---

[4] See page 221 for how to measure CPU time.

Based on this simple optimization example, and our experience with scientific software development in general, we can provide some guidelines on how to implement numerical methods:

1. Use an algorithm that is easy to understand and that can easily be looked up in the literature.
2. Create a computer code that is as close to the algorithm as possible, such that a "one-to-one correspondence" can be checked by reading the code and the algorithm side by side.
3. Test the implementation on a simple problem where the exact answer is known (such as $f_2(x) = 1 + x$ in the previous example).
4. Be careful with hand optimizations before the code is verified.
5. As soon as some hand optimizations are implemented, compile the code with compiler optimization turned on, run the original and the hand-optimized code in a relevant application, check that the results are equivalent, and then compare timings.

Our recommended rule of thumb is to avoid optimization in the early stages of software development since non-optimized code is usually easier to understand and hence easier to debug. If CPU time consumption of the implemented algorithm is a great concern, one can proceed with rewriting the algorithm for optimization. Having the safe, non-optimized version at hand, it is much easier to verify the usually trickier optimized version. We agree with the famous computer scientist Donald Knuth, who said that "premature optimization is the root of all evil".

We should emphasize that there are numerous examples, the first one appearing already in Sect. 6.1.4, especially in advanced scientific computing problems, where a human is better than a compiler to perform optimizations. Most often, however, the human's contribution to faster code is to choose a more efficient numerical method, whereas the benefit of the compiler is to arrange the sequence of arithmetic operations in an optimal manner.

6.1.4 Developing Algorithms for Simpson's Rule

Let us discuss another example regarding the derivation and implementation of algorithms for numerical integration. Simpson's rule for numerical integration reads (see Project 1.7.2)

$$\int_a^b f(x)dx \approx h \sum_{i=1}^n \left\{ \frac{1}{6}f(x_{i-1}) + \frac{4}{6}f(x_{i-\frac{1}{2}}) + \frac{1}{6}f(x_i) \right\}, \qquad (6.3)$$

where

$$x_i = a + ih, \quad x_{i-\frac{1}{2}} = \frac{1}{2}(x_{i-1} + x_i), \quad h = \frac{b-a}{n}.$$

Simpson's rule is more accurate than the trapezoidal rule, in the sense that it requires less computer power to calculate an integral with a certain accuracy. The task now is to implement Simpson's rule.

Mathematical Pseudo Code

As we have learned, the mathematical description of the method in (6.3) should first be turned into a mathematical pseudo algorithm. Then we can proceed with translating the algorithm into computer language. With our recent experience in writing algorithms for the trapezoidal rule, it should be quite easy to come up with Algorithm 6.3. Note that we have not used the notation x_{i-1}, $x_{i-\frac{1}{2}}$, and x_i, but x^-, x, and x^+ instead. This is due to the fact that mathematical symbols with indices, such as x_i, are often translated to arrays in computer codes. In the present case, we do not need to store the x_i values in an array, we just need to compute the x value inside the loop and then call $f(x)$. Therefore, to avoid indices, we work with x^-, x, and x^+. In a computer code these variables could be given names such as xm, x, and xp.

Algorithm 6.3

Simpson's Rule.

Simpson (a, b, f, n)
  $h = \frac{b-a}{n}$
  $s = 0$
  for $i = 1, \ldots, n$
    $x^- = a + (i-1)h$
    $x^+ = a + ih$
    $x = \frac{1}{2}(x^- + x^+)$
    $s \leftarrow s + \frac{1}{6}f(x^-) + \frac{4}{6}f(x) + \frac{1}{6}f(x^+)$
  end for
  $s \leftarrow hs$
  return s

Algorithm 6.3 and a computer implementation that follows the algorithm line by line using similar symbols represent a safe development from description (6.3) of the numerical method. It should not be difficult to get things right and obtain working code.

Optimization of the Algorithm

The downside of Algorithm 6.3 is that we perform more evaluations of the function $f(x)$ than necessary. Computing $f(x)$ is probably the most expensive part of such algorithms, so avoiding too many function evaluations will have a significant impact on CPU time. Even very intelligent compilers will probably have a hard time detecting that we are performing too many function evaluations, so this is work for a human programmer.

The main observation is that we evaluate $f(x_{i-1})$ and $f(x_i)$ in the ith term of the sum in (6.3) and recalculate $f(x_i)$ in the next term of the sum (where we need $f(x_i)$ and $f(x_{i+1})$). We should avoid the recalculation of $f(x_i)$. How to do this is easy to see if we write out $\sum_{i=1}^{n} \left(\frac{1}{6} f(x_{i-1}) + \frac{4}{6} f(x_{i+\frac{1}{2}}) + \frac{1}{6} f(x_{i+1}) \right)$ as

$$\frac{1}{6} f(x_0) + \frac{4}{6} f(x_{\frac{1}{2}}) + \frac{1}{6} f(x_1) +$$

$$\frac{1}{6} f(x_1) + \frac{4}{6} f(x_{2-\frac{1}{2}}) + \frac{1}{6} f(x_2) +$$

$$\frac{1}{6} f(x_2) + \frac{4}{6} f(x_{3-\frac{1}{2}}) + \frac{1}{6} f(x_3) +$$

$$\cdots +$$

$$\frac{1}{6} f(x_{n-1}) + \frac{4}{6} f(x_{n-\frac{1}{2}}) + \frac{1}{6} f(x_n) .$$

This expression can be expressed as two sums plus a contribution from the end points:

$$\frac{1}{6} f(x_0) + \frac{4}{6} \sum_{i=1}^{n} f(x_{i-\frac{1}{2}}) + \frac{2}{6} \sum_{i=1}^{n-1} f(x_i) + \frac{1}{6} f(x_n) .$$

A more efficient formula for Simpson's rule is therefore

$$\int_a^b f(x)dx \approx \frac{1}{6} h \left(f(a) + f(b) + 2 \sum_{i=1}^{n-1} f(x_i) + 4 \sum_{i=1}^{n} f(x_{i-\frac{1}{2}}) \right) . \quad (6.4)$$

The corresponding algorithm appears in Algorithm 6.4. Of course, we could perform the hand optimization of incrementing x_i and $x_{i-\frac{1}{2}}$ instead of using $a + ih$ and $a + (i - \frac{1}{2})h$. This saving of one multiplication in each loop can be detected automatically by a compiler or simply drown in the work required by evaluating $f(x)$.

Algorithm 6.4

Optimized Simpson's Rule.

Simpson (a, b, f, n)
  $h = \frac{b-a}{n}$
  $s_1 = 0$
  for $i = 1, \ldots, n - 1$
    $x = a + ih$
    $s_1 \leftarrow s_1 + f(x)$
  end for
  $s_2 = 0$
  for $i = 1, \ldots, n$
    $x = a + (i - \frac{1}{2})h$
    $s_2 \leftarrow s_2 + f(x)$
  end for
  $s \leftarrow \frac{1}{6} h(f(a) + f(b) + 2s_1 + 4s_2)$
  return s

Algorithm 6.5

Simpson's Rule with One Loop.

Simpson (a, b, f, n)
  $h = \frac{b-a}{n}$
  $s_1 = 0$
  $s_2 = 0$
  for $i = 1, \ldots, n$
      if $i < n$
          $x = a + ih$
          $s_1 \leftarrow s_1 + f(x)$
      end if
      $x = a + (i - \frac{1}{2})h$
      $s_2 \leftarrow s_2 + f(x)$
  end for
  $s \leftarrow \frac{1}{6}h(f(a) + f(b) + 2s_1 + 4s_2)$
  return s

Compiler Optimization Issues

Why do we split the computations in Simpson's rule into two sums, expressed as two loops in Algorithm 6.4? We could get away with one loop as demonstrated in Algorithm 6.5. The single loop now needs an if-test to avoid adding contributions to s_1 when $i = n$. Unfortunately, if-tests inside loops tend to reduce the compiler's possibilities for optimization. Two plain loops are therefore normally much more efficient than one loop with an if-test. The loss of speed introduced by if-tests inside loops depends on the actual code, the programming language, the compiler, problem-dependent parameters, and so on. In the present case, one can argue that function calls also reduce compiler optimizations in the same way as an if-test, so Algorithm 6.5 may not be much less efficient than Algorithm 6.4. However, we can hardly avoid calling $f(x)$ in numerical integration, but the if-test is easy to avoid. We should mention here that one way to avoid the if-test in Algorithm 6.5 is to remove the test and simply subtract the extra term $f(a + nh)$ from s_1.

Another aspect of the current discussion is that some smart compilers will avoid function calls as a part of the optimization; the body of the code for $f(x)$ is inserted directly in the loop as an expression. For example, if we integrate $f(x) = x \sin x$, a smart compiler can translate the code, e.g., in the following way:

```
...
for i = 1, ..., n
    ...
    x = a + (i - ½)h
    s₂ ← s₂ + 1 + x sin x
    ...
```

Of course, hand coding of such an optimization would make the implementation valid for only a specific function. Calling $f(x)$ instead makes the implementation valid for any f.

Testing the computational efficiency of integrating $f_1(x) = e^{-x^2} \log(1+x \sin x)$ and $f_2(x) = 1 + x$ shows that Algorithm 6.4 clearly runs faster than Algorithm 6.3. The difference depends on the amount of compiler optimization and the complexity of $f(x)$. With full optimization and $f_2(x)$, Algorithm 6.4 reduced the CPU time by about 20% in comparison with Algorithm 6.3. The theoretical saving is 33%, since we reduce the number of function evaluations from $3n$ to $2n + 1$.

Computer implementation and debugging can often be very time consuming. When preparing a problem in scientific computing for implementation, one should start with the simplest version of the algorithm. Smart rewriting and hand optimizations in the code can easily result in hunting errors for hours. On the other hand, training the ability to rewrite algorithms in more efficient forms is an important part of scientific computing. Real-life applications of numerical computing in science and technology make such harsh demands on computer power that serious work with optimization is crucial. The bottom line is that such optimization work should be performed *after* safe and easy-to-understand algorithms are implemented and thoroughly tested. Thinking "constantly" of optimization when carrying out the mathematics and programming has been a tradition in scientific computing, and we believe the result has been too many codes that compute wrong numbers simply because the "smart" expressions increased the complexity of the problem and the code at too early a stage in the development.

6.1.5 Adaptive Integration Rules

A basic problem when applying the trapezoidal or Simpson's rule is to find a suitable value of n. Two factors influence the choice of n: the desired accuracy of the integral and the shape of f. A simple strategy is to apply the integration rule for a sequence of n increasing values, say, $n_0, n_1, n_2, \ldots$, and stop when the difference between two consecutive integral values becomes negligible. This is called an *adaptive* algorithm, because the number of points will adapt to the desired accuracy and the shape of f.

Define $I(a, b, f, n)$ as an approximation to $\int_a^b f(x)dx$ using a specific numerical integration rule with n points, and let ϵ be some specified tolerance. The following pseudo code expresses adaptive integration:

```
r = I(a, b, f, n_0)
for n = n_1, n_2, n_3, ...
    s = I(a, b, f, n)
    if |s - r| ≤ ε then
        return s
    else
        r = s
    end if
end for
```

A typical choice of n_i is $(b - a)/2^i$. If ϵ is too small, the adaptive algorithm can run for an unacceptable long time (or forever if ϵ is so small that round-off errors destroy the convergence of the integral as n grows). We should therefore return when n exceeds a prescribed large value. Algorithm 6.6 lists the pseudo code.

Algorithm 6.6

Adaptive Integration: Repeated Evaluations.

adaptive_integration $(a, b, f, I, \epsilon, n_{max})$
   $r = I(a, b, f, n_0)$
   for $n = n_1, n_2, n_3, \ldots$
      $s = I(a, b, f, n)$
      if $|s - r| \leq \epsilon$ then
         return s
      else
         $r = s$
      end if
      if $n > n_{max}$ then
         print error message, return s
      end if
   end for

Another strategy for developing adaptive rules is to base the integration on an unequal spacing of the evaluation points x_i in the interval $[a, b]$. In the case of the trapezoidal rule, we introduce evaluation points x_i, where $x_0 = a$, $x_n = b$, and $x_i < x_{i+1}$, $i = 0, n - 1$. The trapezoidal rule for unequal spacing can be expressed as

$$\int_a^b f(x)dx \approx \sum_{i=1}^{n} \frac{1}{2}(x_i - x_{i-1})(f(x_{i-1}) + f(x_i)). \qquad (6.5)$$

This can be directly expressed in pseudo code. In case f is expensive to evaluate, we could develop an optimized version where f is evaluated only once at every point:

$$\int_a^b f(x)dx \approx \sum_{i=1}^{n} \frac{1}{2}(x_i - x_{i-1})(f(x_{i-1}) + f(x_i)) \qquad (6.6)$$

$$= \cdots + \frac{1}{2}(x_i - x_{i-1})(f(x_{i-1}) + f(x_i)) + \qquad (6.7)$$

$$\frac{1}{2}(x_{i+1} - x_i)(f(x_i) + f(x_{i+1})) + \cdots \qquad (6.8)$$

$$= \frac{1}{2}(x_1 - x_0)f(x_0) + \frac{1}{2}\sum_{i=1}^{n-1} f(x_i)(x_{i+1} - x_{i-1}) + \quad (6.9)$$

$$\frac{1}{2}(x_n - x_{n-1})f(x_n). \quad (6.10)$$

The corresponding pseudo code is given as Algorithm 6.7. One still needs however, to devise a strategy for choosing the unequal spacing, i.e., determine the values of $x_0, \ldots, x_n$. An intuitive approach is to cluster evaluation points where $f(x)$ is rapidly varying, perhaps in a way that makes the error approximately the same on every interval $[x_{i-1}, x_i]$.

Algorithm 6.7

Optimized Trapezoidal Integration: Unequal Spacing.

trapezoidal $(a, b, f, x_0, \ldots, x_n)$
    $s = 0$
    for $i = 1, \ldots, n-1$
        $s \leftarrow s + f(x_i)(x_{i+1} - x_{i-1})$
    end for
    $s \leftarrow s + f(a)(x_1 - x_0) + f(b)(x_n - x_{n-1})$
    return $0.5 \cdot s$

The error when applying the trapezoidal rule on an interval $[x_{i-1}, x_i]$ can be shown to be

$$\hat{E} = -\frac{1}{12}(x_i - x_{i-1})^3 f''(\xi), \quad (6.11)$$

where ξ is some point in $[x_{i-1}, x_i]$. Using a finite difference to approximate f'' at the midpoint of the interval,

$$f''(\frac{1}{2}(x_{i-1} + x_i)) \approx \frac{1}{4(x_i - x_{i-1})^2} \left(f(x_{i-1}) - 2f(\frac{1}{2}(x_{i-1} + x_i)) + f(x_i) \right),$$

we can replace (6.11) by an approximate error quantity

$$E(x_{i-1}, x_i, f) = -\frac{x_i - x_{i-1}}{48} \left(f(x_{i-1}) - 2f(\frac{1}{2}(x_{i-1} + x_i)) + f(x_i) \right).$$
$$(6.12)$$

We could now start with an equally spaced coarse distribution $x_i = a + ih$, compute the error in each interval, and divide an interval into two new equally spaced segments if the error is greater than ϵ, where ϵ is the largest acceptable error on an interval. This procedure can be repeated until no more refinement is necessary. Of course, we should also stop the refinement if the number of points x_i becomes unacceptably large. Algorithm 6.8 presents the pseudo code for one level of refinement. We provide $x_0, \ldots, x_n$ as input and obtain a new set of points $y_0, y_1, y_2, \ldots$, where each original interval is either preserved or divided into two new intervals.

Algorithm 6.8

Interval Refinement based on Error Estimate.

refine1 $(a, b, f, E, \epsilon, n_{max}, x_0, \ldots, x_n)$
$\quad j = 0$
$\quad$ for $i = 1, \ldots, n$
$\quad\quad y_j = x_{i-1}; \; j \leftarrow j + 1$
$\quad\quad e = |E(x_{i-1}, x_i, f)|$
$\quad\quad$ if $e > \epsilon$ then
$\quad\quad\quad$ if $j > n_{max} - 2$ then
$\quad\quad\quad\quad$ error: too many refinements
$\quad\quad\quad$ end if
$\quad\quad\quad y_j = \frac{1}{2}(x_{i-1} + x_i); \; j \leftarrow j + 1$
$\quad\quad$ end if
$\quad\quad y_j = x_n$
$\quad$ end for
$\quad$ return $y_0, \ldots, y_j$

What we want is actually a *recursive* procedure; we want to call Algorithm 6.8 repeatedly until all interval error estimates are less than ϵ or until we reach n_{max} points. The recursive version appears as Algorithm 6.9. We notice that the algorithm is applicable to any integration rule as long as we provide a function E for computing the error in an interval. If you are not familiar with recursive algorithms, you should definitely work out Exercise 6.5.

Algorithm 6.9

Recursive Interval Refinement based on Error Estimate.

refine $(a, b, f, E, \epsilon, n_{max}, x_0, \ldots, x_n)$
$\quad j = 0$
$\quad$ refined $=$ false
$\quad$ stop $=$ false
$\quad$ for $i = 1, \ldots, n$
$\quad\quad y_j = x_{i-1}; \; j \leftarrow j + 1$
$\quad\quad e = |E(x_{i-1}, x_i, f)|$
$\quad\quad$ if $e > \epsilon$ then
$\quad\quad\quad$ if $j > n_{max} - 2$ then
$\quad\quad\quad\quad$ stop $=$ true; jump out of i-loop
$\quad\quad\quad$ end if
$\quad\quad\quad$ refined $=$ true
$\quad\quad\quad y_j = \frac{1}{2}(x_{i-1} + x_i); \; j \leftarrow j + 1$
$\quad\quad$ end for
$\quad\quad y_j = x_n$
$\quad$ if stop or not refined then
$\quad\quad$ return $y_0, \ldots, y_j$
$\quad$ else
$\quad\quad$ refine $(a, b, f, \epsilon, n_{max}, y_0, y_1, \ldots, y_j)$

6.1.6 Ordinary Differential Equations

Writing algorithms in mathematical pseudo code represents a considerable portion of the software development work, at least for the (quite simple) type of problems we deal with in the present book. Before looking at the translation of pseudo code to specific computer languages it can therefore be instructive to go through another example on pseudo code development. Consider an ODE,

$$u'(t) = f(u), \quad u(0) = U_0,$$

solved approximately by *Heun's* method:

$$u_{n+1} = u_n + \frac{\Delta t}{2}[f(u_n) + f(u_n + \Delta t f(u_n))]. \tag{6.13}$$

Here, u_n is an approximation to the exact solution evaluated at the time point $n\Delta t$. A direct translation of (6.13) to an algorithmic form leads to Algorithm 6.10.

Algorithm 6.10

Heun's Method.

heun $(f, U_0, \Delta t, N)$
    $u_0 = U_0$
    for $n = 0, \ldots, N - 1$
        $u_{n+1} = u_n + \frac{\Delta t}{2}[f(u_n) + f(u_n + \Delta t f(u_n))]$
    end for
    return u_N

We have previously mentioned that mathematical symbols with indices normally translate to arrays in a program, which in the present case means that we store $u_1, u_2, \ldots, u_N$. An optimization-oriented person who is very concerned about reducing computer memory might claim that to compute the return value u_N, we can get away with storing only one u value at a time, as expressed in Algorithm 6.10.

Algorithm 6.11

Memory-Optimized Heun's Method.

heun $(f, U_0, \Delta t, N)$
    $u = U_0$
    for $n = 0, \ldots, N - 1$
        $u \leftarrow u + \frac{\Delta t}{2}[f(u) + f(u + \Delta t f(u))]$
    end for
    return u

Computing u at some given time $t_k = k\Delta t$ is now a matter of just calling heun($f, U_0, \Delta t, k$). Suppose we want to plot $u(t)$ as a function of t for $0 \le t \le T$, using N points, we can simply implement the following pseudo code:

```
Δt = T/N
for i = 0, ..., N
    u_i ← heun(f, U_0, Δt, i) (Algorithm 6.11)
    write (iΔt, u_i) to file
end for
```

The result is typically a two-column file, with t and u values in columns 1 and 2, respectively, which can be plotted by (probably) any plotting program.

The outlined implementation is *very inefficient*, yet common among newcomers to scientific computing. The implementation is very efficient in reducing the need for computer memory, unfortunately at the significant cost of CPU time: Every time we call the heun($f, U_0, \Delta t, i$) function, we recompute what we did in the previous call, plus compute a single new u value. We should, of course, integrate the differential equations up to time T and return all the $u_1, \ldots, n_N$ values computed in Algorithm 6.10. After the call to the heun function we can dump the returned array to a file for plotting and data analysis.

Looking more closely at Algorithm 6.10, we realize that the function evaluation $f(u_n)$ is carried out twice. We should avoid this. A more efficient computational approach, which should be considered by the reader for implementation, appears in Algorithm 6.12. This implementation is flexible; we can call it to advance the solution one time step at a time, we can compute the complete solution up to the desired final point of time, or we can call the function several times, computing a chunk of u_n values in each call.

Algorithm 6.12

Optimized Heun's Method.

heun $(f, U_0, \Delta t, N)$
   $u_0 = U_0$
   for $n = 0, \ldots, N - 1$
      $v = f(u_n)$
      $u_{n+1} = u_n + \frac{\Delta t}{2}[v + f(u_n + \Delta t\, v)]$
   end for
   return $u_0, u_1, \ldots, u_N$

6.2 About Programming Languages

There are hundreds of programming languages. Some of them are well suited for scientific computing, others are not. We will present the characteristics of the most popular languages for scientific computing. Our aim is to demonstrate that as soon as you have expressed an algorithm as mathematical pseudo code, the step to a real programming language is quite small. The implementation of an algorithm in a particular language will of course depend on the syntax and built-in functionality of that language. Even for the simple algorithms from Sect. 6.1, the realizations of the algorithms in different languages (see Sects. 6.3 and 6.4) vary considerably. Deciding upon the most convenient programming tool is greatly subject to personal taste and experience. Nevertheless, modern computational scientists need to master several programming languages and have the ability to quickly jump into a new language. The reason is obvious: Different tools have different strengths and weaknesses and hence are suitable for different type of problems.

We shall discuss four basic issues that influence the choice of programming language for scientific computing tasks:

– Static typing versus dynamic typing
– Computational efficiency
– Built-in numerical high-performance utilities
– Support for user-defined objects

The languages we bring into the discussion are Fortran 77, C, C++, Java, Maple, Matlab, and Python. All of these languages are widely used in scientific computing. There are different versions of Fortran, e.g., Fortran 95 and Fortran 2000, which have some of the features of C++, Java, and Python. Fortran 77, on the other hand, is a small and compact language with fewer features than C. We will normally specify the type of Fortran language, and use the term Fortran (without a number) when we speak of the Fortran family of languages in general.

6.2.1 Static Typing Versus Dynamic Typing

Computer languages such as Fortran, C, C++, and Java are said to be *strongly typed*, or statically typed. This means that the programmer must explicitly write the type of each variable. If you want to assign the value 3 to a variable called a, you must first specify the type of a. In this case a can be an integer or a real (single precision or double precision) variable, since both types can hold the number 3. Say you specify a to be an integer. Later on in the program you cannot store a string in a, because a can only hold an integer; a is related to a memory segment consisting of four bytes, which are sufficient to store an integer, but too large or (more likely) too small to store a string. Even if we could store a four-character string in four bytes, the program would interpret these four bytes as an integer, not as four characters.

Maple, Matlab, and Python are examples of programming languages where one does not need to specify the type of variables. If we want to store the number 3 in a variable a, we usually write a = 3 (or a:=3 in Maple) to perform the assignment. We may well store a string in a afterward, and then a real number, if desired. The variable a will know its content, i.e., it will know whether it contains an integer, a real number, or a string and adjust its behavior accordingly. For example, if we multiply a by 15, this is a legal mathematical operation, provided a is a number; otherwise a will know that it does not make sense for it to participate in a multiplication. The type information store in a is there, but we can say that the typing is dynamic, in the sense that the programmer does not explicitly write the type of a and a can dynamically change the type of its content during program execution. Dynamic typing gives great flexibility and a syntax that is closer to the mathematical notation of our pseudo code language.

Static typing is often said to be very useful for novice programmers. Incompatible types are quickly detected and reported before the program can be run. Suppose you hit the s key instead of the a key when writing the statement a = 3. If you do not introduce a variable s elsewhere, s is not declared with a type, and a typing error is detected. With dynamic typing nothing about types is known before the program is run. Incompatible types are detected when operations become illegal, but a writing error such as s = 3 instead of a = 3 is not detected; s is as good as any variable name, and it is brought into play by the assignment. However, as soon as the program works with a, this variable is now not initialized, and an error will result.

The authors have programmed in languages with both static and dynamic typing. The conclusion is that the debugging styles are a bit different, but we cannot claim that one of the approaches generally leads to fewer errors than the other.

6.2.2 Computational Efficiency

Some computer languages are faster than others. For example, the loop over i in Algorithm 6.4 will be much faster in Fortran and C/C++ than in Maple, Matlab, or Python, no matter how much we try to optimize the implementation. The reason that Fortran, C, and C++ are so much faster stems from the fact that these languages are *compiled into machine code*. We can classify computer languages as either *interpreted* or *compiled*. This distinction reflects the potential speed of the language and is therefore of particular interest in scientific computing. Normally, compiled languages have static typing, whereas interpreted languages have dynamic typing. Java is something in between: It has static typing, was originally interpreted, but is being equipped with compiler techniques.

Fortran, C, and C++ are examples of compiled languages. A compiler translates the computer program into machine code, i.e., low-level, primitive instructions tied to the hardware. Since most programs call up functionality in some external libraries, the machine code of a program must be *linked* with a machine code version of the required libraries. Compilation and linking are normally two distinct steps. The machine code of the program and the libraries are merged in one file, the *executable*. To run the program, you simply write the name of the executable.

Maple, Matlab, and Python are examples of interpreted languages. A program is read (line by line) by an *interpreter* that translates statements in the code into function calls in a library. The translation takes place while the program is running. Hence, inside a loop the translation of a particular statement is repeated as many times as iterated in the loop. This causes the program to run much more slowly than in the case where we could just execute pre-made machine instructions. The benefit is that we get a high degree of flexibility; statements and data structures can change dynamically at run time in a safe and convenient way from a user's point of view. Interactive computing is one application of this flexibility.

Comparing implementations of Algorithm 6.4 in different languages shows that the speed of the various programs varies greatly. Using compiled languages results in significantly faster programs than interpreted languages. Some numerical applications require so much computing power that one is forced to use the fastest algorithms and languages in order to solve the problem at hand. Other applications demand only minutes or hours of execution time on a desktop computer, and in these cases one can often trade computational efficiency in favor of increased human efficiency, i.e., quicker, safer and more convenient programming.

6.2.3 Built-in High-Performance Utilities

Programming in Fortran, C, C++, and Java can be very different from programming in Maple, Matlab, and Python, especially when it comes to algorithms with loops. In the former four languages you can implement a loop as in Algorithm 6.4 using basic loop constructions in those languages, and the performance will be very good. This means that there will often be small differences between your implementation and a professional implementation of the algorithm made available through a numerical library. Many researchers and engineers like to create their own implementation of algorithms to have full control of what is going on. This strategy often leads to only a minor loss of performance.

The situation is different in Maple and Python, and also to some extent in Matlab. Plain loops in these languages normally lead to slow code. A Maple implementation of Algorithm 6.4 can easily be more than 150 times slower than a similar implementation in Fortran 77. However, this comparison is not fair. Plain, long loops in Maple are to some extent a misuse of the language, especially if we try to implement algorithms that are available in libraries or as a part of the language. Numerical integration is an example; Maple has a rich functionality for numerical integration, where accuracy and performance can be controlled. Calling up a built-in numerical integration rule in Maple (which means an algorithm of much higher sophistication than the trapezoidal rule) produces a result in about the same time as required by our hand-made Fortran 77 function. In other words, you should not consider implementation of basic numerical algorithms in Maple, because this will give low performance; you should utilize built-in functionality. Clearly, not all algorithms are available in Maple, and sometimes you have to make your own implementations. It is then important to try to break up the steps in an algorithm into smaller steps, which may be supported by built-in functionality.

The comments about Maple are also relevant for Matlab and Python. All these three programming platforms have a large collection of highly optimized functions in numerical libraries. Algorithms to be implemented in Maple, Matlab, and Python should utilize the languages' high-performance built-in functionality to as high a degree as possible. Hence, programmers using these languages need to be familiar with the libraries, otherwise they will both reinvent the wheel (by re-implementing basic algorithms) and decrease performance. While loops play an important role in compiled languages, loops are very slow in interpreted languages and should hence be avoided. Instead, one needs to express the algorithms in terms of built-in basic vector operations, where all loops are solely executed in libraries written in compiled languages. The process of rewriting an algorithm in terms of basic vector operations is usually referred to as vectorization and constitutes the subject of Sect. 6.3.7.

We should mention that Matlab has a just-in-time compiler that can automatically turn loops over arrays into as efficient code as vectorization can offer. However, the just-in-time compilation strategy works only in certain situations, and in particular not if there are function calls inside the loops, as we have in the numerical integration algorithms.

6.2.4 Support for User-Defined Objects

Fortran 77 has only a few different types of variables: integer, real, array, complex, and character string. One can build any code with these types of variables, but as soon as the complexity of the program grows, one may want to group a collection of primitive variables into a new, more advanced variable type. Many languages therefore allow the programmer to define new variable types. In C, one can define a struct to be a collection of basic C variables and previously defined structs. This construction acts as a method for creating user-defined variable types.

When people start to create their own types with structs and similar constructions in other languages, it becomes apparent that a user-defined variable should not only store information, but also be able to manipulate and process that information. This results in more intelligent variable types, but requires the struct to contain functions as well as data. C++ is an extension of C that offers user-defined variable types containing both functions and data. Such a user-defined type is normally referred to as a *class*. Fortran 90 and Fortran 95 are extensions of Fortran 77 (and earlier versions known as Fortran IV and Fortran 66) where the programmer can collect data and functions in *modules*. Modules can do many of the same things as classes, but classes in C++ are much more advanced than the modules in Fortran 90/95. The Fortran 2000 initiative aims at equipping Fortran with modules having the functionality of classes that programmers expect today.

Java supports classes, but the flexibility of the classes is less than in C++. On the other hand, Java has functionality that makes the language much more user-friendly and easier to program with than C++. Python has classes that are more advanced than those in C++. Matlab originally had only matrices and strings, but recent extensions to Matlab allow the definition of classes. Maple has many variable types, but no way to create user-defined types. Having said this, we should add that languages with a rich collection of built-in data types and dynamic typing, such as Maple and Python, often offer the ability to create heterogeneous lists or hash structures that replace the traditional use of classes in C++ and Java.

A class can be viewed as a definition of a new type of variable. To create variables of this type, one creates *objects* of the class. The program will therefore work with objects in addition to variables of built-in types. Programming with objects is a bit different from programming with fundamental built-in types, such as real, integer, and string. In the latter case, collections of primitive variables are shuffled in and out of functions (subprograms) and the software is organized as a library of functions. With classes, the functions are often built into the classes, and the software is a collection of classes (i.e. data types) instead of a function library.

The real power of class-based programming comes into play when classes can be related to each other in a kind of family tree, where children classes inherit functionality and data structures from parent classes. Although the implementational and functional details can differ between members in such a tree of classes, one can often hide these differing details and work solely with the topmost parent class. This is called object-oriented programming and constitutes a very important programming technique in computer science. The Java language is very tightly connected to object-oriented programming. Object-oriented programming is highly valuable in scientific computing, especially for administrating large and complicated codes, but the traditionally popular languages Fortran and C do not support this style of programming.

6.3 The Trapezoidal Rule in Different Languages

On the following pages we shall give a glimpse of some widely used programming languages: Fortran 77, C, C++, Java, Matlab, Maple, and Python. All three languages are popular in scientific computing communities. The exposition is strongly example-oriented. The simple trapezoidal rule from Algorithm 6.2 on page 201 is implemented in each of the mentioned languages. The code segments are briefly explained and accompanied by some comments related to topics such as static typing and computational efficiency. In Sect. 6.4 we shall implement Heun's method from Algorithm 6.12 in the aforementioned computer languages and thereby learn more about arrays and file handling in different languages. When solving ODEs, we also need to display the solution graphically, and this is shown by plotting the solution using Gnuplot and Matlab.

6.3.1 Code Structure

The main structure of the code for performing numerical integration with the aid of the trapezoidal rule will be quite independent of the programming tool being utilized. Our mathematical model is an integral,

$$I = \int_a^b f(x)dx,$$

and the purpose is to compute an approximation of I. In a program we need to

1. Initialize input data: a, b, and n
2. Specify a function $f(x)$
3. Call a program module that implements Algorithm 6.2
4. Write out the value of I

The input data are normally provided by the user of the program at run time, but for simplicity, we shall explicitly set $a = 0$, $b = 2$, and $n = 1,000$ in our sample codes.

The function $f(x)$ is selected as $e^{-x^2} \log(1 + x \sin x)$ and implemented with the name f1. Most codes also have a simple linear function $1 + x$, called f2 and used for checking that the numerical results are correct. Since $1 + x$ is integrated without numerical errors, when using the trapezoidal rule, we should expect $I = \int_0^2 (1 + x)dx = 4$ to machine precision. If the program works for f2, we expect it to work for f1.

Algorithm 6.2 is more or less directly translated to a function. The main difference between the various implementations is how $f(x)$ is technically treated as an argument in this function.

6.3.2 Fortran 77

The Fortran language was designed in the 1950s as a high-level alternative to assembly programming. With Fortran, scientists could express mathematical formulas in a program using a syntax close to the syntax of mathematics. Since the 1950s, Fortran has been improved several times, resulting in Fortran IV, Fortran 66, Fortran 77, Fortran 90, Fortran 95, and Fortran 2000. The latter three versions are very different from the former ones, so it makes sense to talk about two families of Fortran, before and after Fortran 90.

Fortran 77 is a quite primitive language in the sense that there are few variable types (only integer, real, complex, array, and string), few keywords, few control structures, and a strict layout of the code. These limitations are not significant until one tries to build fairly large codes. Fortran IV and Fortran 66 are even more primitive and are of no significance today, except that there is lots of legacy code written in these languages.

Fortran 77 is without competition the most widespread programming language for scientific computing applications. The reasons for this are simplicity, tradition, and high performance. The newer versions, Fortran 90, 95, and 2000, support programming with objects (called modules), and have modernized Fortran, to some extent, in the direction of C++ and Java. However, Fortran 77 is normally faster than the newer versions and therefore preferred by many computational scientists.

Fortran 77 is very well suited for implementing algorithms with CPU-intensive loops, especially when the loops traverse array structures. Applications involving sophisticated data structures, text processing, or large amounts of code will normally benefit from being implemented in more modern languages such as C++ or Fortran 90/95/2000.

Algorithm 6.2 can be implemented in a Fortran 77 function as illustrated below:

```
real*8 function trapezoidal (a, b, f, n)
real*8 a, b, f
external f
integer n

real*8 s, h, x
integer i
h = (b-a)/float(n)
s = 0
x = a
do i = 1, n-1
   x = x + h
   s = s + f(x)
end do
s = 0.5*(f(a) + f(b)) + s
trapezoidal = h*s
return
end
```

Fortran 77 has some limitations on the layout of the statements, which seem a bit old-fashioned now:

– No statement must begin before column 7.
– There can only be one statement per line.
– Comments start in the first column, usually with the character c, and extend to the end of the line.

Some other rules are as follows:

– Fortran is case-insensitive, so whether you write `trapezoidal` or `TraPEZoidal` does not matter.
– All variables, i.e., arguments and local variables, must be declared before the first computational statement in the function.
– Real variables can be of different lengths; `real*4` denotes a four-byte single-precision variable, and `real*8` denotes an eight-byte double-precision variable.
– For loops are implemented with the do–end do construction (and hence are referred to as "do loops").
– The `external` keyword is used to indicate an external function to be called.

Note that we compute $\frac{b-a}{n}$ by converting n to a real variable (the `float` function performs this operation in Fortran). The reason is because division by an integer will in some languages imply `integer` division,[5] which yields $h = 0$ in the present case. It is a good habit to always convert an integer in division operations to real, regardless of how the computer language handles the division operation.

A test program calling up the `trapezoidal` function can look like the following:

```
C       test function to integrate:

        real*8 function f1 (x)
        real*8 x
        f1 = exp(-x*x)*log(1+x*sin(x))
        return
        end

C       main program:

        program integration
        integer n
        real*8 a, b, result
        external f1
        a = 0
        b = 2
        n = 1000
        result = trapezoidal (a, b, f1, n)
        write (*,*) result
        end
```

Compiling, Linking, and Executing the Program

Suppose the code segments above are stored in a file `int.f`. Fortran programs must be compiled and linked. On a Unix system this can take the form

```
unix> f77 -O3 -c int.f      # compilation
unix> f77 -o int int.o      # linking
```

The compilation step translates `int.f` into object (machine) code. The resulting file has the name `int.o`. The linking step combines `int.o` with the standard libraries of Fortran to form an executable program `int`. The `-O3` option means optimization level 3 when compiling the code. What a certain optimization level means depends on the compiler. On Linux systems the common Fortran compiler is the `g77` GNU compiler, not `f77`. You will normally find `g77` on other Unix systems too.

To run the program, we write the name of the executable, in this case `int`. On Unix systems you probably need to write `./int`, unless your `PATH` environment variable contains a dot[6] (the current working directory).

---

[5] The result of dividing the integer p by the integer q is the largest integer r such that $rq \leq p$. Note that if $q > p, r = 0$.

[6] There is a significant security risk in having a dot in the `PATH` variable, so this is not recommended.

Measuring the CPU Time

The forthcoming subsections contain implementations of the trapezoidal algorithm in many other languages. The relative computational efficiency of these implementations is of significant interest. The CPU time consumed by an algorithm is a good measure of computational efficiency. Unix systems have a command `time` that can be used to infer the timings of a program. In the present case one can write

```
unix> time ./int
```

and the output will be something like

```
real      0m4.367s
user      0m4.310s
sys       0m0.050s
```

"Time" on a computer is not a unique term. The time it takes for the `int` program to execute, measured on a wall clock or wrist watch, is called *wall clock time*. Sometimes this time measurement is called *elapsed time* or *real time*. This measure of time may not reflect the efficiency of the program well if there are many other processes and users running concurrently on the machine. Therefore, other time measurements are introduced. The *user time* is the time it takes to execute the statements in the program, minus the time that is spent in the operating system on, e.g., reading and writing files. The time consumed by the latter operations is called *system time*. The sum of the user and system time constitutes the *CPU time*, i.e., the time it takes to execute all statements in the program.[7] In the example above, the CPU time is $4.310 + 0.05 = 4.360$ s. The accuracy of such timings varies with the computer system. To be on the safe side, the program should run for several seconds (say, at least 5 s).

The current Fortran 77 program with 10,000 repetitive calls to the function `trapezoidal`, with $n = 1,000$, requires about 4 s on an IBM X30 laptop running Linux and using the `g77` compiler. (Note that a "smart" compiler may be so smart that it throws away repetitive calls unless we insert statements that require all the calls!) For comparison purposes it is advantageous to use relative time measures. We will hence report the timings of the other implementations as the actual CPU time divided by the CPU time required by the Fortran 77 code.

Remark

The CPU time spent by a set of statements can in most languages also be measured inside the program by a suitable function call. Unfortunately, Fortran has no unified standard for doing this, although many Fortran compilers offer a timing function (`g77` offers `dtime`). We therefore rely on the Unix command `time` for timing the complete execution of a program.

---

[7] This is almost correct: Child processes, typically launching a new program inside another program, are not included in the CPU time, but this is of no relevance in the present simple program.

6.3.3 C and C++

The C language was invented in the beginning of the 1970s. Its main application was implementation of the Unix operating system. The combination of Unix and C became a dominating programming environment in the late 1980s. C is clearly a more advanced programming language than Fortran 77, and some computational scientists switched from Fortran 77 to C. C gives the program developer more flexibility, normally at the cost of a slight loss in computational performance. C has the same fundamental data types as Fortran 77, but offers a struct type, which can hold any composition of fundamental data types and other structs. This offers the possibility of creating more advanced data types than in Fortran 77.

C++ was designed to offer classes and object-oriented programming as in the famous SIMULA 67 language, but with C syntax and C as a subset. Several features of languages with classes were removed to improve performance. Classes in C++ extend C's struct such that user-defined types contain both data and functions operating on the data. This provides much more flexible data types than the struct in C. Applications that can benefit from sophisticated user-defined data structures, and perhaps also the style of object-oriented programming, will normally be much simpler to implement in C++ than in Fortran 77 or C. The applications will also often be simpler to maintain and extend. However, C++ code can seldom compete with Fortran 77 when it comes to computational speed, although the difference is small. With some tweaking, C and C++ can on many occasions be as fast as, or even slightly faster, than Fortran 77, but this depends on the type of algorithms and the involved data structures.

C and C++ programs for Algorithm 6.2 look very similar since this is a very simple problem. C++ has a slightly more readable syntax, so we present the algorithm in that language first:

```
typedef double (*fptr) (double x);

double Trapezoidal (double a, double b, fptr f, int n)
{
   double h = (b-a)/double(n); double s = 0; double x = a;
   for (int i = 1; i <= n-1; i++) {
     x = x + h;
     s = s + f(x);
   }
   s = 0.5*(f(a) + f(b)) + s;
   return h*s;
}
```

C and C++ have, unfortunately, some constructs that make it somewhat difficult to explain even very simple programs to the novice. For example, the Trapezoidal function needs a function $f(x)$ as an argument, and this argument must be a *function* pointer, here called fptr and defined through a typedef statement. The rest of the code should be easier to understand. Real variables with double precision (real*8 in Fortran 77 terminology) are called double in C and C++. The Trapezoidal function returns a double (the approximation to the integral). Contrary to Fortran 77, the type of each argument precedes the argument name, providing a more compact code. Every statement ends with a semi-colon, and one can have as many statements as desired on the same line. In C++, variables can be declared wherever they are needed. The loop is implemented via the for construction, which is adopted in a wide range of programming languages.

A test program, calling up the Trapezoidal function to perform the numerical integration of a specific function, can take the form

```
#include <iostream>
#include <cmath>

/* the Trapezoidal function: */
...

/* test function to be integrated */
double f1 (double x)
{
  return exp(-x*x)*log(1+x*sin(x));
}

int main()   // main program
{
  double a = 0, b = 2; int n = 1000;
  double result = Trapezoidal (a, b, f1, n);
  std::cout << result << std::endl;
}
```

The #include statements include information about functions in libraries (here input/output functionality from iostream and mathematical functions from cmath); the compiler needs to see a declaration of a function, i.e. the type of return value and the type of arguments, before a call to the function can be compiled. For example, calling the exponential math function exp requires us to include the cmath file where exp is declared. If you call the function with the wrong types of arguments, a compiler error will be issued. Fortran 77 has no information about the functions one calls, so if you provide the wrong type or wrong number of arguments, the call will compile normally, but the run-time behavior is unpredictable. Very often strange things happen, and it requires some experience to find such bugs. Many programmers switched from Fortran 77 to C or C++ just because the C and C++ compilers find so many more coding errors.

The code segments above are written in C++, but the C version is quite similar. Since C++ has C as a subset and is a more flexible computer language, the authors recommend using C++ instead of plain C. Some constructs in C++, especially related to programming with classes, can result in a slight performance loss. One can, nevertheless, often rewrite time-critical parts in C++ programs using the C subset in order to take advantage of the performance of C.

A very popular C compiler is gcc (GNU's C compiler). On most Unix systems, the vendor's compiler has the name cc. The C++ counterpart to gcc is g++. This is a good all-round compiler, but some commercial compilers, especially the KCC compiler, can give significantly better performance in scientific computing applications.

A typical session for compiling and linking a C++ program in a file `int.cpp` on a Unix machine, using the `g++` compiler, goes like this:

```
unix> g++ -O3 -c int.cpp      # compiling int.cpp to int.o
unix> g++ -o int int.o        # linking; form executable int
```

These steps are the same as for Fortran 77 and will be similar for a C program; only the name of the compiler differs. Executing the program follows the description for the corresponding Fortran 77 program; just name the executable file.

The CPU time can be measured in C and C++ programs by calling the `clock` function. We have not done this. Instead we rely on the Unix `time` command. The performance test, involving calling the `Trapezoidal` function 10,000 times with $n = 1,000$, ran at the same speed as our Fortran 77 code. Normally, C and C++ run at the same or slightly lower speed than Fortran 77.

6.3.4 Java

Java was constructed by Sun Microsystems in the beginning of the 1990s and can be viewed as a simpler and more user-friendly version of C++. At the time Java was released, C++ had recently become a very popular and dominating language, but Java appeared to be more convenient and productive, so many programmers converted quickly to Java. Java is now the most popular programming language in the world.

Java programs are not compiled in the classic sense of Fortran, C, and C++ so performance was a significant negative feature for computational scientists. Despite numerous efforts to improve performance (and it is now indeed possible to obtain execution times close to those of C++ in some applications), the initial interest in Java as a language for scientific computing has declined. However, Java has become the dominant language for teaching introductory computer programming, so in the early stages of student programs it makes great sense to use Java for numerical computing. In the professional world of scientific computing, Fortran 77 dominates, but C++ and Matlab have a significant share.

Let us look at a Java implementation of Algorithm 6.2. At first sight, this implementation is significantly less straightforward than the implementations in Fortran or C/C++. The reason is because one has to work with classes, because all functions in Java must belong to a class. The bottom line is, therefore, that we implement the functions from our Fortran and C/C++ programs as functions in otherwise empty Java classes. We also note that functions in Java are referred to as *methods*.

```
import java.lang.*;

interface Func {  // base class for functions f(x)
    public double f (double x);  // default (empty) impl.
}

class Trapezoidal {
    public static double integrate (double a, double b,
                                    Func f, int n)
```

```
    {
    double h = (b-a)/((double)n);
    double s = 0;
    double x = a;
    int i;
    for (i = 1; i <= n-1; i++) {
        x = x + h;
        s = s + f.f(x);
    }
    s = 0.5*(f.f(a) + f.f(b)) + s;
    return h*s;
    }
}
```

The `integrate` method is very similar to Algorithm 6.2 and the implementations in most other languages. Java is statically typed and much of the syntax is inspired by C and C++. Calling $f(x)$ is a bit different than in Fortran and C/C++, because we need to pass f as an object of type `Func` to class `Trapezoidal`'s integrate method. To evaluate f, we call the `f` method in class `Func`. An actual function to be integrated must be implemented as a method `f` in a subclass of `Func`. Here is an example:

```
class f1 implements Func {
    public double f (double x)
    { return Math.exp(-x*x)*Math.log(1+x*Math.sin(x)); }
}
```

That is, to send a function $f(x)$ as an argument we actually need to implement the function as a class and make use of object-oriented programming. Many readers will find the Java code unnecessarily comprehensive in simple numerical problems because of this fact.

A main program in Java is actually a method `main` in some class, here a test class called `Demo`:

```
class Demo {
    public static void main (String argv[])
    {
    double a = 0;
    double b = 2;
    int n = 1000;
    double result = 0;
    int i;
    f1 f = new f1();
    //f2 f = new f2();
    for (i = 1; i <= 10000; i++) {
        result = Trapezoidal.integrate(a, b, f, n);
    }
    System.out.println(result);
    }
}
```

Since `integrate` is a static method in the `Trapezoidal` class, we can call the method without creating an object of type `Trapezoidal`. To pass the $f(x)$ function defined in the `f` method in class `f1`, we must create an object of type `f1` and pass it to the `integrate` method. When developing larger computer applications, the idea of encapsulating everything in classes helps to modularize the code and ease extensions, but in this simple example it may appear as unnecessary "overhead" compared to most of the other programming languages we address.

Suppose the shown code segments are located in a file `Trapezoidal.java` (the `.java` extension is required). The programs must first be compiled to bytecode:

```
unix> javac Trapezoidal.java
```

Now we can run the code by specifying the name of the class containing the main program, i.e., the `main` method:

```
unix> java Demo
```

There is unfortunately no built-in method for measuring the CPU time in Java programs so the easiest way to measure the efficiency is to use the Unix `time` command (`time java Demo`).

The CPU time ratio relative to the Fortran 77 and C/C++ codes was 2.1, i.e., Java ran at approximately half the speed of Fortran 77 and C/C++ in this example. The test was performed with Java Development Kit 1.1 for Linux. The speed of Java is likely to improve in future releases.

6.3.5 Matlab

Matlab was originally a user-friendly front end to efficient Fortran 77 libraries for numerical linear algebra computing. During the 1980s and especially the 1990s the Matlab environment became more and more user-friendly, and manifested Matlab as a leading development platform for at least simpler scientific computing applications. Working with Matlab is not too different from programming in Fortran 77, but some differences are striking:

- Statements can be issued in an interactive way, i.e., you can type a command and immediately view the result.
- There is no need to declare variables (what we called dynamic typing in a previous discussion).
- Matlab has many high-level commands that replace the need for writing detailed loops.
- If statements go wrong, Matlab normally issues easy-to-understand error messages.
- Matlab offers advanced, integrated visualization. This makes it easy to compute something and immediately display the results graphically.

These features improve the user's productivity. Matlab is considerably simpler to work with than programming code segments in Fortran, C, C++, or Java and calling up some visualization program.

As a programming language, Matlab is convenient, but somewhat primitive. The code is interpreted and runs slowly, unless the computations are performed solely in the built-in Fortran and C libraries. This means that one should avoid long loops in Matlab and instead try to perform the computations by combining various Matlab commands whose loops are implemented in Fortran or C (more about this in

Sect. 6.3.7). Recent extensions include the possibility to define classes and to translate Matlab code to C++. Well-trained users will often be able to generate highly efficient Matlab code. For large or advanced scientific computing applications, Matlab can seldom compete with tailor-made programs in the compiled languages (Fortran, C/C++), but for simpler calculations and testing ideas, Matlab is a very popular and productive tool.

The trapezoidal integration method is to be implemented in a Matlab function. Any Matlab function we want to call from the interactive Matlab environment or from a Matlab script must have its source code written in a file with extension .m. Such files are called M-files. In the present example, we create a file Trapezoidal.m, containing more or less a direct translation of Algorithm 6.2:

```
function r = Trapezoidal(a, b, f, n)
% TRAPEZOIDAL Numerical integration from a to b
% by the Trapezoidal rule
f = fcnchk(f);
h = (b-a)/n;
s = 0;
x = a;

for i = 1:n-1
    x = x + h;
    s = s + feval(f,x);
end
s = 0.5*(feval(f,a) + feval(f,b)) + s;
r = h*s;
```

The return parameter is called r and must be assigned before the end of this function. Comment lines start with %. The first comment segment after the heading of the function is taken as a documentation of the function and its usage if the first word is the name of the function. Writing help Trapezoidal in Matlab will then print out this documentation.

In Matlab, functions can be passed as arguments to other functions in various ways. The statement f=fcnchk(f) makes a unified function representation f out of different types of supplied function representations (names of M-files, strings with function expressions, inline Matlab function objects).

The next three statements should be trivial to understand. The semicolon at the end of each statement is not strictly required, but if you omit it, Matlab will echo the statement to the screen. This is inconvenient (and takes much time) if the program executes many statements (which is the case if you have loops).

For loops in Matlab are of the form for variable = expr, where expr is a loop expression. Typical loop expressions are

- a:b, which generates the sequence a, a+1, a+2, ..., b, and
- a:s:b, which generates the sequence a, a+s, a+2*s, ..., b.

In the present example we need an expression with step 1 from 1 to n. To call the function f with argument x, we need to use the Matlab construction feval(f,x).

A function to be integrated is naturally placed in another M-file, say, f1.m:

```
function y = f1(x)
y = exp(-x*x)*log(1+x*sin(x));
```

To integrate `f1`, we can call the `Trapezoidal` function in the interactive Matlab environment as follows:

```
a = 0; b = 2; n = 10;
result = Trapezoidal(a, b, @f1, n);
disp(result);  % print result
```

Notice that a function `f1` in an M-file `f1.m` is transferred as an argument by adding the @ prefix to the name `f1`.

Adding timing functionality is easy with the Matlab `cputime` function. However, we need to run a long loop over the `Trapezoidal` function call to obtain some seconds of CPU time:

```
a = 0; b = 2; n = 1000;
t0 = cputime;
for i = 1:10000   % repetitions to obtain some seconds CPU time
    result = Trapezoidal(a, b, @f1, n);
end
disp(result);
t1 = cputime - t0;
disp(t1);
exit
```

Matlab allows a flexible assignment of functions, as demonstrated next:

```
a = 0; b = 2; n = 10;

% function f1 defined in f1.m (function handle):
result = Trapezoidal(a, b, @f1, n);
disp(result);

% inline object f:
f = inline('exp(-x*x)*log(1+x*sin(x))');
result = Trapezoidal(a, b, f, n);
disp(result);

% string expression:
result = Trapezoidal(a, b, 'exp(-x*x)*log(1+x*sin(x))', n);
disp(result);
```

Running the simple main program in Matlab on a laptop resulted in a CPU time ratio of 85 relative to the Fortran 77 and C/C++ codes. This can, however, be dramatically improved by vectorizing the code, see Sect. 6.3.7.

6.3.6 Python

Python is a very flexible and convenient programming language that supports much more advanced concepts than C or Fortran, and also more powerful constructions than in C++ or Java. The nature of Python allows one to build libraries with an interface that gives the programmer access to powerful high-level statements. Application codes therefore tend to be smaller, more compact, and easier to read than their counterparts in Fortran, C, C++, and Java. Because Python programs are interpreted, some constructions (especially loops) run much more slowly than in

Fortran, C, C++, or Java. However, for many applications Python is fast enough, in scientific computing contexts as well [19]. Only a glimpse of Python is given below. The reader is referred to the book [20] for a more comprehensive treatment of Python's applicability in a numerical context.

Matlab and Python have much in common: Both are easy to learn, they have a very clean syntax, and they have high-level tools for performing compound operations in a few statements. Matlab has more built-in functionality for scientific computing, but Python is a more advanced and flexible programming environment. A good strategy is to use Matlab for tasks where Matlab is strong, mainly linear algebra-related problems, and build your own, tailored Matlab-like environment via Python for more advanced tasks. Of course, Python can also be used for linear algebra problems.

Variables in Python can readily be brought into play, without explicitly mentioning their type. Since the Python syntax (like Matlab's) is close to mathematical notation, the difference between Algorithm 6.2 and Python code is small:

```python
#!/usr/bin/env python
from math import *

def Trapezoidal(a, b, f, n):
    h = (b-a)/float(n)
    s = 0
    x = a
    for i in range(1,n,1):
        x = x + h
        s = s + f(x)
    s = 0.5*(f(a) + f(b)) + s
    return h*s

def f1(x):
    f = exp(-x*x)*log(1+x*sin(x))
    return f

def f2(x):           # simple function for verification
    return 1 + x     # integral from 0 to 2 should be 4

a = 0; b = 2; n = 1000
for i in range(10000):
    result = Trapezoidal(a, b, f1, n)
import time  # measure time spent in the program
t1 = time.clock()  # CPU time so far in the program
print result, t1
```

This is the complete code; the `trapezoidal` function, the test function `f1` to be integrated, and a main program calling up the integration 1,000 times such that measurements of the CPU time become reliable. As in most other languages, we need to import library functionality, here through the `import` statement. Most of the syntax should be self-explanatory, even for the novice[8] with the exception, perhaps, that functions are prefixed by `def` and for loops from `p` to `q` in steps of `r` are written `for`

---

[8] The intuitive treatment of functions as arguments in functions, cf. the `f` variable in function `Trapezoidal`, is considerably simpler and clearer than the corresponding treatment of function arguments in the other languages we treat herein.

i in range(p,q+1,r), or just for i in range(q) if the indices from 0 up to
and including q-1 are wanted (these for loops can only work with integer counters).

Python codes are interpreted. Hence, there is no need to compile the file containing the code; just write the name of the file, int.py, to execute it under Unix, or write

```
python int.py
```

which works under all operating systems where Python is installed.

Python also allows interactive computing. You can write python on the command line to enter Python's interactive mode, but we recommend using the more user-friendly IDLE shell that comes with the Python source code distribution.[9] Start with

```
from int import *
```

where int.py is the complete Python program shown above. The import statement causes all the statements in int.py to be executed, including the integral computations at the end of the file.[10] The nice thing is that we have, through the import statement, defined the functions Trapezoidal and f1, so we can easily experiment with, e.g., the n value as follows:

```
>>> from int import *
0.250467798247 70.6
>>> n=10; Trapezoidal(a, b, f1, n)
0.25021765112667527
>>> n=1000; Trapezoidal(a, b, f1, n)
0.25046779824710441
>>> n=3; Trapezoidal(a, b, f1, n)
0.24752392127618572
```

The Python program executed on a laptop machine resulted in a CPU time ratio of 14 relative to the Fortran 77 and C/C++ codes. The next section explains how to significantly improve the CPU time.

6.3.7 Vectorization

We have seen that both Matlab and Python enable implementation of the trapezoidal rule close to the syntax used in Algorithm 6.2, but the code is interpreted and runs much more slowly than in Fortran, C, C++, and Java. Loops in interpreted languages will run slowly; actually, such languages are not meant for intensive numerical computations in explicit for loops – this is a kind of misuse of the

---

[9] If PYTHONSRC is the root of the Python source code tree, the IDLE shell is launched by executing PYTHONSRC/Tools/idle/idle.py. You should make an alias for this (long) path in your working environment.

[10] The CPU time of the computations is significant, so you would probably get rid of the repetitive calls to Trapezoidal before using int.py in interactive mode.

language. Instead, interpreted languages often offer alternative programming techniques for achieving efficiency in computations with large data structures or long loops.

Basic Ideas of Vectorization

Algorithms with long for loops in Matlab and Python should be *vectorized*. Vectorization means expressing mathematical operations in terms of vector operations instead of loops with element-by-element computations. A simple example may be summing up the square of the elements of a vector $(v_1, \ldots, v_N)$:

$$s = 0$$
$$\text{for } i = 0, \ldots, N$$
$$\quad s \leftarrow s + v_i^2$$
$$\text{end for}$$

In Matlab this translates to

```
s = 0;
for i = 1:N
  s = s + v(i)*v(i);
end
```

Similar Python code reads

```
s = 0
for i in range(N):
  s = s + v[i]*v[i]
```

These loops normally run slowly. To vectorize the code, we need to express the loops in terms of vector operations. This is very simple in the present case, because s is nothing but the square of the norm of v.

Matlab has a function norm that we can use:

```
s = norm(v); s = s*s
```

Python has a special package, Numerical Python, for efficient array storage and computation. This module does not provide a norm function, but the norm of v is the square root of the inner product (also called the scalar product) of v and itself. Since Numerical Python offers an innerproduct function, we can compute s as

```
s = innerproduct(v,v)
```

The major difference between computing s through a for loop and through a norm or innerproduct function is that the latter functions are implemented as a for loop *in C or Fortran*. Therefore, norm in Matlab and innerproduct in Python run as fast as highly optimized C or Fortran code.

The idea of vectorization is to avoid explicit loops and instead invoke C or Fortran functions where the loop can be much more efficiently implemented. However, vectorizing an algorithm can require quite some rewriting of a loop-based algorithm. The rewriting also depends on what vector operations are offered by the programming language. The linear algebra field in mathematics defines some vector operations, such as inner (scalar) products, matrix-vector products and so on. Programming languages support these mathematical operations, but also offer additional functionality. Unfortunately, there is no standard naming convention for this additional functionality. The vectorization of an algorithm is therefore more tightly coupled to its actual implementation in a particular programming language.

The Trapezoidal Rule in Vectorized Form

Vectorization of Algorithm 6.2 requires rewriting the instructions such that the loop is replaced by built-in operations working on vectors. Let us start with a vector x, going from a to b in steps of $h = (a - b)/n$:

$$x_1 = a, \quad x_2 = a + h, \ldots \quad x_j = a + (j - 1)h, \ldots \quad x_{n+1} = b.$$

We then apply f to each element in x such that $f(x)$ is a new vector v of the same length as x. This application of f to x should be performed in Fortran, C, or C++ for efficiency. Languages with support for vector operations normally supply a function for summing all the elements of the vector $f(x)$: $s = sum_{i=1}^{n+1} f_i$. This operation is often named reduce or sum. The quantity s is not exactly what we want, since the end points f_1 and f_{n+1} should have weight $\frac{1}{2}$ and not 1, as implied in the sum over all elements f_i. The correction is easily performed by subtracting $\frac{1}{2}(f_1 + f_{n+1})$. Finally, we need to multiply s by h. Algorithm 6.13 summarizes the various steps (sum(v) here is our notation for adding all elements in v).

Algorithm 6.13

> *Vectorized Trapezoidal Integration.*
>
> Trapezoidal_vec (a, b, f, n)
> $\quad h = \frac{b-a}{n}$
> $\quad x = (a, a + h, \ldots, b)$
> $\quad v = f(x)$
> $\quad s = h \cdot (sum(v) - 0.5(v_1 + v_{n+1}))$
> $\quad$ return s

Vectorized Numerical Integration in Python

Translating Algorithm 6.13 into Numerical Python is straightforward. Generation of the sequence of evaluation points $(a, a + h, \ldots, b)$ is performed by the statement

```
x = arrayrange(a, b, h, Float)
```

Because of round-off errors the upper limit can end up as b-h instead of b. It is therefore smart to set the upper limit as b+h/2; this will guarantee that b becomes the last element in x. The sum operation on a vector v is called add.reduce(v) in Numerical Python. Knowing about arrayrange and add.reduce, it should be quite easy to understand the implementation of the vectorized algorithm in Python:

```
def trapezoidal(a, b, f, n):
    h = (b-a)/float(n)
    x = arrayrange(a, b+h/2, h, Float)
    v = f(x)
    r = h*(sum(v) - 0.5*(v[0] + v[-1]))
    return r
```

Indexing Python arrays is a slow process, but we do it only twice, which is not critical if n is large. Arrays in Python start with index 0, and -1 is a notation for the last index.

Numerical Python redefines the multiplication operator * such that h*f(x) is a scalar h multiplied by a vector f(x). The expression f(x) means applying a function f to a vector x, element by element. However, we must avoid implementing f(x) as an explicit loop in Python. To this end, Numerical Python supports the efficient application of standard mathematical functions and operators on vectors. For example, sin(x) computes the sine of each element in a vector x. We can therefore implement a function $f(x) = e^{-x^2} \ln(1 + x \sin x)$ applied to each element of a vector x as

```
def f1(x):
    f = exp(-x*x)*log(1+x*sin(x))
    return f
```

This is exactly the same syntax as that used for applying a scalar function on a scalar value. Sending, e.g., a number 3.1 to f1, results in the computation $e^{-3.1^2} \ln(1 + 3.1 \sin 3.1)$. Sending a vector x to f1 results in more complicated computations "behind the curtain":

1. temp1 = sin(x), i.e., apply the sine function to each entry in x
2. temp2 = 1 + temp1, i.e., add 1 to each element in temp1
3. temp3 = log(temp2), i.e., compute the natural logarithm of all elements in temp2
4. temp4 = x*x, i.e., square each element in x
5. temp5 = exp(temp4), i.e., apply the exponential function to each element in temp4
6. f = temp5*temp3, i.e., multiply temp5 and temp3 element by element, as in a scalar multiplication

As we see, f is built of several vector operations, requiring temporary arrays, and hence additional storage, compared with a straightforward loop. However, a straight loop computing f element by element required 40 times as long CPU time as the vectorized version (on an Intel laptop computer with Linux and Python compiled by GNU's C compiler gcc).

The main program calling up the vectorized version of the trapezoidal function is exactly the same program as we used to test the scalar version of Trapezoidal:

```
a = 0; b = 2; n = 1000
for i in range(10000):
    result = trapezoidal(a, b, f1, n)
import time
t1 = time.clock()
print result, t1
```

We note that in Python you can say print a, where a is any built-in object (including an array) and get nicely formatted output. This is very convenient during program development and debugging.

The reader should notice how convenient a language using dynamic typing is; many of the code segments developed for scalar computations can be directly reused for vectorized computations. (The same functionality is obtained in C++ by the use of templates and overloaded operators, though at the cost of increased language and technical complexity.)

The vectorized version of the trapezoidal method uses one-seventh of the CPU time required by the loop-based counterpart. This means that vectorized Python runs slightly faster than Java and at about half the speed of Fortran 77, C, or C++.

Vectorized Numerical Integration in Matlab

Algorithm 6.13 can be implemented directly in Matlab:

```
function r = Trapezoidal_vec(a, b, f, n)
% TRAPEZOIDAL Numerical integration from a to b
% by the Trapezoidal rule
f = fcnchk(f);
h = (b-a)/n;
x = [a:h:b];
v = feval(f, x);
r = h*(sum(v) - 0.5*(v(1) + v(length(v))));
```

The construction of the sequence of evaluation points having a uniform distance h between a and b is created by the expression [a:h:b]. The feval(f,x) expression we know from the scalar (non-vectorized) version of the function can also be used when x is a vector. As in Python, the return value from f is then a vector. Summing up all elements in a vector v is performed with sum(v), and indexing the last element in v can be written as v(length(v)) (if the first index in v is set to 1, which is the default value).

The function to be integrated must now work for a vector. In the previous Python example we could reuse a function implemented for a scalar input and return value. Such reuse is not possible in Matlab when the function expression $f(x) = e^{-x^2} \log(1 + x \sin x)$. The reason is because the expression exp(-x*x) for a vector x implies a matrix multiplication between x and itself. What we need is the .* operator such that an element in the answer is the product of the corresponding two elements in the operands. Hence, the vectorized version of the scalar f1 function is

```
function y = f1_vec(x)
y = exp(-x .* x) .* log(1 + x .* sin(x));
```

We store this function in an M-file f1_vec.m. The main program used to test the efficiency of the scalar version of the trapezoidal rule implementation in Matlab can be reused in the vectorized case if the name of the integration routine and the function to be integrated are changed:

```
function y = f1_vec(x)
y = exp(-x .* x) .* log(1 + x .* sin(x));
```

The vectorized version above reduced the CPU time to 1/42nd of the CPU time required by the loop-based Matlab scripts from Sect. 6.3.5. The CPU times of the vectorized versions of the Python and Matlab codes were approximately equal, running at half the speed of the Fortran 77 and C/C++ implementations.

Discussion

It might seem complicated to use languages that require vectorization for computational efficiency. Straightforward implementations in terms of loops are indeed simpler, but vectorization has one great advantage: The vector operations can utilize a library particularly tuned for the hardware in question. For example, smart tricks to speed up array operations can be implemented in the library, but more importantly, the library can implement parallel versions of the array operations such that one can utilize parallel computers with only minor competence in parallel programming. At least in principle, it may be hard to write a straightforward loop that can compete with highly tuned basic array operations in libraries. The present example is sufficiently simple such that straightforward loops inside a library yield high (and close to optimal) efficiency.

Vectorization became an important programming technique with the advent of the first supercomputers in the late 1970s. We just argued that in a compiled language like Fortran there is no need to transform loops into calls to vector operations. The situation was a bit different for the first supercomputers. The point then was to break up more complicated nested loops into a series of simpler loops performing basic vector operations. These simpler loops could easier be translated by the compiler to a suitable form for very efficient computations on the super computers. This type of vectorization can still be beneficial on modern RISC architectures found in PCs and workstations. More importantly, basic vector operations have been implemented with maximum efficiency, tuned to the hardware, and made available in libraries (BLAS and ATLAS are examples). To get the ultimate speed out of code, it may be advantageous to rewrite the algorithms and implementations such that they can utilize very efficient libraries for vector operations.

In the 1980s, supercomputers were based on parallel computations, and algorithms had to be recast again, this time from serial into parallel versions. The importance of vectorization disappeared, but due to the popularity of interpreted environments such as Matlab and Python, RISC-based computers, and standardized libraries for vector operations, the vectorized programming style remains highly relevant.

6.3.8 Maple

Maple is a complete problem-solving environment for *symbolic* calculations. Symbolic in this context means manipulating mathematical expressions, not just numbers. For example, Maple can calculate the integral of $\cos x$ to be $\sin x$ and differentiate $\sinh x$ to get $\cosh x$. Maple is also equipped with the widely used NAG libraries for efficient numerical computations.

A particularly attractive feature of Maple is that you can write *interactive* scientific reports with formulas, text, and plots. The report can be read as is, but the reader can also change Maple commands (formulas, equations, etc.) and view the consequences of such changes. This encourages explorative and interactive investigations of scientific problems.

Maple is a complete programming language. As in Matlab and Python, variables are not explicitly typed and there are many high-level commands, making Maple programs compact.

The bridge between Algorithm 6.2 and a Maple function is short:

```
Trapezoidal := proc(a,b,f,n)
local h, s, x, i:
h := (b-a)/n:
s := 0:  x := a:
for i from 1 to n-1 do
  x := x + h:
  s := s + f(x):
od:
s := s + 0.5*f(a) + 0.5*f(b):
s := h*s:
s;
end:
```

Maple performs symbolic computations. That is, (b-a)/n is not a number unless a, b, and n are assigned numbers; if these arguments are assigned mathematical expressions, (b-a)/n remains an expression, i.e., a mathematical formula instead of a number. In the present case, a, b, and n are numbers, so the operations in the trapezoidal function are numerical computations.

A test function f1 to be integrated is defined by

```
f1 := x -> exp(-x*x)*log(1+x*sin(x));
```

The numerical integration of f1 is now carried out by

```
q := Trapezoidal(0, 2, f1, 1000)
```

This code runs very slowly. Repeating it 10 times and multiplying the CPU time by 1,000 (corresponding to 10,000 repetitions as we have used for all previous CPU time comparisons) showed that Maple required more than 600 times the CPU time of the Fortran 77 implementation! We should add that the speed of Maple varied greatly (up to a factor of 4) between different executions.

The reason why the implementation of Algorithm 6.2 runs so slowly in Maple is that Maple code is interpreted. Long for loops are, in some sense, a misuse of Maple. One should either implement such loops with numerical computations in

C or Fortran code to be called from Maple, or (better) use a built-in Maple function to perform the task. In the case of numerical integration, Maple offers the user many computationally efficient functions. The plain trapezoidal rule is not available. Maple instead offers sophisticated methods with user interfaces for specifying the desired precision and a particular rule rather than the detailed parameters of a method. Maple then calls up highly optimized Fortran 77 functions, which automatically find a rule compatible with the user's accuracy requirement.

In the present case, we just write

```
q:=evalf(Int(f1,0..2,'method'=_NCrule));
```

to evaluate the integral of f1 numerically by the simplest rules in Maple. The answer is returned in a fraction of a second. Maple computes our integral faster than the Fortran 77 program and with higher precision. The reason is because Maple applies a much more efficient numerical integration rule.

To summarize, we can easily implement Algorithm 6.2 in Maple, but the execution time will be long if n is large. The "right" solution is to use the built-in numerical integration functions, but this invokes much more sophisticated and efficient methods than Algorithm 6.2.

6.3.9 Summary

In the previous sections we have presented implementations of the trapezoidal rule in different computing environments. Although the trapezoidal rule is among the very simplest of all numerical algorithms, quite significant differences in performance and convenience of the various programming tools have come to be. These differences are much larger and more serious when addressing more complicated numerical algorithms.

It is almost impossible to draw firm conclusions about the preferred computing environment, since this depends heavily on personal taste and previous experience. We have tried to define some characteristics of the different programming tools in Table 6.1 when implementing the trapezoidal rule. The purpose is to provide a quick comparison of some features that programmers of numerical applications may emphasize when choosing a language.

Some trends are clear when comparing Fortran 77, C/C++, Java, Python, Matlab, and Maple. If computational speed is the most important factor, one should choose one of the compiled languages: Fortran 77, C, C++, or Java. If programming convenience is regarded as most important, the interpreted and dynamically typed environments Maple, Matlab, and Python constitute the preferred choice. Speed versus convenience can also be expressed as computational versus human efficiency. The former is very important if the total CPU time is hours, days, or weeks, or if the program is to be executed "as is" in thousands of production runs by many people. Students or scientists will often be most concerned with human efficiency. Some programmers regard C++ as a "middle of the road" choice: It provides speed and a fair amount of programming convenience.

Maple is an obvious choice if it is important to integrate symbolic and numerical computing. However, Matlab has a communication link to Maple, and Python can call up symbolic mathematics engines such as SymPy or Ginac, so integration of symbolic functionality can be performed in all three environments.

Even if a particular feature is supported in several languages, the usage of the feature may be considerably more favorable in one particular language. Classes constitute an example; the authors find class programming easiest and most convenient in Python, followed by Java and C++, while we find Matlab classes less convenient. These finer points are not easy to express in a tabular set-up such as Table 6.1.

The most obvious conclusion is that there is no single preferred scientific computing tool. You need to gain experience with several and learn how to pick the most productive tool for a given task. We recommend you gain this knowledge from "learning by doing", so that you can build up your own experience rather than trying to deduce the conclusion from theoretical considerations.

Finally, we mention that only some relevant programming languages are compared here. We have said little about C and not mentioned Python's "brothers" Perl and Ruby; there are also excellent computing environments such as Mathematica, Gauss, and R (or S-Plus) that are not mentioned here. We have also neglected the modern Fortran versions: 90, 95, and 2000. These represent a smooth transition from the simple Fortran 77 language to a rich language such as C++. For many purposes, Fortran 2003[11] shares the same characteristics as C++, with some convenient additional built-in features for numerical computing.

Table 6.1 Comparison of different programming languages. The numbers given in the "speed" column equal the actual CPU time of Algorithm 6.2, implemented in the particular language, divided by the CPU time required by the associated Fortran 77 implementation. For Python and Matlab we have reported the CPU time of the vectorized version of Algorithm 6.2. The numbers in the "program length" column correspond to the number of lines in our implementation, including comment lines, and depend significantly on the personal tastes and habits of the programmer. The dashes indicate that it is difficult to give a precise yes or no answer: Maple's speed is improved by calling built-in numerical functionality, but vectorization as a programming technique is not required; Java and Python (and C++ to some extent) can easily incorporate visualization in programs, but this is not a part of standard libraries

Language	Program length	Speed	Vectorization required	Dynamic typing	Classes	Interactive computing	Built-in visualization
Fortran 77	41	1.0	No	No	No	No	No
C++	38	1.0	No	No	Yes	No	No
Java	50	2.1	No	No	Yes	No	–
Python	24	1.9	Yes	Yes	Yes	Yes	–
Matlab	21	2.0	Yes	Yes	Yes	Yes	Yes
Maple	36	613.6	–	Yes	No	Yes	Yes

---

[11] At the time of this writing, Fortran 2003 compiler are just starting to emerge.

6.4 Heun's Scheme in Different Languages

The purpose of this section is to implement Heun's method from Algorithm 6.12 on page 212 in the following programming languages: Fortran 77, C, C++, Java, Matlab, Maple, and Python. Section 6.3 gives a glimpse of these languages applied to the trapezoidal integration technique. You should therefore be familiar with Sect. 6.3 before proceeding with the implementation examples regarding Heun's method. These new examples involve array and file handling in the various computer languages.

6.4.1 Code Structure

Before diving into the details of various programming language, we should have a clear view of the structure of the complete program. It would be convenient to have Δt and the total simulation time $t_{\text{stop}} = N\Delta t$ as input parameters to the main program. Algorithm 6.12 must then be called and the solution should be displayed graphically. The latter task can in some environments easily be done inside the program, while in other situations we need to store the solution in a file and use a separate plotting program to view the graph of $u(t)$. We have decided to always write the solution to a file (to demonstrate file writing) and in addition display the solution directly when this is feasible. The code must also implement a right-hand side function $f(u)$. For testing purposes, we choose $f(u) = -u$ and $u(0) = 1$ such that the exact solution reads $u(t) = e^{-t}$.

6.4.2 Fortran 77

Algorithm 6.12 can be directly translated to Fortran 77:

```fortran
      subroutine heun (f, u0, dt, n, u)
      integer n
      real*8   u0, u(0:n), dt, f
      external f

      integer i
      real    v
C     initial condition:
      u(0) = u0
C     advance n steps:
      do i = 0, n-1
         v = f(u(i))
         u(i+1) = u(i) + 0.5*dt*(v + f( u(i) + dt*v ))
      end do
      return
      end
```

There are two new issues (compared with the `trapezoidal` function explained in Sect. 6.3.2):

1. We now have a subroutine, i.e., a subprogram that does not return a value.
2. One of the arguments is an array. The indices in this array range from 0 up to (and including) n, specified by the declaration `real*8 u(0:n)`. The array is indexed[12] using the syntax `u(i)`.

All arguments in Fortran 77 subroutines and functions are both input *and* output variables, in the sense that any changes in these variables are visible outside the function. That is, if we change u, which we intend to do, since u is supposed to hold the computed function values, the final u is immediately available in the calling program:

```
C       calling code:
        integer n
        real*8 v(0:n), v0, some_f, dt
        external some_f

        dt = 0.01
        v0 = 1.0
        call heun (some_f, v0, dt, n, v)
```

Note that subroutines are called using the `call` statement (in most other languages it is sufficient to write the name of the subprogram). After the return from the `call heun` statement, the v array will contain the computed discrete $u(t)$ values.

If we accidentally change u0 inside the `heun` subroutine, the new value will be reflected in the v0 variable in the calling program. This is not a feature we want. Most other programming languages have mechanisms for taking a copy of the arguments (usually referred to as call-by-value). Using copies of pure input variables reduces the danger of unintended changes to the variables in the calling code.

Having called the subroutine `heun` to compute discrete function values in an array, we may want to dump this array to file for later plotting. A suitable file format is to have two numbers on each line, the time point and the corresponding value of u, e.g.,

```
0.  1.
0.004  0.996008
0.008  0.992031936
0.012  0.988071745
0.016  0.984127362
0.02   0.980198726
0.024  0.976285772
0.028  0.97238844
```

Most plotting programs can read such data pairs: t_i, $u(t_i)$, and draw straight line segments between the points to visualize the curve. In our code we include a subroutine `dump` that creates such a file. Some key statements for illustrating writing in Fortran 77 are given next.

---

[12] Some compilers have flags for checking the validity of the index (with an associated efficiency penalty).

```
      integer iunit
      real*8  v
      ...
      iunit = 22
      open(iunit, name='sol.dat', status='unknown')
      write(iunit, *) 'some text and variables', v, ' and text'
      ...
      close(iunit)
```

Note that a file is referenced by an integer, iunit here. Our dump subroutine assumes that a file has already been opened, such that we can pass iunit as an input argument and work with it inside the subroutine:

```
      subroutine dump (iunit, u, n, t0, dt)
      integer n, iunit
      real*8  u(0:n), t0, dt

      integer i
      real*8  time
      time = t0
      do i = 0, n
         write(iunit,*) time, u(i)
         time = time + dt
      end do
      return
      end
```

The main program is listed next and commented upon afterward:

```
      program ode
      integer nmax, n, i, iunit
      parameter (nmax=1000)
      real*8 dt, tstart, tstop, u0, u(0:nmax)
      external f1

      tstart = 0.0
      write(*,*) 'stop time:'
      read(*,*) tstop
      write(*,*) 'time step:'
      read(*,*) dt
C     set a reasonable time step if dt<0:
      if (dt .lt. 0.0) then
         dt = 0.004
      end if

C     check that the u array is sufficiently large:
      n = tstop/dt
      if (n .gt. nmax) then
         write(*,*) 'ERROR: too small time step'
      end if

C     time integration:
      u0 = 1.0
      call heun (f1, u0, dt, n, u)
      write(*,*) 'end value =', u(n), ' error = ', u(n)-exp(-dt*n)

C     write solution to file (for plotting):
      iunit = 21
      open(iunit, name='sol.dat', status='unknown')
      call dump (iunit, u, n, tstart, dt)
      close (iunit)
      end
```

Arrays in Fortran 77 must be allocated at compile time. That is, we cannot ask the user for Δt and t_{stop}, compute $N = t_{\text{stop}}/\Delta t$, and create an array u of length N. Instead, we must allocate a sufficiently long u array in the main program. All subroutines or functions can, however, receive arrays with lengths specified by variables. In our case, subroutine heun works with the first n+1 elements of the u array allocated at compile time in the main program. The `parameter` statement enables setting the value of constants at compile time.

To get user information about Δt and t_{stop}, we use `write` and `read` statements and communicate with the user in terms of questions and answers. In other programming languages we will get input from the command line instead. This is normally possible in Fortran by calling up functionality in a C library, but the feature is not a part of the Fortran standard. We also mention that comparisons of numbers in if-statements apply the operators `.lt.`, `.le.`, `.eq.`, `.ge.`, and `.gt.` rather than <, <=, == or =, >=, and >, respectively.

Checking that n is not greater than nmax is very important. Without this test, subroutine heun will index the u array outside the allocated chunk of numbers. This may or may not produce strange error messages (usually just `segmentation fault`) that are hard to track down. Some compilers have flags for detecting if the array index is outside valid values. However, in subroutine heun u is declared as `u(0:n)`, and the compiler-generated check is just that the index i does not exceed the value n, not whether i is outside the physical bounds of the array.

The Fortran 77 code is compiled, linked, and run as explained in Sect. 6.3.2.

6.4.3 Graphics

Here we explain how we can plot curve data stored in files. The data consists of points (x_i, y_i), $i = 0, \ldots, N$, and the goal is to draw line segments between these points in an xy coordinate system. We assume that the (x_i, y_i) points are stored in two columns in a file, the x_i values in column 1 and the y_i values in column 2. This file is given the name sol.dat in the forthcoming demonstrations of plotting programs. In the present example with the numerical solution of ODEs, the x_i values correspond to points in time, and the y_i values are the associated values of the unknown function u in the differential equation.

Gnuplot

A widely available plotting program is Gnuplot. Assuming that the name of the file containing the data points in two columns is sol.dat, the Gnuplot command

```
plot 'sol.dat' with lines
```

displays the graph. To try it out: Start Gnuplot by typing gnuplot. This enters a command mode where you can issue the command above.

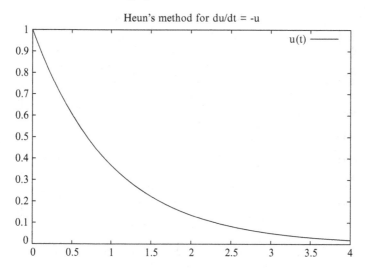

Fig. 6.1   Plot made by Gnuplot

Normally, we want to set a heading in the plot and also produce a hardcopy of the plot in PostScript format (for inclusion in reports). The following set of Gnuplot commands perform these tasks:

```
set title "Heun's method for du/dt = -u"
# plot on the screen:
plot 'sol.dat' title 'u(t)' with lines
# hardcopy in PostScript format:
set term postscript eps monochrome dashed 'Times-Roman' 20
set output 'tmp.ps'
plot 'sol.dat' title 'u(t)' with lines
```

We recommend putting such commands in a file, say `plot`, and executing Gnuplot with the name of this file as the argument:

```
unix> gnuplot -persist plot
```

The `-persist` option makes the plot stay on the screen after the Gnuplot program has terminated. The PostScript version of the plot is displayed in Fig. 6.1.

Matlab

Having the x_i values as a vector x and the y_i values as a vector y, plotting in Matlab is easily done through the standard command `plot(x,y)`. In our example with $u(t)$ data, the following Matlab commands create a plot of the data in the file `sol.dat` and annotate a title and curve label, in addition to making a hardcopy of the plot:

```
data = load('sol.dat');   % load file into two-dim. array data
t = data(:,1); u = data(:,2);   % extract columns as vectors
plot(t, u);
```

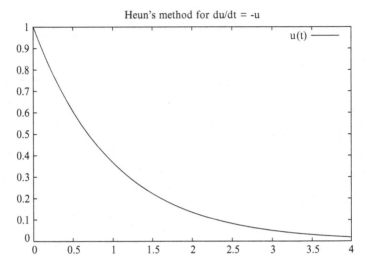

Fig. 6.2   Plot made by Matlab

```
title('Heun method for du/dt = -u');
legend('u(t)');    % attach curve label
print('-deps', 'tmp.eps');  % make Encapsulated PostScript plot
```

Suppose these commands are stored in a file `plotmatlab.m`. Inside Matlab we can then write `plotmatlab` to execute the commands in `plotmatlab.m` and thereby create the plot. Alternatively,

```
unix> matlab -nodesktop -r  plotmatlab
```

creates the plot and invokes Matlab in a terminal window. Type `exit` to terminate the session. The PostScript version of the plot, as created by Matlab, is displayed in Fig. 6.2.

6.4.4 C++

The Computational Algorithm

The implementation of Algorithm 6.12 in the C++ programming language can take the following form:

```
void heun (fptr f, double u0, double dt, int n, double u[])
{
  u[0] = u0;  // initial condition
  // advance n steps:
  for (int i = 0; i <= n-1; i++) {
    double v = f(u[i]);
    u[i+1] = u[i] + 0.5*dt*(v + f( u[i] + dt*v ));
  }
}
```

The `heun` function is like a subroutine in Fortran, i.e., it does not return any value, which in C++ is indicated by declaring the return type as `void`. The `f` argument is used to transfer the right-hand side of the ODE to this function (see page 222 for details). For our test program, the following `f1` function will be supplied as the `f` argument to the `heun` function:

```
double f1 (double u)   { return -u; }
```

Plain Arrays in C++

The argument `u` in the `heun` function is a basic C++ (or C) one-dimensional array, with one index starting at 0. The array argument is declared as `double u[]` or `double* u`. The latter notation actually declares a pointer to a double, i.e., the memory address of a double. Arrays and pointers in C and C++ are very closely related; array variables are represented by pointers to (i.e. the address of) the first array element in memory. Indexing arrays is done with square brackets: `u[0]`, `u[1]`, and so on. The array length (here `n+1`) must be transferred as a separate argument, as in Fortran 77. Another similarity with Fortran is that there is no built-in check of the validity of the array indices. Indexing arrays beyond their bounds is one of the most common errors in these programming languages.

Arrays in C and C++ can be created at run time with an appropriate length after some input data are read.[13] The syntax is as follows:

```
double* u = new double[n+1];
```

The `new` operator allocates a chunk of memory of sufficient size for storing `n+1` double-precision real numbers and returns a pointer to the first array element. We can assign the returned pointer to a pointer variable `u` and index the array as `u[i]`. When there is no further use for the array, it must be deallocated (deleted from memory) by the syntax

```
delete [] u;
```

For every `new` there should a corresponding `delete` action. Issuing a `delete` twice is a common error, and the result is just strange program behavior and abort messages such as `segmentation fault`. Matching all the `new` and `delete` statements correctly appears to be one of the greatest challenges when developing C and C++ code.

Array Objects

Fortunately, C++ has safer array variables. The class `valarray` in the standard C++ library offers a vector type (array with a single index). Here is a declaration of an array `u` of length `n+1`, where each array element is a `double`:

---

[13] Recall that Fortran arrays are declared with a fixed size at compile time.

```
std::valarray<double> u(n+1);
```

The u array is automatically deleted when it is no longer in use. The declaration of arrays of type `valarray` in a function such as `heun` is as follows:

```
void heun (fptr f, double u0, double dt, int n,
          std::valarray<double>& u)
```

The ampersand `&` indicates that a *reference* to the u object is transferred; without the ampersand, u will be a copy of the transferred array. This is not what we want, since our goal is to compute values in u and view these values outside the `heun` function. The syntax of the body of the `heun` function is independent of whether we use `std::valarray` or plain C/C++ arrays.

Multi-dimensional arrays in C and C++ can be constructed, but the allocation and use of `new` is more complicated. Details can be found in books on the C or C++ programming language. Unfortunately, the standard C++ library does not offer a class such as `valarray` for multi-dimensional arrays. One can, nevertheless, create a class for multi-dimensional arrays with minor effort. There are also several open source and commercial libraries in C++ that offer sophisticated multi-dimensional arrays for numerical computing.

File Writing

Basic file writing in C++ is illustrated next[14]:

```
std::ofstream out("sol.dat");  // open a file sol.dat
out << "some text " << object << " more text\n";
out.close()
```

A `dump` function for writing the $u(t)$ values to a file can typically be coded as follows:

```
void dump (std::ostream& out, double u[], int n, double t0, double dt)
{
  double time = t0;
  for (int i = 0; i <= n; i++) {
    out << time << " " << u[i] << '\n';
    time += dt;
  }
}
```

Although the file is of type `std::ofstream`, it is common to declare the corresponding function argument as `std::ostream`, because this type is a common name for all types of output media. For example, the `out` argument above can be a file (`std::ofstream`) or the terminal window (`std::cout`), or even a string.

---

[14] This is C++ file writing. One can also use the C functionality for file writing. This can be convenient if format control via the `fprintf` function is desired.

The Main Program

The main program (in the case where we use a primitive C/C++ array for u) is listed below.

```cpp
int main (int argc, const char* argv[])
{
    // check that we have enough command-line arguments:
    if (argc < 3) {
        std::cout << "Usage: " << argv[0] << " tstop dt\n";
        exit(1);
    }
    // read tstop and dt from the command line:
    double tstop = atof(argv[1]); // get 1st command-line arg.
    double dt = atof(argv[2]);    // get 2nd command-line arg.

    if (dt < 0.0) { dt = 0.004; } // set a suitable time
    step if dt<0
    int n = int(tstop/dt);        // no of time steps

    double* u = new double[n+1];  // create solution array

    // time integration:
    double u0 = 1.0;              // initial condition
    heun (f1, u0, dt, n, u);      // advance solution n steps
    printf("end value=%g  error=%g\n", u[n], u[n]-exp(-dt*n));

    // write solution to file (for plotting):
    std::ofstream out("sol.dat");
    dump (out, u, n, 0.0, dt);
    out.close();

    delete [] u;   // free array memory
    return 0;      // success of execution
}
```

Command-Line Arguments

In this main program we get input data, such as tstop and dt, from the command line. All command-line arguments in C and C++ programs are transferred to the main function as an array of strings. For example, if the name of the executable file is ode and we run

```
unix> ./ode 4.0 0.04
```

there are two command-line arguments: 4.0 and 0.04. These arguments are stored in an array of strings argv. This array has the name of the program (here ode) as its first argument (argv[0]). The first command-line argument is argv[1], the second is argv[2], and so on. Observe that all command-line arguments are available as strings, so if we want to compute with tstop, we need to convert the string to a floating-point number. This is performed by the atof (ASCII to float) function. The number of array elements in argv is provided as argc. We can thus test if the user has provided a sufficient number of arguments (two in this case). The handling of command-line arguments in C and C++ is reflected in many other languages, including Java and Python.

Graphics

Plotting the solutions is performed as in the Fortran case; i.e., an external plotting program is used to plot the data stored in the file sol.dat.

There exist plotting programs written in C or C++ that can be used as libraries for the present program such that the u array can be sent directly to a plotting routine, without using a file for intermediate data storage. However, there are no "standard" plotting libraries for C++.

6.4.5 Java

The Java program for solving an ODE by Heun's method is quite similar to the corresponding C++ program, except that the stand-alone functions heun and dump must be methods of a class in Java.

The implementation of the heun function, as described in Algorithm 6.12, is performed via the advance method in class Heun:

```
class Heun {
  public static double[] advance (Func f, double u0,
                    double dt, int n)
  {
    double u [] = new double[n+1];
    u[0] = u0; // initial condition

    // advance n steps:
    for (int i = 0; i < n; i++) {
      double v = f.f(u[i]);
      u[i+1] = u[i] + 0.5*dt*(v + f.f( u[i] + dt*v ));
    }
    return u;
  }
}
```

The right-hand side of the ODE is represented as function objects in a Func hierarchy, as explained in Sect. 6.3.4. Arrays in Java are created in the same manner as in C++. However, there is no need to delete the array, since Java will do this when the array is no longer in use. This makes it possible to create our array wherever it is convenient, e.g., in the Heun.advance function, and return it from the function to the calling code, if desired. (In C++ this is also possible, but it is usually a good habit, from a programming safety point of view, to perform both the new and delete statements in the calling code. Declaring a variable u of type std::valarray inside the C++ heun function can of course be performed, but C++ requires us to *copy* this array if we want to return it, whereas with plain C/C++ and Java arrays we avoid this copy, and only references/pointers are returned.)

File writing in Java is typically illustrated by the following statements:

```
PrintStream f = new PrintStream(new FileOutputStream("sol.dat"));
f.println(time + " " + u[i]);
f.close()
```

Dumping the solution array u to a file is performed by a dump method in a class IO:

```
class IO {
  public static void dump (PrintStream f, double u[], double t0,
    double dt)
  {
    int n = u.length;
    double time = t0;
    for (int i = 0; i < n; i++) {
      f.println(time + " " + u[i]);
      time += dt;
    }
  }
}
```

The main program can take this form:

```
class Demo {
  public static void main (String argv[])
    throws IOException
  {
    double tstop = Double.valueOf(argv[0]).doubleValue();
    double dt = Double.valueOf(argv[1]).doubleValue();
    int n = (int) (tstop/dt);
    Func f = new f1();
    double u0 = 1.0;
    double u [] = Heun.advance (f, u0, dt, n);
    double error = u[n] - Math.exp(-dt*n);
    System.out.println("end value=" + u[n] + " error=" + error);
    PrintStream file = new PrintStream(new FileOutputStream
      ("sol.dat"));
    IO.dump (file, u, 0.0, dt);
    file.close();
  }
}
```

Plotting of the solution is subjected to the same comments as we made for C++: Calling a Java plotting library is possible, but there are no standard plotting tools to be used.

6.4.6 Matlab

In Matlab, we implement Algorithm 6.12 as a function heun in an M-file heum.m:

```
function u = heun (f, u0, dt, n)
% HEUN numerical integration of ODEs by Heun's method

f = fcnchk(f);
u = zeros(n+1,1);  % make n+1 zeros (as initialization)
% initial condition:
u(1) = u0;
% advance n steps:
for i = 1:n  % i <-> i-1 compared to the original algorithm
    v = feval(f, u(i));
    u(i+1) = u(i) + 0.5*dt*(v + feval(f, u(i) + dt*v));
    %fprintf('u(%d)=%g\n',i+1,u(i+1))
end
```

As in Java, it is convenient to create arrays wherever we need them and return them from functions, if desired. Matlab will automatically delete arrays when they are no longer in use. Transfer of arrays in and out of functions is performed very efficiently by transferring references to the arrays.[15]

Transferring functions (here the right-hand side, $f(u)$, in the ODE $u'(t) = f(u)$) to the heun function follows the set-up from Sect. 6.3.5. We find it convenient to create the u array inside the heun function. The syntax [0:n]*0 first creates an array with values 0, 1, up to and including n, and then all the elements in this array are multiplied by zero. There is actually no need to initialize the array elements, since this will be done in the subsequent loop. Arrays in Matlab always start with 1 as the index so we need to rewrite Algorithm 6.12. Although this is easy (just replace i by $i + 1$), the rewrite is often an error-prone process.

File writing in Matlab is very similar to file writing in C. The basic statements are typically

```
fid = fopen('sol.dat', 'w');
fprintf(fid, 'some text, u[%d]=%g\n', i, u(i));
fclose(fid);
```

Here we have used an output function fprintf, which is very convenient when we want to control the format of the output. The function follows the syntax of the printf-family of functions originating from C. The first argument is a string, where variables can be inserted in "slots" starting with %. The variables are listed after the format string (here i and u(i)). The character after % determines the formatting of integers, reals, or strings. Here we specify that an integer and a real are to be written as compactly as possible: %d and %g, respectively. More sophisticated formatting, e.g., writing a real number in scientific notation with six decimals in a field 12 characters wide can be specified as %12.6e-4. Complete documentation of the printf-syntax is obtained by typing man sprintf or perldoc sprintf on Unix/Linux systems (in general one can look up the online documentation of Perl's sprintf function).

A dump function for writing u to an already opened file reads

```
function dump (fid, u, t0, dt)
% DUMP write array, containing the solution of an ODE, to file

n = length(u);
time = t0;
for i = 1:n
    fprintf(fid, '%g %g\n', time, u(i));
    time = time + dt;
end
```

One should note that strings in Matlab are always enclosed in single quotes.

In the main program there is no natural reading of input data from the command line, since one usually operates Matlab interactively and sets variables directly in the command interpreter. We therefore "hardcode" the input parameters tstop and dt:

---

[15] This is not an issue for the programmer to worry about. On the other hand, when applying C++, the programmer is actually required to gain the necessary knowledge and control the low-level details of transferring arrays between functions.

```
% main program for solution of an ODE:
tstop = 2.0;
dt = 0.004;
n = tstop/dt;
u0 = 1.0;
u = heun(@f1, u0, dt, n);
m = length(u);
fprintf('end value=%g  error=%g\n', u(m), u(m)-exp(-dt*n));
fid = fopen('sol.dat', 'w');
dump(fid, u, 0.0, dt);
fclose(fid);
```

Matlab has a wide collection of visualization tools. In particular, it is straightforward to plot Matlab arrays without going through a data file. The commands listed in Sect. 6.4.3 can be issued right after u is computed:

```
% visualization:
% plot t and u array, i.e., (t(i),u(i)) data points:
t = [0:dt:tstop];
plot(t, u);
title('Heun method for du/dt = -u');
legend('u(t)');   % attach curve label
```

6.4.7 Python

The Python code for solving the present ODE problem is quite similar to the Matlab code. In general, these two languages usually lead to the same nature of compact, easy-to-understand programs.

Algorithm 6.12 can be implemented directly as a Python function called heun:

```
def heun (f, u0, dt, n):
    """numerical integration of ODEs by Heun's method"""
    u = zeros(n+1)   # solution array
    u[0] = u0        # initial condition
    # advance n steps:
    for i in range(0,n,1):
        v = f(u[i])
        u[i+1] = u[i] + 0.5*dt*(v + f(u[i] + dt*v));
    return u
```

In Python, we create arrays when we need them. Arrays are efficiently transferred to and from functions using references, and Python automatically deletes arrays that are no longer in use (these two features are shared with Matlab and Java). Python arrays have zero as the base index, as in C++. The programmer can, however, create her own arrays with arbitrary base indices. We emphasize that the arrays we use in Python for numerical computing are created by the Numeric module; the built-in Python lists (also sometimes called arrays) are not well suited for scientific computations.

File writing in Python follows this setup:

```
file = open("sol.dat", "w")  # open file for writing
file.write("some text, u[%d]=%g\n" % (i,u[i])
file.close()
```

Here we have used Python's way of formatting strings with `printf`-syntax (see page 250 for a brief introduction to `printf`-style formatting). The Python syntax is a string, containing slots for variables, followed by % and a list of the variables to be inserted in the slots.

The `dump` function for writing u to a file is now straightforward:

```
def dump (file, u, t0, dt):
    """write array, containing the solution of an ODE, to file"""
    time = t0
    for i in range(len(u)):
        file.write("%g %g\n" % (time, u[i]))
        time += dt
```

The main program, fetching input data from the command line, can be in this form:

```
if len(sys.argv) <= 2:
    print "Usage: %s tstop dt" % sys.argv[0]
    sys.exit(1)
tstop = float(sys.argv[1])    # 1st command-line arg
dt = float(sys.argv[2])       # 2nd command-line arg
if dt < 0:   dt = 0.004
n = int(tstop/dt)

def f1(u): return -u    # right-hand side: du/dt = f(u) = -u
u0 = 1.0
u = heun(f1, u0, dt, n)
print "end value=%g   error=%g\n" % (u[-1], u[-1]-exp(-dt*n))
f = open("sol.dat", "w")
dump(f, u, 0.0, dt)
f.close()
import time; print "CPU-time=", time.clock()
```

The indexing `u[-1]` means the last element in u, i.e., the same as `u[len(u)-1]` (`u[-2]` is the second last element, and so on).

The solution can, of course, be visualized by invoking a plotting program such as Gnuplot or Matlab with the `sol.dat` file. On the other hand, there is a package called SciTools (developed at the authors' institution) that offers a Matlab-like set of commands for plotting in Python. Executing

```
from scitools.std import *
```

imports all of the array and plotting functionality. The relevant Python statements for visualizing the computed u as a curve of $(t, u(t))$ points then read

```
# visualize solution:
t = linspace(0, tstop, u.size)
plot(t, u, axis=[t[0], t[-1], 0.1, 1.1], title='u(t)',
     hardcopy='tmp.ps')
```

6.4.8 Summary

Section 6.3.9 provides a comparison of Fortran 77, C/C++, Java, Matlab, Python, and Maple for implementing simple functions and loops. The present section adds comments regarding the handling of arrays and graphics. The other new program feature in the present ODE example, file writing, is handled in a similar way in the various languages.

Array creation and processing is particularly efficient in Fortran. First, arrays are allocated at compile time, with a fixed size (although arrays *seemingly* have a dynamic size in subroutines and functions). Second, Fortran 77 compilers have been developed over five decades to a very sophisticated level with respect to optimizing (long) loops with array traversal.

Java, Matlab, and Python offer a user-friendly handling of arrays. Arrays can be created wherever convenient, arrays are efficiently transferred in and out of functions, and arrays are deleted when they are no longer in use.[16] There is, of course, some efficiency penalty to be paid for having such flexible and easy-to-use arrays. C++ arrays fall somewhere in between; they are as primitive as in Fortran 77, but wrapping them in a class opens up the possibility to quickly create much more user-friendly array objects, and with some significant effort one can implement arrays that mimic those found in Java, Matlab, and Python. C++ also allows you to have good (low-level) control of efficiency, which you do not have in Java, Maple, Matlab,[17] or Python.[18]

The convenient array handling and simple plotting capabilities found in Maple, Matlab, and Python make these languages very attractive for explorative scientific computing. One should bear in mind, however, that programs written in Fortran or C/C++ can run much faster.

For exploration of the physical problems and numerical methods covered in most of this book, we think both Matlab and Python represent very productive programming environments. In our opinion, Python has some advantages over Matlab:

- The Python programming language is more powerful
- Several global functions can be placed in a single file (a complete toolbox/module can be contained in one file)
- Transferring functions as arguments to functions is simpler
- Nested, heterogeneous data structures are simple to construct and use
- Object-oriented programming is more convenient
- Interfacing C, C++, and Fortran code is better supported
- Scalar functions works with vectors to a larger extent (without modifications)
- The source is free and runs on more platforms

Having said this, we must add that Matlab has a significantly more comprehensive numerical functionality than Python (linear algebra, ODE solvers, optimization,

---

[16] All this functionality can in fact be implemented in Fortran 77, but the language does not support the functionality, so one has to write lots of low-level code that simulates memory management in a long fixed-size array, reflecting the computer's memory.

[17] In Matlab, most of the array processing functionality takes place in highly optimized Fortran 77 and C functions. Just avoid explicit Matlab loops, see Sect. 6.3.7.

[18] With Python it is very easy to migrate array handling code to Fortran 77, C, or C++ to achieve this low-level control of efficiency.

time series analysis, image analysis, etc.) The graphical capabilities of Matlab are also more convenient than those of Python, since Python graphics rely on external packages that must be installed separately.

There is an interface pymat that allows Python programs to use Matlab as a computational and graphics engine. Thus, we prefer to use Python as the programming platform, calling up Matlab when needed.

6.5 Numerical Software Engineering

Numerical software engineering deals with methods for structuring a large piece of numerical software such that it is easy to reuse and maintain the code. Software engineering in general is a huge area of computer science. Many of the methods developed in software engineering are, of course, useful for numerical software as well, but numerical software has certain features that make special demands on software engineering techniques. Contrary to most common non-numerical codes, one often encounters huge data structures, very long execution times, parallel computations, and large amounts of computed numbers whose accuracy can be difficult to verify. The codes can also be very large and complicated. It therefore makes sense to use the term *numerical software engineering* for software engineering adapted to the world of scientific computing.

6.5.1 Function Libraries

The classic way of organizing numerical codes is to implement algorithms as subroutines in Fortran and collect these subroutines in libraries. An application solving a particular scientific computing problem would then define a set of arrays and other variables and call the appropriate subroutines to get the job done.

The term subroutine is associated with Fortran. When using most other programming languages one would probably refer to a *function* library.

Numerical Integration Library

Suppose you need to write a piece of code to integrate a function, say $f(x) = x \tanh x$ from 0 to 10, by a suitable numerical integration rule. Choosing the trapezoidal rule as the method, the minimum amount of code would look like

$n = 200$
$s = 0$
$x = 0$
for $i = 1, \ldots, n - 1$
$\quad x \leftarrow x + h$

$$s \leftarrow s + x \tanh x$$
$$s \leftarrow h(s + 0.5 \cdot 0 \tanh 0 + 0.5 \cdot 10 \tanh 10)$$

Although the code is simple and to the point, it is not reusable on other occasions, i.e., when you want to integrate a new function by the trapezoidal rule. You will then need to write and test the integration method once more. This is a waste of effort – perhaps not that much in the present example, but this is the simplest of all examples. We believe that adopting sound programming styles is a good habit for obtaining a reliable and efficient way of solving problems, even in simple problems where any approach can work well.

The trapezoidal rule should, naturally, be implemented as indicated in all our algorithms from Sect. 6.1.2, namely as a Trapezoidal function (or subroutine) where as many parameters as possible are represented by variables. This means that you should never hardcode the function $f(x)$ and the integration limits a and b as we did above. As soon as the function is well tested, it can be reused as a working "black box" later. If you call this piece of software to integrate a function and see that the result is wrong, you can be quite sure that the error is outside the Trapezoidal function; it must be in the calling code. This principle of breaking up a code into reusable and well-tested pieces is crucial for the efficient and reliable development of software for more complicated problems.

As you need more sophisticated methods for numerical integration, you can implement these as reusable functions. After some time you will have a collection of functions for numerical integration. At that point you need to organize these functions such that you can call the methods you want in an application in a convenient way. The interface, i.e. the argument list, to the functions should be the same, if possible. Furthermore, the functions should be available in a library such that there is no need to copy the complete function codes to your application. The latter point has two important advantages: It keeps the application code small, and there is only one version (code) of the numerical integration function.

Suppose your collection of functions for numerical integration counts four methods:

- Trapezoidal (a, b, f, n)
- Simpson (a, b, f, n)
- GaussLegendre (a, b, f, n)
- GaussLobatto (a, b, f, n)

These functions have the same list of arguments. More sophisticated integration methods, e.g. adaptive schemes, may need more input than a, b, f, and n (see page 260).

To make a library, we collect the code associated with the functions in a single file or a collection of files (the file structure can depend on the programming language). None of these files should contain a main program, only functions or subroutines.

Integrating a function numerically, utilizing this library, consists of writing a program doing three things:

1. Enabling access to functionality in the library
2. Implementing the function $f(x)$ to be integrated

3. Setting *a*, *b*, and *n*, calling the numerical integration function, and printing the result

We will present small applications running a loop over different *n* values and calling up the four methods for each value of *n*.

Libraries and Compiled Languages

How you technically create the library depends on the programming language used to implement the algorithms. In Fortran, C, and C++ you must compile the library file and place it in a directory that acts as a repository for your libraries. Most real-life libraries are made up of many files, since it would naturally be inconvenient to put all the library code in one big file. These individual files must be compiled, one by one, and then a tool for merging the compiled files into one library file must be invoked. Here is a typical manual procedure, in a Unix environment, for compiling three Fortran 77 files file1.f, file2.f, and file3.f and making a library mylib out of them:

```
unix> f77 -O3 -c file1.f file2.f file3.f
unix> ar libmylib.a file1.o file2.o file3.o
```

The first line compiles the files, resulting in three object files file1.o, file2.o, and file3.o. The second line runs the ar utility to merge the object files into a library file libmylib.a. The name of the library file must always start with lib, and this is not a part of the library name (just mylib is the library name in this case). There are many variants of this theme; the purpose here is to give a brief description of how classic subroutine libraries are created. The library file libmylib.a must be located in a directory for libraries. On a Unix system, /usr/lib is such a directory, but only system administrators have write access to this directory, so one probably needs to create one's own library directory, e.g., $HOME/lib. In the following we assume that libmylib.a is located in $HOME/lib.

An application can now be written that calls up functionality in the library:

```
real*8 function g(x)
real*8 x
g = x*tanh(x)
return
end

real*8 a, b, I
integer n

program test
a = 0
b = 10
n = 200
I = trapezoidal(a, b, g, n)
write(*,*) I
end
```

This code is placed in a file, say, `gint.f`. The file must be compiled, resulting in the object file `gint.o`. This object file must thereafter be linked with the library file. The typical manual steps on a Unix computer read

```
unix> f77 -O3 -c gint.f
unix> f77 -L$HOME/lib -o app gint.o -lmylib
```

The first command is the compilation step. The second command is the linking step. The `-L` option tells `f77` where to search for library files, and `-o` specifies the name of the executable (here `app`, for application). The rest of the arguments are object files and library files that should be linked together to form the executable. Observe that library files are written with a special syntax, a prefix `-l` (for library) followed by the name of the library.

C and C++ libraries are created in the same way as explained for Fortran libraries.

Java Libraries

Java libraries consist of classes, and each publicly available class is placed in a file. These files are conveniently organized in directory hierarchies. In the present example with numerical integration, we would typically implement the integration methods as classes `Trapezoidal`, `Simpson`, etc., stored, respectively, in files `Trapezoidal.java`, `Simpson.java`, and so on. The files are naturally located in a directory, say `/home/me/java/int`. The application code must import one or more of these methods, and Java searches for classes to import in the directories contained in the `CLASSPATH` environment variable. In a Bash Unix environment, the `CLASSPATH` variable is typically initialized in a Bash start-up file `.bashrc` in your home directory. The initialization statement reads

```
export CLASSPATH=$CLASSPATH:/home/me/java/int
```

There is a utility called `jar` for packing a set of Java source code files in a single file (`jar` can be viewed as Java's counterpart to the Unix `tar` utility).

Matlab Libraries

Matlab functions are stored in so-called M-files with extension `.m`. One M-file defines a single function to be used outside the file. For example, the file `x.m` defines a function `x`, with as many input arguments and return arguments as desired. Other functions can be placed after the function `x` in the file `x.m`, but these functions are only local to the file, so they cannot be accessed from the calling Matlab environment (or script).

In the present example, we would put the various integration methods in separate M-files, i.e., `Trapezoidal.m`, `Simpson.m`, and so on. These M-files are conveniently stored in a directory whose name reflects the numerical integration, say `/home/me/matlab/int`. The name of this directory must be contained in Matlab's

PATH variable, either by executing

```
path(path, '/home/me/matlab/int')
```

inside Matlab or by setting the MATLABPATH environment variable in the start-up file. In case your start-up file is .bashrc (Unix Bash environment), the relevant statement becomes

```
export MATLABPATH=$MATLABPATH:/home/me/matlab/int
```

The current working directory is always in Matlab's search path.

To summarize, one collects each Matlab library function to be offered to users in a separate M-file, one organizes these M-files in a suitable directory structure, and Matlab's search path must be updated so that Matlab searches these directories for M-files.

Python Libraries

Python libraries are composed by *modules* and *packages*. A module is a file with Python functions and classes, whereas a package is a collection of modules, often with a tree structure (actually reflecting the organization of modules files in a directory tree).

To make a Python module, just place the desired functions in a file. In our example on numerical integration functions, we collect the functions in one file, say, numint.py, and place the file in a directory where Python looks for modules.

The main program for experimenting with different values of n and different methods can look like this:

```
#!/usr/bin/env python
import numint  # give access to library for numerical integration
from math import *

def f1(x):
    return exp(-x*x)*log(1+x*sin(x))

a = 0; b = 2
n_values = (10, 100, 1000, 10000)
for n in n_values:
    for method in (Trapezoidal, Simpson, GaussLegendre,
                   GaussLobatto):
        result = method(a, b, f1, n)
        print method.__name__, ":", result
```

In this code example, method is a function. We can run through a list of functions, call an entry, and write the name of an entry. This flexibility is not found in the other languages we address in this chapter. As a result, Python makes code organization and experimentation very simple and convenient.

The import statement causes Python to search for a file numint.py in a set of directories specified by the internal Python variable sys.path or the environment variable PYTHONPATH. Suppose you stored numint.py in the directory

/home/me/python/modules. To notify Python about this directory, you can either add the directory name to sys.path in the application code[19]

```
sys.path = sys.path + ['/home/me/python/modules']
```

or you can set PYTHONPATH appropriately in your start-up file. If you work in a Unix Bash environment, the start-up file is normally .bashrc, and in this file you can write

```
export PYTHONPATH=$PYTHONPATH:/home/me/python/modules
```

Maple Libraries

A collection of Maple procedures can be turned into a *package*, which is the Maple terminology for what we previously have referred to as library. The procedure for creating a package is a bit more comprehensive than for the other programming languages we touch upon here, so we refer to the online help facility in Maple for complete information. Try first help – topic search – package, structure. A complete example on creating a package is found under the main chapter Example Worksheets in the help menu. Go further to Language and System and then to examples/binarytree.

Say the name of your package is numint. Here is a typical use of the package:

```
with(numint); # import functions like Trapezoidal, Simpson, ...
f1 := x -> exp(-x*x)*log(1+x*sin(x));  # func to be integrated
qt := Trapezoidal(0, 2, f1, 1000):
qs := Simpson(0, 2, f1, 1000):
printf("Trapezoidal=%e,  Simpson=%e", qt, qs);
```

6.5.2 Motivation for Object-Oriented Libraries

The function libraries presented in the previous section are easy to apply; a user just needs to get access to the library and call the desired function. Application codes often require the user to specify at run time the numerical methods to be used for, e.g., integration. This is easily accomplished by reading information about the method and calling the appropriate function in an if–else test:

if method eq. "Trapezoidal" then
    Trapezoidal (a, b, f, n)
else if method eq. "Simpson" then
    Simpson (a, b, f, n)

and so on.

With Python we can build code at run time and execute a string of code with the exec function. This allows reading both the method and the function as user input,

---

[19] Here sys.path is a list of strings (directory names), and the statement adds two lists.

and performing the integration with the desired method and function with just one line of code (!):

```
exec("r = %s(a, b, lambda x: %s, n)" % (method, function_expression))
```

The integration result is stored in the variable r. The lambda construction is a way of defining a type of inline function. More straight and less compact code could be a multi-line string such as

```
exec("""
def myf(x):
    return %s

r = %s(a, b, myf, n)
""")
```

Of course, this construction of code at run time works only if all integration functions have the same set of parameters, here a, b, f, and n. This may not be the case when we extend the libraries with more sophisticated integration methods. For example, we can think of methods producing a result with an error less than a specified tolerance. Such methods can take the tolerance and additional parameters, specifying numerical details (such as the order of the integration scheme) as input arguments. A flexible library for numerical integration could have a method collection such as the following:

- Trapezoidal (a, b, f, n)
- Simpson (a, b, f, n)
- AdaptiveTrapezoidal $(a, b, f, n_{max}, \epsilon)$
- AdaptiveSimpson $(a, b, f, n_{max}, \epsilon)$
- AdaptiveGaussLegendre $(a, b, f, n_{max}, \epsilon, p_1, p_2)$
- GeneralGaussLobatto $(a, b, f, n, q_1, q_2, q_3)$

Here we have just listed some functions for illustrating that the number and type of arguments differ among the functions.

In the if–else approach it is easy to cope with differing argument lists in the function calls, since each call is written explicitly the way it is required. However, in large codes if–else tests tend to be long, and the same tests are often repeated in many places in the code. This assertion may be less easy to realize from the present simple integration example, but the authors' experience with building large numerical codes points to repeated if–else lists as a major problem. When a new integration method is added to the library, the programmer must remember to add a call to the new function in all relevant if–else tests throughout the code. If the numerical software counts some hundred thousand lines of code, this is a tedious and error-prone task. We shall therefore present a better way of implementing flexible choices of numerical algorithms.

Every time we need to integrate a function in the code, we would like to issue a statement where we do not need to see what type of method that is being used or what type of parameter is needed. The advantage of such an approach is obvious: All calls to numerical integration routines will work with a new method when we have added that new method to the library. This is a key point in handling large pieces of numerical software. One solution is to employ *object-oriented programming*. In the following we assume that the reader is familiar with object-oriented programming from a basic course in Java or C++.

In object-oriented programming we represent the numerical algorithms by *objects*. Objects contain both data and functions (operating on the object data and external data). We introduce a pseudo-code notation for objects:

```
class Trapezoidal
  data:
    a, b, f, n
  methods:
    integrate()
    init(a, b, f, n)
```

This pseudo code defines the contents of objects of type `Trapezoidal`. Each such object will contain the data `a`, `b`, `f`, and `n` (with obvious meaning in the present example) and a function called `integrate`, for carrying out the numerical integration, and another function `init`, for initializing the data members.

We create a class for each integration method, say, `Trapezoidal`, `Simpson`, and so on. Each class is *derived* from a general base (or super) class called `Integration`. The base class defines the set of legal methods that can be applied to all integration methods and it also declares the set of data that can be shared among various integration methods (typically a, b, f, and n in our example). As soon as we have created an object representing a particular method, say, `Simpson`, we can treat this object as being of type `Integration` in the rest of the code. This means that we hide the information that we are working with Simpson's rule; all we can see is a general numerical integration object. When we want to integrate a function, we have an object (say) `i` and just call `i`'s `integrate` method, without arguments. The magic of object-oriented programming is that the program knows the type of integration object we actually have (say, `Simpson`) and calls the `integrate` method in the appropriate subclass (here `Simpson`'s `integrate` method). Since the object type can be treated as "identical" for all methods, and since all `integrate` methods take the same argument list (empty in this case), the application code for integrating a function is independent of the method we use.

At this stage it might be encouraging to realize that the outlined object-oriented programming approach has wide applications. Numerical software is basically a collection of different types of numerical methods (integration, the solution of ODEs, the solution of nonlinear equations, the solution of linear systems of equations, etc.) and a "main program" calling up a combination of methods to solve a specific problem. A certain type of numerical method has many different algorithms. Employing the idea of object-oriented programming, each type of numerical method is implemented as a class hierarchy, where the different algorithms are realized as different subclasses. In this way, we will have a class hierarchy for numerical integration, ODE solvers, nonlinear equation solvers, and so on. When an ODE solver needs a nonlinear equation solver as part of the algorithm, it can work with a general (base class) interface to all methods for nonlinear equations. The code applying these methods does not need to distinguish between bisection or Newton's method; just the nonlinear equation solver base class and its generic (say) `solve` method are seen. This way of programming hides details in numerical algorithms and reduces complexity. This is of course very important for the reliability of large numerical codes.

6.5.3 Design of a Numerical Integration Hierarchy

Class `Integrate` is the base class of our numerical integration methods hierarchy. Specific integration rules are implemented as subclasses. Common to all methods are the integration limits and the function to be integrated. Other parameters can differ among the subclasses. The idea of an `init` method with differing argument lists is not good. We should look for a strategy that handles input data and the initialization of integration objects in the same way for all subclasses of `Integration`. To achieve a unified treatment of input data for all subclasses, it may be convenient to collect all the parameters in a so-called *parameter class*, here called `Integration_prm`:

```
class Integration_prm
  data:
    a, b, f
    npoints
    method
    tolerance
    ...
  methods:
    constructor()
    create()
    default()
    read()
    write()
```

This outline of the class shows some possible parameters and (at least) five methods:

- `constructor` for creating and initializing an object
- `create` for creating an integration rule object (as a subclass instance in the `Integration` hierarchy), based on the data in this `Integration_prm` instance
- `default` for assigning appropriate default values to data members
- `read` for reading values from some kind of user interface
- `write` for dumping the contents of the parameters (for a check)

Observe that our pseudo code does not coincide with a particular programming language.[20]

The base class `Integration` for all integration methods typically has a constructor that takes the bag of parameters (i.e., an `Integration_prm` object) and stores it internally in the class. The base class also offers access to the `integrate` method

---

[20] For example, the `constructor` method will have different names in different languages (`Integration_prm` in Java and C++, `__init__` in Python).

common to all subclasses, where the numerical integration rule is to be implemented
(in the subclasses):

```
class Integration
  data:
    prm
  methods:
    constructor(Integration_prm p)
      prm = p
    integrate()
```

Whenever appropriate, we insert the body of a function as indented statements under
the function heading.

The trapezoidal method is our first subclass candidate:

```
class Trapezoidal, subclass of Integration
  data:
  methods:
    constructor(Integration_prm p)
      Integration.constructor(p)

    integrate()
      h = (prm.b-prm.a)/(prm.npoints-1)
      x = a
      s = 0
      for i = 2,...,n-1
        x = x + h
        s = s + f(x)
      s = s + 0.5*(f(prm.a) + f(prm.b))
      return h*s
```

This class apparently has data on its own, but it inherits the prm object from its base
class Integration, and no additional data are needed. The constructor passes
the Integration_prm input object onto the base class constructor, which stores the
parameter object as the data member prm. In the integrate method, the integration
parameters are accessed through the prm object. For example, prm.a means the a
data member of object prm. This is the same notation as used in common languages
supporting classes, such as C++, Java, and Python.

6.5.4 A Class Hierarchy in Java

Readers who are familiar with the Java programming language will quite quickly
convert the previous pseudo code to Java. The following text is written for this group
of readers.

The base class Integration is realized as

```
class Integration {
    public Integration_prm prm;

    public Integration (Integration_prm p)
    { prm = p; }

    public double integrate ()
    {
```

```
        System.out.println("integrate: not impl. in subclass");
        return 0;
        }
    }
```

A specific integration method, such as the trapezoidal rule, is implemented in the subclass `Trapezoidal`:

```
class Trapezoidal extends Integration {
    public Trapezoidal (Integration_prm p)
    { super(p); }

    public double integrate ()
    {
    double a = prm.a; double b = prm.b; int n = prm.n;
    Func f = prm.f;

    double h = (b-a)/((double)n);
    double s = 0;
    double x = a;
    int i;
    for (i = 1; i <= n-1; i++) {
        x = x + h;
        s = s + f.f(x);
    }
    s = 0.5*(f.f(a) + f.f(b)) + s;
    return h*s;
    }
}
```

The `Integration_prm` class holds a, b, n, and f. In addition, the class provides a function `read` for extracting information about a, b, and n from command-line arguments. A function `write` dumps the contents of a, b, and n on the screen. Here is the code:

```
class Integration_prm {
    public double a, b;
    public Func f;
    public int n;
    public String method;

    public Integration_prm ()
    {   /* default values */
    a = 0; b = 1; n = 10; f = new f2();
    method = "Trapezoidal";
    }

    public void read (String argv[])
    {
    int i;
    for (i = 0; i < argv.length; i=i+2) {
        if (argv[i].compareTo("-a") == 0) {
        a = Double.valueOf(argv[i+1]).doubleValue();
        }
        if (argv[i].compareTo("-b") == 0) {
        b = Double.valueOf(argv[i+1]).doubleValue();
        }
        if (argv[i].compareTo("-n") == 0) {
        n = Integer.valueOf(argv[i+1]).intValue();
        }
        if (argv[i].compareTo("-m") == 0) {
```

```
                    method = argv[i+1];
                 }
          }
    }

    public void write ()
    {   System.out.println("a=" + a + " b=" + b + " n=" + n +
                 " method=" + method);
    }

    public Integration create ()
    {
    Integration i;
    if (method.compareTo("Trapezoidal") == 0) {
          i = new Trapezoidal(this);
    }
    // else if (method.compareTo("Simpson") == 0)   and so on...
    else {
          i = new Trapezoidal(this);
    }
    return i;
    }
}
```

The specification of functions to be integrated is exactly the same as in Sect. 6.3.4; that is, $f(x)$ must be programmed as a subclass of Func, since Java does not allow stand-alone functions.

The final piece of the Java code is the main program, here a part of a class called Demo for testing the implementation:

```
class Demo {
    public static void main (String argv[])
    {
    Integration_prm p = new Integration_prm();
    p.read(argv);
    p.write();
    f1 f = new f1();
    Integration i = p.create();
    double result=0; int j;
    for (j = 1; j <= 10000; j++) {
          result = i.integrate();
    }
    System.out.println(result);
    }
}
```

Notice that the variable i is of type Integration and that i.integrate is correctly interpreted as a call to the integrate method in the Trapezoidal class.

6.5.5 A Class Hierarchy in Python

Let us show how the Integration hierarchy can be implemented in the Python programming language. We exemplify one subclass, Trapezoidal, and with this information, the reader should be able to construct other subclasses (Simpson, for instance). The text here assumes that you are familiar with several Python topics:

class programming, overloaded operators in classes, type checking, and eval/exec.
A suitable background can be gained by looking up these keywords in the index of
H. P. Langtangen's *Scripting Tools for Scientific Computations*.

We suppose that we have created a class Integration_prm (details follow later).
The base class for all integration methods can then be implemented as the Python
class Integration.

```
class Integration:
    """
    Base class for numerical methods for integrating
    f(x) over [a,b].
    """

    def __init__(self, p):
        self.prm = p   # Integration_prm object

    def integrate(self):
        """Perform integration."""
        # to be implemented in subclasses
        return "Integration.integrate; not impl. in subclass"
```

The trapezoidal rule is conveniently realized as a subclass:

```
class Trapezoidal(Integration):
    def __init__(self, p):
        Integration.__init__(self, p)

    def integrate(self):
        p = self.prm; a = p.a; b = p.b; n = p.npoints; f = p.f
        h = (b-a)/float(n)
        s = 0
        x = a
        for i in range(1,n,1):
            x = x + h
            s = s + f(x)
        s = 0.5*(f(a) + f(b)) + s
        return h*s
```

Other subclasses implement other integration rules.

The Integration_prm class just holds a, b, n, and f, but it is convenient to
supply the class with functionality for setting default values and for reading information from the command line. To set the function f, we use a utility Function, to
be explained later. For now it is sufficient to know that

```
Function('sin(x)+1', independent_variable='x'
```

defines a function of x equal to sin(x)+1. In other words, the Function utility
allows us to specify Python functions directly from text expressions (!).

Convenient reading of command-line parameters is done with a function
cmldict, which returns all command-line options as a dictionary (say) p, such
that p[option]=value. That is, -a 0.1 on the command-line results in p['a']
being 0.1. Alternatively, we can traverse the array of command-line strings as we
did in the Java code. Here is the code for the parameter class:

```
class Integration_prm:
    """
    Holds all parameters needed to initialize objects
```

```
    in the Integration hierarchy.
    """

    def __init__(self):
        self.default()
        return

    def default(self):
        """Assign appropriate default values to all parameters."""
        self.a = 0
        self.b = 1
        self.npoints = 10
        # default f(x)=1+x:
        self.f = Function('1+x', independent_variable='x')
        self.method = "Trapezoidal"
        # alternative: could use a dictionary and __getitem__

    def read(self, argv=sys.argv[1:], short_opt="", long_opt=""):
        """Read from the command line."""
        from cmldict import cmldict
        defaults = { 'a' : 0, 'b' : 1, 'n' : 10,
                     'f' : Function("1+x"), 'm' : "Trapezoidal" }
        p = cmldict(argv, cmlargs=defaults, validity=0)
        self.a = p['a']
        self.b = p['b']
        self.f = Function(p['f'])
        self.npoints = p['n']
        self.method = p['m']

        # alternative manual parsing by traversing argv:
        for i in range(0,len(argv),2):
            if argv[i] == "-a":  self.a = float(argv[i+1])
            if argv[i] == "-b":  self.b = float(argv[i+1])
            if argv[i] == "-n":  self.n = int(argv[i+1])
            if argv[i] == "-f":  self.n = Function(argv[i+1])
            if argv[i] == "-m":  self.method = argv[i+1]

    def write(self):
        print "a=%g b=%g f='%s' n=%d method=%s" % \
              (self.a, self.b, self.f.__name__, \
               self.npoints, self.method)

    def create(self):
        """Create subclass of Integration"""
        code = "i = %s(self)" % self.method
        exec(code)  # turn string into Python code
        return i
```

Note that the `create` function is very simple, because the string in `self.method` is supposed to coincide with a subclass name in the `Integration` hierarchy. We can therefore construct a string `code` containing the construction of the proper subclass instance and use `exec` to execute this code segment. Alternatively, we could have used an if–else test, as we did in the corresponding `create` method in the Java code.

Here is an example on a typical main program using an `Integration_prm` instance and an `Integration` subclass instance to evaluate an integral:

```
p = Integration_prm()
p.read(sys.argv[1:])
p.write()
i = p.create()
```

```
result = i.integrate()
print result
```

Running such a type of Python program with

```
-a 0 -b 2 -m Trapezoidal -f '1+x^x^x'
```

as command-line arguments gives the result 6.04.

The details of the `Function` class are not important, but the code is listed here for reference and to demonstrate how easy it is with Python to build a convenient utility for representing functions. This utility can be used to hold either Python functions or string definitions of functions.[21]

```
class Function:
    """
    Unified treatment of functions; strings or function objects.
    Examples on usage:

        def myfunc(x):
            return 1+x

        f = Function(myfunc, 'x')   # attach Python function
        v = f(1.2)

        f = Function('1+t', 't')    # specify function by string
        v = f(1.2)
    """
    def __init__(self, f, independent_variable='x'):
        self.f = f   # expression or function object
        self.var = independent_variable   # 'x', 't' etc.
        if type(f) == type(""):
            self.f_is_string = 1
            self.__name__ = self.f
        else:
            # function object:
            self.f_is_string = 0
            self.__name__ = self.f.__name__

    def __call__(self, x):
        if self.f_is_string:
            exec("%s = %g" % (self.var, x))
            return eval(self.f)
        else:
            return self.f(x)
```

6.5.6 Object-Oriented Programming in Matlab

The Matlab language supports object-oriented programming. Nevertheless, these authors find it is more convenient to perform object-oriented programming in languages such as Java, C++, and Python. These languages were originally designed

---

[21] Class `Function` implements the flexibility of the `fcnchk` function in Matlab, but results in nicer call syntax (functions represented as `Function` instances are called like any other Python functions).

for object-oriented programming, while Matlab offers this programming technique as a more recent add-on to the language. On the other hand, Matlab allows the use of Java classes mixed with Matlab statements. This is a convenient way of introducing object-oriented programming in Matlab code. In our example, we could reuse the Java classes `Integration_prm`, `Integration`, `Trapezoidal` and so on, from Sect. 6.5.4, but we shall not go further into this topic here.

6.6 Exercises

Exercise 6.1. Write a mathematical pseudo code for solving the quadratic algebraic equation
$$ax^2 + bx + c = 0.$$
The algorithm should take a, b, and c as input, and return one or two roots.   ◇

Exercise 6.2. A straightforward mathematical pseudo code from Exercise 6.1 can fail if $a = 0$. Also, if $a = 0$ and $b = 0$, c must equal zero. Extend the pseudo code with appropriate tests on $a = 0$ and $a = b = 0$.   ◇

Exercise 6.3. The roots in the algorithm in Exercise 6.1 can be real or complex, depending on the sign of $b^2 - 4ac$. Many programming languages handle the square roots of negative numbers correctly, i.e., compute the roots as complex numbers. However, some languages do not support complex numbers well, and the algorithm should in such cases ensure that only real arithmetic is involved. Modify the pseudo code in Exercise 6.1 such that the roots are represented by four real numbers, the real and imaginary parts of the two roots. Make sure that all operations involve real numbers only.   ◇

Exercise 6.4. Implement Algorithm 6.6 in a computer program. Find a suitable test problem for verifying the implementation.   ◇

Exercise 6.5. Work through Algorithm 6.9 by hand for the choices $a = 0$, $b = 1$, $f(x) = x^3$, and an initial set of points between $x_0 = 0$ and $x_1 = 1$. Let E correspond to the error in the trapezoidal rule (6.12), and set $\epsilon = 0.01$.   ◇

Exercise 6.6. Implement Algorithm 6.9 in a computer program. Find a test problem that can be used to verify the implementation.   ◇

Exercise 6.7. Calling the heun function in Algorithm 6.11 N times, integrating from 0 to some time level in each call, was pointed out as a very inefficient way of producing a set of data points $(i\,\Delta t, u(i\,\Delta t))$. Find out how you can produce these data points more efficiently, still using the heun function in Algorithm 6.11 (hint: perform one step at a time). Count the number of f-function evaluations in the two approaches.   ◇

Exercise 6.8. Suppose we have a differential equation on the form

$$u'(t) = f(u, t), \quad u(0) = U_0,$$

that is, f can be an explicit function of t. A simple example is $f(u, t) = -2t * u$, with the corresponding solution $u = e^{-t^2}$.

Understand the idea behind the derivation of Heun's method, and use this understanding to generalize Algorithm 6.12 such that it works for the right-hand side functions $f(u, t)$ in the above differential equation.   ◇

6.7 Projects

6.7.1 Computing the Volume of a Cylindrical Container

The purpose of this project is to compute the volume of water in a cylindrical container. An example of such a container is given in Fig. 6.3. The radius of the container can change along the center axis, and is given by a function $R(x)$. Water is stored in the container up to a height H. The goal is to develop a program that can compute the water volume in such a container, given the function $R(x)$ and the parameter H as input.

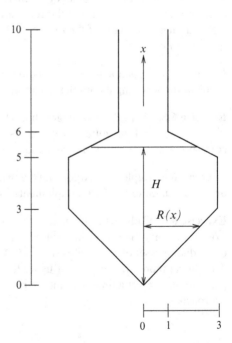

Fig. 6.3 Sketch of a cylindrical container with circular cross section and radius $R(x)$. The container contains water up to a height H

(a) Show that the water volume is given by the expression

$$V = \int_0^H \int_0^{2\pi} \int_0^{R(x)} r\,dr\,d\theta\,dx\,. \qquad (6.14)$$

Here, (r, θ, x) denotes cylindrical coordinates. Show that the integral (6.14) reduces to

$$V = \pi \int_0^H [R(x)]^2 dx\,. \qquad (6.15)$$

(b) Write the function $R(x)$ in Fig. 6.3 as mathematical pseudo code.
(c) Write a mathematical pseudo code for calculating V,

 volume (R, H, n)

given $R(x)$, H, the number of evaluation points n, using the trapezoidal rule.
(d) Implement the algorithm from (c) with the aid of a programming language.
(e) Show how an implementation of Algorithm 6.2 in a specific programming language can be reused as is for computing V. (Hint: Define a function $f(x)$ as $\pi[R(x)]^2$, with $R(x)$ as a global function.) Discuss the pros and cons of this approach and the tailored solution in (d).

6.7.2 A Class Hierarchy for Scalar ODE Solvers

This project concerns applying the ideas of Sect. 6.5.2 to the solution of scalar ODEs. We introduce a base class ODESolver for all solver algorithms, and a class ODESolver_prm holding all parameters needed for initializing and running an ODESolver subclass instance.

(a) Write a pseudo code for class ODESolver_prm, specifying data members and methods.
(b) Write a pseudo code for class ODESolver.
(c) Write a pseudo code for a specific ODE algorithm, such as Heun's method, as a subclass of ODESolver.
(d) Implement the class hierarchies in Java, C++, and Python.
(e) Find at least three different solution algorithms for the ODEs and implement these algorithms as subclasses of ODESolver.
(f) Make libraries out of the implementation in (e). The C++ library can be named libODE.a, the Java library ODE.jar, and the Python module ODE.py. (See Sect. 6.5.1 for information on how to create such libraries.)
(g) Construct a test problem for verifying the implementation. Write associated main programs and check that the code works. The main programs should make use of the libraries from (f).

6.7.3 Software for Systems of ODEs

This project extends Sect. 6.4 and Project 6.7.2 to *systems* of ODEs.

(a) Generalize Algorithm 6.12 to systems of ODEs. Make two versions of the algorithm: one that does not store values in an array, except at the final step, and one that stores all the computed values in an array.

(b) Implement the algorithm for solving a system of ODEs as a function in a programming language. Consider only the algorithm that stores the computed values at the final step.

The other algorithm would need to return a two-dimensional array, with one index for the step numbers and one index for the components in the ODE system. Since two-dimensional arrays are not covered in Sect. 6.4, we drop the corresponding implementation. In the version of the algorithm to be implemented, return the solution as an array, where each element contains the corresponding component in the solution (at the final, N-th, step of the algorithm). Note that calling the algorithm repeatedly with $N = 1$ (single step at a time) makes it easy to load the solution at many time steps into some array in the calling code. This array might be convenient for visualizing the solution.

(c) Extend the ODESolver_prm class and the ODESolver hierarchy to systems of ODEs. Use the code from (b) in a subclass of ODESolver.

(d) Construct a test problem and verify the implementations.

Chapter 7
The Diffusion Equation

This chapter treats the numerical simulation of *diffusion* processes. Diffusion takes place in space and time simultaneously and is an important phenomenon in nature and technology. Scientific computations of physical quantities such as temperature, pollution, and velocity are frequently based on models for diffusion. Diffusion is often coupled with other processes in physical problems, but in this chapter we will only consider diffusion as an isolated process.

A central message to communicate in the present chapter is that diffusion arises in widely different physical contexts, and the derivations of appropriate mathematical models differ correspondingly, but all the models we look at can be expressed in the same form. Therefore we can derive common algorithms for simulating different diffusion processes, and implement these algorithms in a program that can be executed to study various diffusion phenomena in different physical contexts. For example, we can use the program to study both heat conduction and the diffusion of bacteria. What we learn about heat conduction can be transferred to knowledge on how bacteria populations move in space and time.

The idea that the same mathematical model can describe different kinds of physics is quite simple, but it is non-trivial to get all the details right such that you can really write a single program for a collection of application areas, run the program, and present results in the context of a particular physical case. That is why we explain all the details regarding how a single model and program can be used for three widely differing physical applications: heat conduction, diffusive (molecular) transport, and thin-film fluid flow.

The main purpose of the chapter is to learn about mathematical models involving partial differential equations (PDEs). The PDE arising in diffusion phenomena is one of the simplest PDEs, but to formulate and solve it, we encounter the major topics about PDEs and the numerical methods used to solve them. To keep the mathematical and numerical details at a minimum, we mainly work with one-dimensional diffusion models.

Sections 7.1.1–7.1.4 explain the basic physical features of diffusion processes with examples from heat conduction and diffusive transport (fluid mixing). We also argue why one-dimensional models are relevant in a three-dimensional world. Section 7.2 is of particular importance since here we state the governing PDE model for diffusion and introduce basic mathematical quantities.

A. Tveito et al., *Elements of Scientific Computing*, Texts in Computational Science and Engineering 7, DOI 10.1007/978-3-642-11299-7_7,
© Springer-Verlag Berlin Heidelberg 2010

Section 7.3 is devoted to deriving mathematical PDE models for diffusion. Three different physical applications of increasing complexity are presented in Sects. 7.3.1–7.3.3. These models, in complete form, are summarized in Sect. 7.3.4. Section 7.3.5 discusses scaling, which is a useful tool for simplifying PDE models.

Simple (explicit) numerical methods for one-dimensional diffusion problems are treated in Sect. 7.4. Basic finite difference methodology is introduced and applied to a model problem in Sects. 7.4.1–7.4.3. How to verify that a computer implementation of the algorithms works is the subject of Sect. 7.4.4. Some numerical extensions to handle heterogeneous media are covered in Sect. 7.4.7.

The methods introduced in Sect. 7.4 are only useful if the time step is below a certain critical limit. Unconditionally stable methods, without any restriction on the time step, are introduced in Sect. 7.5. We first deal with the backward Euler scheme, then we make extensions to the Crank–Nicolson scheme, and finally we summarize all our schemes for diffusion problems in terms of the so-called θ scheme.

As will be evident, simulation of diffusion processes is a wide scientific topic, including physics, mathematics, numerics, and programming. The mathematical problem and the basic solution tools are described in Sects. 7.2–7.5. The remaining chapters aim at providing some understanding of how diffusion models are formulated and some intuition of what kinds of solutions are expected when simulating such phenomena. All these aspects are crucial for serious simulations of diffusion problems on a computer.

The information on modeling diffusion processes in Sects. 7.1 and 7.3 is not required to understand how to simulate diffusion. It is therefore possible to jump directly to Sect. 7.2, where the mathematical model of diffusion is listed and then continue reading about the numerics in Sect. 7.4, and perhaps continue with Sect. 7.5 if one is interested in more advanced numerical methods for diffusion problems.

7.1 Basics of Diffusion Processes

In this section we give some examples of diffusion phenomena, and we justify why we later work almost exclusively with one-dimensional models.

7.1.1 Heat Conduction

Suppose you have two pieces of metal. The two pieces have different constant temperatures, so that we can refer to one as hot and the other as cold. Then we bring the pieces in contact with each other. At the time of contact, the temperature in the combined piece is discontinuous, since it makes a jump when going from one piece into the other. Heat will then flow from the hot piece to the cold one. The amount of heat flow per unit time, the "velocity of heat", depends on the temperature difference between the two pieces. The initial jump in the temperature is smoothed, and as time goes by, the hot piece gets cooler and the cold one gets hotter. After a long time, the temperature is the same in both pieces, i.e., the temperature is constant throughout both metal pieces. The physical process where heat flows from hot to cold regions is called *heat conduction* and is caused by the motion of the molecules in the metal pieces.

If the two metal pieces are surrounded by air, there will also be heat exchange with the air. Heat will flow into the air if the metal is hotter than the air. On the contrary, the air will heat the metal if the metal is initially cooler than the air. Let us imagine a physical experiment where we can neglect the heat exchange with the surrounding air. Each metal piece is then a rod with some outer isolating material such that hardly any heat can flow through the isolating material. At one end of the rod, there is no isolation, and this end is brought in contact with the similar end of the other rod. In this situation there will be a very small temperature difference throughout the cross sections of the rods. We can therefore assume that the temperature, called T, only varies along the rods, and in time. We then set $T = T(x,t)$, where x is a coordinate along the center axis of the rods. If the lengths of the left and right rods are L_L and L_R, respectively, we let x go from 0 to the total length $L = L_L + L_R$. The $T(x,t)$ function is then defined from $x \in [0, L]$ and $t > 0$. Initially, the metal pieces have different temperatures: $T = T_L$ in the left rod, $x \in [0, L_L]$, and $T = T_R$ in the right rod, $x \in (L_L, L]$. Say $T_L > T_R$. Then heat will flow from the left to the right piece. The initial jump in T is smoothed and the $T(x,t)$ function becomes a constant T_∞ as $t \to \infty$. Because no heat can escape from this isolated system, $T_\infty = (T_L + T_R)/2$. Later, we will compute the $T(x,t)$ function in detail and see how it evolves in space and time.

7.1.2 Diffusive Transport of a Substance

Another diffusion process occurs when you drop some dark blue ink into a glass of water. At once, you can observe the ink in a small, localized region of space. As time goes by, the ink diffuses into the water, i.e., the ink molecules mix with the water molecules. The water gets bluer as the initial dark blue ink spreads throughout the glass. After quite some time, all the water is light blue. The color is reflected by the concentration of ink molecules. We can use a function c of space and time to represent the concentration of ink molecules. When $c(x, y, z, t) = 1$, we have only ink molecules inside a small volume[1] surrounding the spatial point (x, y, z) at time t. Initially (i.e., when $t = 0$), $c = 1$ in a small region corresponding to the droplet of ink. In the water there are no ink molecules and thus $c = 0$. The significant jump from $c = 0$ in the water to $c = 1$ in the ink is smoothed out in time. After some time, c might end up at a value larger than 0, but significantly less than 1, since there is much more water than ink in the glass. The final light blue color we

---

[1] A small volume in the present context can be a cube with a width of, say, 0.1 mm.

observe may correspond to $c = 0.1$. We refer to this phenomenon as the *diffusive transport* of a substance. The substance, ink in this case, flows from regions of high concentration to regions of low concentration. As in the heat conduction example, the flow tends to smooth out variations. An example related to ink in water is that of sugar in coffee: If the coffee is not stirred, the sugar will dissolve and the sugar molecules will distribute throughout the coffee by diffusion.[2] Cooking spaghetti is yet another example: The water molecules diffuse into the spaghetti strands and make them thicker and more flexible.

Widely different physical problems, such as heat conduction in metals, dissolving sugar in coffee, or cooking spaghetti can all be described by the same mathematical model. This is the great power of using mathematics to model the physical world: Seemingly very different phenomena can be described by the same equations, and these equations can be solved in a single program, which then can be used to simulate diverse physical phenomena.

7.1.3 Diffusion Versus Convection

Getting back to the problem of putting a cube a sugar in the coffee, one is often too impatient to wait for the diffusion process to finish; it takes some time before the sugar cube has dissolved and spread throughout the coffee. Normally, one grabs a spoon and furthers the mixing by moving the spoon around in the coffee. From a physical point of view, one creates a fluid flow in the coffee, and the sugar molecules are then transported more efficiently with the aid of the flow. This accelerates the mixing of sugar and coffee. Transport due to fluid flow is often called *convection*. Transport due to diffusion only is much slower. Most transport processes in nature and technology are a combination of convection and diffusion. For example, heat distribution in a building makes use of ventilation systems, imposing fluid flows, to make the transport of heat more efficient than if we only waited for diffusion to heat the building.

A problem of particular interest in our times is the transport of pollution, e.g., the depositing of some polluting substance in an ocean, a river, a lake, or the ground. In the case of a lake at rest, the transport of pollution has great similarities to dropping ink in water: Both are diffusion processes. However, in a river or in an ocean, the transport of pollution due to fluid flow is essential, so convection contributes much more than pure diffusion to the distribution of the pollution. In fact, one can neglect diffusion in such transport phenomena.

Even when you put a sugar cube in a cup of coffee, or ink in water, the mixing will be influenced by both diffusion and convection. The convection arises because the sugar or ink introduces density differences in the fluid mixture. Light fluid flows

---

[2] Dissolving a substance in a fluid may involve convection as well as diffusion, as explained in the forthcoming section.

upward, while denser and heavier fluid sinks. That is, gravity and density differences set up a flow, leading to transport of the dissolved substance by convection. This convection is usually small in liquids, but can often be substantial in gases, which flow much more easily. In air, for example, there is always some minor flow, either due to temperature and thereby density differences, or due to some external forcing. When you take your shoes off, you can sometimes quickly smell your feet. If the smell molecules were brought to your nose by diffusion only, it would take a very long time (about a year!) to notice the smell. Minor air flow will introduce convection of the scent molecules and contribute to a transport that takes just a second. Similarly, heating up a room by diffusion would take a very long time, but the temperature differences introduced by a heating device set up density differences and a corresponding flow that efficiently distributes the warmed air throughout the room – by convection. The diffusion process can in such cases be neglected.

Contrary cases also exist: Convective transport can be approximated by diffusion. The ground, consisting of media such as soil and rock, is porous. That is, there is a network of very small pores in which fluid can flow. The transport of pollution in the pores is dominated by convection. However, this network, when viewed from a *macroscopic* level (containing a large number of pores), makes the transport look like a diffusion process. Hence, mathematical models for the diffusive transport of a substance are highly relevant for a wide range of transport problems in the ground.

Sediment transport and the formation of geological layers involve the convective transport of sediments in water. Nevertheless, on large spatial and temporal time scales, the cumulative effect of the transport can well be described as a diffusion process. Diffusion equations are therefore used to simulate the geological evolution of sedimentary basins that may contain oil and gas.

The above discussion reveals that building mathematical models and applying them correctly to solve problems in nature and technology can be complicated. Many consider this topic as beyond the scope of scientific computing, meaning that scientific computing only deals with the numerical treatment of a given model. However, the formulation of a sound numerical method is tightly connected to the formulation of the model, which is connected to the derivation of the model and the physics of the problem. The boundaries of scientific computing as a discipline are therefore (also) diffuse.

7.1.4 The Relevance of One-Dimensional Models

Mathematicians, physicists, and computer scientists are often accused of having a one-dimensional life. What is definitely true, is that they prefer to work in a one-dimensional world whenever possible. We will follow this line of reasoning and concentrate on one-dimensional diffusion problems in this book. Since the world is definitely three-dimensional, this restriction deserves some justification.

It turns out that many three-dimensional physical problems are well described by one-dimensional models, and since one space dimension is much easier to deal with than three, reduction to one dimension is a common task. Even if the one-dimensional model reveals itself to be a crude quantitative approximation of the real-world problem, we can gain a lot of qualitative insight and understanding about the physics and mathematical description of the problem by studying the one-dimensional model.

One-Dimensional Models are Convenient

There are two major practical advantages of one-dimensional models: Solutions can be computed in a short time, allowing many experiments to be done, and visualization of the solution is easy, since a simple curve plot will be sufficient. Another advantage from a teaching point of view is that one-dimensional models contains fewer numerical details than higher-dimensional models, making it easier to focus on the principal ideas and to get a computer code to work.

Long and Thin Geometries

Some examples of reducing the three-dimensional world to a one-dimensional approximate model will now be discussed. Our first example concerns "long and thin geometries", i.e., the geometry has a dominant dimension, which makes the problem one-dimensional. Think of a long, thin tube filled with water and imagine that we inject some ink at one end. If the injected ink expands the cross-section of the tube, the transport of ink by diffusion will be directed along the tube. What we observe is a diffusive front propagating away from the end where the ink was injected. Figure 7.1 depicts this process. The underlying microscopic physics of this problem, i.e., molecular vibrations, is highly three dimensional. The macroscopic physics, i.e., diffusive transport, is also three dimensional, but the transport in the directions perpendicular to the tube walls is very small if the tube is thin, since the whole cross-section is then supposed to have approximately the same concentration. In mathematical terms, this means that only the derivative in the direction of the tube axis needs to be taken into account. This property makes the mathematical model one-dimensional.

What happens if the injected ink does not fill the whole cross section? A spherical droplet at the inlet makes the problem three-dimensional, at least in principle. Figure 7.2 presents simulations of this case. Comparing Figs. 7.1 and 7.2 shows that after a short time, the problem in Fig. 7.2 also seems to be well described by a one-dimensional model, due to the smoothing effect of diffusion and the "thin" geometry the tube. A more accurate assessment of the accuracy of a purely one-dimensional model can be obtained by comparing c plotted along the tube axis and a concentration function computed by a one-dimensional model. Figure 7.3 displays the curves obtained by the two models: Some discrepancies are indeed observed, but the one-dimensional model is clearly a reasonable approximation.

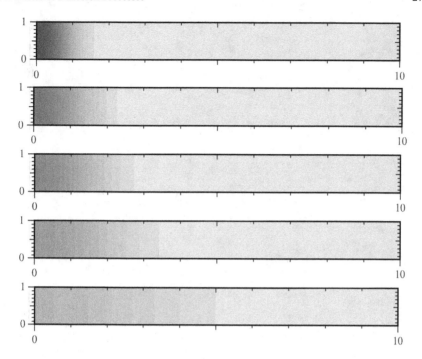

Fig. 7.1 Diffusion of ink in a long, thin tube simulated by a three-dimensional mathematical model. The *top figure* shows the initial concentration (*dark* is ink, *white* is water). The *lower three figures* show the concentration of ink at (*scaled*) times $t = 0.25$, $t = 0.5$, $t = 1$, and $t = 3$, respectively. The evolution is clearly one-dimensional

One-Dimensional Variation Only

A closely related problem is our introductory example of bringing two metal pieces at different temperatures into contact. If we assume that the surfaces of the metal pieces are *insulated*, i.e., covered by some isolation material such that heat cannot escape from the metal, except at the end surface that is in contact with the other piece, see Fig. 7.4, the heat transport vanishes at the surfaces of the combined piece. If the temperature is constant throughout the pieces before they come into contact, the heat will be transported in a normal direction to the contact surface, since this is the only direction where we encounter temperature differences. Figure 7.5 depicts the temperature evolution, which is clearly one-dimensional. The strong smoothing effect of diffusion is evident from this figure, but it is perhaps even more evident from the corresponding curves through the two materials in Fig. 7.6.

On the other hand, if the surfaces are not insulated, there would be a difference between the temperatures inside and outside the piece, resulting in a heat flow out of the surface, and the problem would be three dimensional. A specific example where the surfaces are in contact with air temperature at a fraction 0.8 of the initial temperature difference is shown in Fig. 7.7. This plot, at time $t = 0.1$, shows

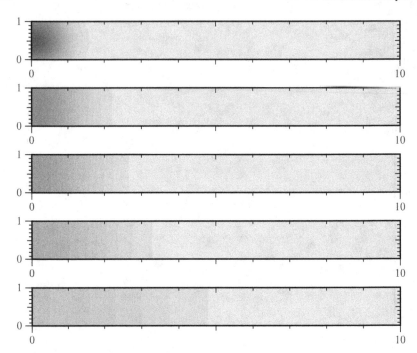

Fig. 7.2 The same diffusion problem and three-dimensional mathematical model as in Fig. 7.1, but the initial concentration of ink at the *left end* does not fill the tube cross-section entirely, thus inducing some small initial three-dimensional effects. How appropriate a one-dimensional model is for the present case becomes evident by comparing the plots with those in Fig. 7.1

that, in comparison with the corresponding plot in Fig. 7.5 (second from the top), there are some visible three-dimensional (two-dimensional in a flat plot) effects. The final temperature will in this case stabilize at the surrounding air temperature (cf. Fig. 7.8). The evolution as a whole looks one-dimensional, except at the early stages (e.g., $t = 0.025$).

Cylindrical and Spherical Symmetry

Replacing the familiar Cartesian coordinates x, y, and z by another coordinate system can lead to one-dimensional models, although the mathematical problem in Cartesian coordinates are two or three dimensional. Consider computing the temperature inside a very long cylinder, where none of the input data varies along the surface of the cylinder. Switching to cylindrical coordinates, we can omit the dependence on all coordinates, except the radial one. This leaves a one-dimensional problem on an interval from zero to the radius of the cylinder in cylindrical coordinates. In Cartesian coordinates, the problem is two dimensional in a circular domain. Similarly, heat conduction in a hollow sphere (e.g., a gas container) is a

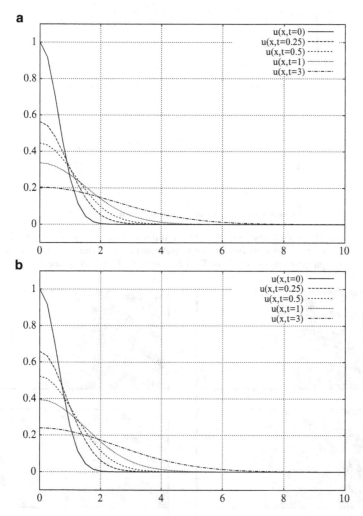

Fig. 7.3 (a) Concentration along the tube axis in Fig. 7.2. (b) Concentration computed by a corresponding purely one-dimensional model

three-dimensional problem in Cartesian coordinates, which in many applications can be reduced to a one-dimensional radial problem in spherical coordinates.

Approximate Variation along a Line

Sometimes we can work with a one-dimensional model even when it cannot be well justified mathematically or physically. Consider putting a can of beer in the refrigerator. If we want to compute how the temperature inside the can decreases

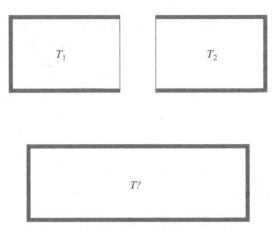

Fig. 7.4 The *top figure* shows two metal pieces, at different temperatures T_1 and T_2, insulated on all sides except for one (*thinner line*), being brought into contact with each other in the *lower figure*. The initial jump in temperature from T_1 to T_2 is smoothed out in time, see Fig. 7.5

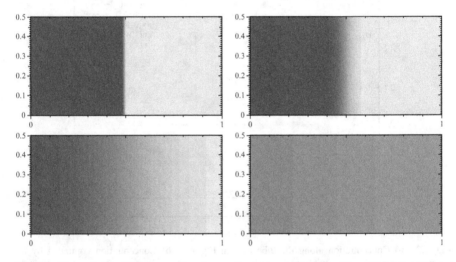

Fig. 7.5 The evolution of the temperature in a medium composed of two pieces of metal, at different initial temperatures. In the *gray-scale plots*, *dark* is hot and *white* is cool. The plots correspond to $t = 0$ (*upper left*), $t = 0.001$ (*upper right*), $t = 0.025$ (*lower left*), and $t = 0.25$ (*lower right*). Since all boundaries are insulated, the temperature approaches a constant value, equal to the average $(T_1 + T_2)/2$ of the initial temperature values. The simulations are based on a three-dimensional mathematical model

in time, we actually deal with a three-dimensional problem. For convenience we may, nevertheless, want to introduce a one-dimensional model. Imagine a horizontal line through the center axis of the can of beer. We could compute the heat transport along this line, disregarding heat transport in the other space directions. Since the can of beer is cylindrical, we could in fact compute the heat transport in the

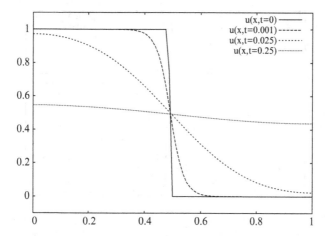

Fig. 7.6 Curve plots of the temperatures along the lines $y = 0.5$ for the same physical problem as in Fig. 7.5

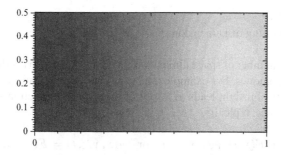

Fig. 7.7 Plot of the temperature at $t = 0.025$ for the same problem as in Fig. 7.5, but the boundaries are now in contact with the surrounding air instead of being insulated. This produces a slight three-dimensional effect

radial direction, which would utilize the radial symmetry of the problem and hence actually model two-dimensional horizontal heat transport, disregarding the transport in the vertical direction only. The actual temperature values we compute from such one-dimensional models may not be very accurate, compared with real-world observations, but the *qualitative features* of the temperature distribution may be well represented. That is, we can use the crude one-dimensional model to achieve a rough description of how the temperature inside the can develops in time. This gives us a better understanding of the features of a diffusion process. Since many different physical phenomena can be described by the same model, our understanding from studying a particular example, even in a very simplified one-dimensional form, reaches far beyond the example itself. In a nutshell, this is the power of mathematical modeling.

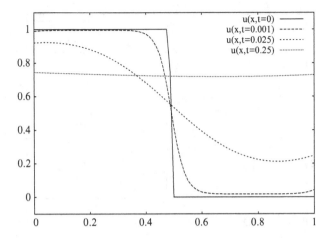

Fig. 7.8 Curve plots of the temperatures along the line $y = 0.5$ for the same physical problem as in Fig. 7.7

Mathematical Splitting of Dimensions

Reduction of the number of space dimensions in a model can also result from purely mathematical techniques. For example, one can think of a situation where physical and/or mathematical insight leads to a split of the unknown function's dependence on space coordinates: typically,

$$u(x, y, z, t) = F(x, t) + G(y, z, t) \quad \text{or} \quad u(x, y, z, t) = F(x, t)G(y, z, t).$$

If one can easily deduce what the G function is, one is left with a one-dimensional problem for finding $F(x, t)$.

Neglecting Variations in Time

When discussing the reduction of the dimensions of a model, we should also mention that the time parameter can sometimes be discarded, if the changes of the unknown function in time are sufficiently small. Such problems are usually referred to as *stationary* or *time-independent problems*. One example is the physical problem corresponding to Figs. 7.4 and 7.5, where two metal pieces at different temperatures are brought into contact. Our everyday experience tells us that the temperature jump will be smoothed out in time. As time increases, the solution approaches a constant state. Mathematically, the solution and thus also the model do not depend on the time parameter in the limit $t \to \infty$. In real computations we reach an approximately stationary (time-independent) problem after a finite time.

Summary and Concluding Remarks

We shall in this chapter deal with one-dimensional models only, because this keeps the amount of mathematical and numerical details at a modest level. One-dimensional models are relevant in a number of contexts. Of course, reducing any three-dimensional phenomena to one dimension implies some approximation. This approximation can be rough, as in the refrigerator example, or it can be very accurate, as when one models diffusion of some substance in a long, thin tube filled with, e.g., water. The results may be qualitatively or quantitatively correct, depending on whether the approximation is rough or very good, respectively.

It requires quite some experience with physical and mathematical modeling to judge the quality of reducing the number of space dimensions from three to one or from three to two. From now on, we just assume that it is meaningful to work with a one-dimensional diffusion model in a physical problem of interest. Our focus will be on how one builds such a model, how one can solve the involved equations, mainly by means of a computer, and the graphical display (visualization) of the solutions. The latter topic will contribute to increasing our understanding of what a diffusion process is, at least from a mathematical point of view. (Most diffusion processes arise from the motion of and interaction between molecules. Our models can only predict the average (macroscopic) result of very detailed microscopic molecular dynamics. If you want to learn about the underlying physics of diffusion, you should consult an introductory book on physics.)

7.2 The Mathematical Model of Diffusion

Let $u(x, t)$ be the temperature or the concentration function in a diffusion process that can be considered one dimensional. At each point x in the one-dimensional space, and at each point t in time, we associate a value of u. The purpose of the present chapter is to compute u, and occasionally quantities that can be derived from a knowledge of u. Computing u means solving equations entering a mathematical model for diffusion. We will briefly present the different type of equations in the following.

7.2.1 The Diffusion Equation

The function $u(x, t)$ is governed by the *diffusion* equation

$$\frac{\partial u}{\partial t} = k \frac{\partial^2 u}{\partial x^2} + f(x, t). \tag{7.1}$$

The parameter k depends on the physical problem we are studying, and we will come back to its physical interpretation in Sect. 7.3, but already here we can state that a large k implies that u spreads quickly in space, whereas a low k leads to the slow evolution (smoothing) of u. Roughly speaking, the function $f(x, t)$ reflects the supply of u (as in internal heat generation or pollution injection).

Equation (7.1) contains partial derivatives of u and is referred to as a PDE. The particular PDE (7.1) has different names: Some refer to it as the *diffusion equation*, while others call it the *heat equation*.

The PDE is supposed to govern a process in some spatial domain Ω for times $t > 0$. The domain Ω is in the one-dimensional case an interval along the x axis, e.g., $\Omega = (a, b)$. Equation (7.1) is used to compute u inside the spatial interval (a, b) for times $t > 0$. In practice, we typically compute $u(x, t)$ for $x \in (a, b)$ and up to a finite t value t_{stop}. Note that the PDE governs u in an open interval (a, b), not including the end points, and for $t > 0$. Other equations, called boundary and initial conditions, are used for u at the end points and at $t = 0$ (see below).

7.2.2 Initial and Boundary Conditions

To solve a diffusion problem, it is not sufficient to just solve the governing PDE. We also need *boundary conditions* at the boundary of Ω and *initial conditions* at $t = 0$. When $\Omega = (a, b)$, the boundary consists of the two end points $x = a$ and $x = b$. The symbol $\partial\Omega$ is commonly used to denote the boundary of a domain Ω.

The boundary and initial conditions are necessary for the solution of our mathematical problem to be unique. Suppose that we have solved (7.1) and found a solution u. Then the function $v = u + C$, where C is any constant, is also a solution of (7.1). In other words, there are infinitely many solutions of (7.1) – since C can take on infinitely many values. In Nature we observe that, in a given problem, there is one and only one temperature or concentration distribution in space and time. Uniqueness of the solution is therefore a fundamental concept. How many boundary and initial conditions one needs to ensure a unique solution is quite a complicated mathematical issue. In the present case one can prove that uniqueness is guaranteed if we specify

- The boundary conditions $u(a, t)$ and $u(b, t)$ for $t > 0$
- The initial condition $u(x, 0)$ for $x \in [a, b]$

Other choices of boundary conditions are also possible, as will be evident when we begin the modeling of diffusion problems. Instead of prescribing u at $x = a$ and/or $x = b$, we could prescribe

$$\frac{\partial}{\partial x} u(a, t), \quad \frac{\partial}{\partial x} u(b, t).$$

The important point is that we need one condition at $x = a$ and one condition at $x = b$, i.e., *one condition at each point on the boundary of the domain Ω*.

7.2.3 The One-Dimensional Initial-Boundary Value Problem

The collection of a PDE together with initial and boundary conditions is called an *initial-boundary value problem*. An example of an initial-boundary value problem for diffusion is

$$\frac{\partial u}{\partial t} = k\frac{\partial^2 u}{\partial x^2} + f(x,t), \quad x \in \Omega, t > 0,$$
$$u(a,t) = U_a, \quad t > 0,$$
$$u(b,t) = U_b, \quad t > 0,$$
$$u(x,0) = I(x), \quad x \in \Omega.$$

These equations constitute the complete set of information for determining the unknown function $u(x,t)$. For this to be true, the u values at the boundary of Ω, U_a and U_b, must be known functions of time or constants. Similarly, the initial value of u in Ω must be known as a function $I(x)$. The parameter k in the PDE must also be known in advance.[3]

7.2.4 The Three-Dimensional Diffusion Equation

Mathematical models for three-dimensional diffusion are quite similar to one-dimensional models. The function u now depends on three spatial coordinates x, y, and z, as well as time t, so we seek a function $u(x, y, z, t)$ (describing temperature or concentration). The governing PDE for $u(x, y, z, t)$ reads

$$\frac{\partial u}{\partial t} = k\left(\frac{\partial^2 u}{\partial x^2} + \frac{\partial^2 u}{\partial y^2} + \frac{\partial^2 u}{\partial z^2}\right) + f(x, y, z, t). \tag{7.2}$$

This PDE is defined in a three-dimensional domain Ω and for times $t \geq 0$. The boundary conditions can be that u is known at the boundary $\partial\Omega$ of Ω. The initial condition is, as in the one-dimensional case, that u is known as some function $I(x, y, z)$ for $t = 0$: $u(x, y, z, 0) = I(x, y, z)$.

The sum of second-order derivatives on the right-hand side of (7.2) appears in numerous applications, not only diffusion. The sum has therefore received its own notation, $\nabla^2 u$:

---

[3] We should add that in real life, some of the input data U_a, U_b, k, $f(x,t)$, and $I(x)$ may be unknown, seemingly making the mathematical model inapplicable. Nevertheless, one can have some other kind of information, e.g., some measurements of u at some points in space and time. One can then use the PDE to estimate the unknown parameters. Such situations are referred to as *inverse problems*. The PDE model can then, in some cases, be combined with other mathematical techniques such that u can be computed (Chap. 9 gives a brief introduction to the topic), but this is a much more complicated topic than calculating u from a PDE with standard boundary and initial conditions.

$$\nabla^2 u = \frac{\partial^2 u}{\partial x^2} + \frac{\partial^2 u}{\partial y^2} + \frac{\partial^2 u}{\partial z^2} \,. \tag{7.3}$$

Equation (7.2) can then be more compactly written as

$$\frac{\partial u}{\partial t} = k\nabla^2 u + f \,. \tag{7.4}$$

The $\nabla^2 u$ term is a meaningful notation also in one-dimensional and two-dimensional problems; if u is not a function of z, i.e. a two-dimensional problem, the z derivative vanishes, and if $u = u(x, t)$, as in one-dimensional problems, both the terms involving the y and z derivatives are omitted in the expression for $\nabla^2 u$.

The diffusion equation without a time derivative is known as the *Poisson*[4] *equation*:

$$-\nabla^2 u = f \,. \tag{7.5}$$

If the source term is also neglected, the resulting equation is referred to as the *Laplace*[5] *equation*:

$$\nabla^2 u = 0 \,. \tag{7.6}$$

This is perhaps the most famous of all PDEs.

Cylindrical Symmetry

Many geometrical objects have the shape of a cylinder (a can of beer being one example). Computations in such geometries can benefit from working with *cylindrical coordinates*. Instead of the Cartesian coordinates x, y, and z, we introduce $(r, \theta, \hat{z})$. The r coordinate measures the distance from the z axis, and the θ coordinate measures the rotation of a point around the z axis. The relation between cylindrical and Cartesian coordinates reads as[6]

$$x = r\cos\theta,$$
$$y = r\sin\theta,$$
$$z = \hat{z} \,.$$

The z axis is supposed to coincide with the central axis in the cylinder.

---

[4] Siméon-Denis Poisson, 1781–1840, French mathematician and physicist, contributed to many topics in mathematics and physics, including celestial mechanics, Fourier series and integrals, probability, and the theory of electricity and magnetism. He also independently derived the Navier–Stokes equations (7.29)–(7.30) governing the flow of fluids.

[5] Pierre-Simon, marquis de Laplace, 1749–1827, was a French mathematician and astronomer. Besides formulating the famous Laplace equation, he worked on a wide range of topics in astronomy and mathematics (including statistics). He is considered one of the greatest scientists of all time.

[6] See your favorite book on vector calculus for more details about cylindrical (and spherical) coordinates.

The real benefit of cylindrical coordinates occurs when we have *radial symmetry*, i.e., no quantities in the problem depend on the angle θ. This is often the case (cooling a can of beer is one example). One can then utilize $\partial/\partial\theta = 0$, which means that the three-dimensional problem in Cartesian coordinates is reduced to a two-dimensional problem in cylindrical coordinates. If, in addition, we have small variations in the z direction, such that $\partial/\partial z = 0$ is a reasonable approximation, we get a one-dimensional problem involving r as the only space coordinate.

The term $\nabla^2 u$ can be shown to be reduced to the simple form

$$\frac{1}{r}\frac{\partial}{\partial r}\left(r\frac{\partial u}{\partial r}\right) \tag{7.7}$$

when u depends on the cylindrical coordinate r and time t only.

Spherical Symmetry

Diffusion in spherical geometries can benefit from switching to spherical coordinates, at least if it is reasonable to assume that the unknown depends only on the spherical radial coordinate r in addition to time. The $\nabla^2 u$ term can in this case be shown to reduce to

$$\frac{1}{r^2}\frac{\partial}{\partial r}\left(r^2\frac{\partial u}{\partial r}\right). \tag{7.8}$$

A nice feature of a spherically symmetric diffusion equation is that it can be reduced to a standard one-dimensional equation by a simple transformation. We introduce a new function v related to u by $u = v/r$. Inserting $u = v/r$ into

$$\frac{\partial u}{\partial t} = k\frac{1}{r^2}\frac{\partial}{\partial r}\left(r^2\frac{\partial u}{\partial r}\right) + f(r) \tag{7.9}$$

yields

$$\frac{\partial v}{\partial t} = k\frac{\partial^2 v}{\partial r^2} + rf(r). \tag{7.10}$$

We can therefore use a simulation program (or known analytical solutions) of the standard one-dimensional diffusion equation to find solutions of spherically symmetric *three-dimensional* diffusion. This is a quite convenient observation, since computing with a Cartesian coordinate is always easier than computing with the radial spherical coordinate.

From One to Three Dimensions

As soon as one has understood the basics of diffusion in one space dimension, and learned the technicalities of the solution procedures and their implementations, the knowledge can be generalized in quite a straightforward manner to two or three space dimensions. Or perhaps we should be frank and say that it is straightforward *in principle*, since there are no new physical, mathematical, or numerical *ideas*, but the book-keeping and the amount of technical details are increased significantly, so getting all details right in a computer program can be quite challenging. The more you work with algorithms and implementations, the easier it will be to meet such challenges.

7.3   Derivation of Diffusion Equations

Although the same diffusion equation can be applied to model phenomena of different physical nature, the *derivation* of the equation from physical principles depends, not surprisingly, on the physical context. We will therefore quickly go through how the diffusion equations arise in three different physical contexts: the diffusive transport of a substance, heat conduction, and viscous fluid flow. These three cases are of increasing complexity.

7.3.1   Diffusion of a Substance

Conservation of Mass

We consider a one-dimensional medium where a substance undergoes diffusive transport. An example can be ink spreading in a long, thin tube filled with water. The motion of the substance will be determined by two physical laws: (a) the conservation of mass and (b) Fick's law relating the velocity (flux) to the concentration. Conservation of mass is a principle that has been verified in physical experiments for centuries, and is hence used without uncertainty. Fick's law, on the other hand, is a simple empirical relation, based on physical reasoning and experiments, and with potentially significant uncertainty.

Deriving a One-Dimensional PDE

In a part $\Omega = [a, b]$ of the medium, we shall express the principle of mass conservation. Now, the world is not one-dimensional, so we imagine a tube with the x axis as the center line and that there are no variations in the y and z directions. Only variations in the x direction, inside the interval $\Omega = [a, b]$, are taken into account. Let $c(x, t)$ be the concentration of the substance, and let $q(x, t)$ be the velocity of the substance. In a small time interval Δt the net mass flow of the substance into Ω must lead to an increase in the total mass in Ω. The mass flow, during the time interval Δt, into Ω can be expressed as $\varrho q(a)\Delta t$, where ϱ is the mass density of the pure substance. (In time Δt, the substance travels a distance $q\Delta t$, which upon multiplication of ϱ gives the mass, since mass is volume times density.) The flow out of Ω during the time Δt is $\varrho q(b)\Delta t$. The net mass flow is hence $\varrho\Delta t(q(a)-q(b))$. The total mass inside Ω is obtained by integrating the density, but in a small space interval dx. Not all of the space is occupied by the substance, only a fraction c. We must hence multiply ϱ by c and dx to get the mass in dx. The change in total mass in Ω is therefore $\int_a^b \varrho\Delta c\,dx$, where Δc is the change in concentration.

The balance of the net inflow of mass and the increase in mass, in an interval $\Omega = [a, b]$, can now be summarized as

$$\varrho q(a)\Delta t - \varrho q(b)\Delta t = \int_a^b \varrho\Delta c\,dx. \tag{7.11}$$

Sometimes we can inject or extract mass by some means, and this will influence the mass balance. Suppose $f(x,t)$ reflects the injected volume of the substance, per unit volume and unit time. In a small time interval Δt and in a small space interval Δx, the injected mass then becomes approximately $f(x,t)\varrho\Delta t\Delta x$, where x is an arbitrary x value inside the interval Δx, and t is an arbitrary t value inside the interval Δt. The total injection of mass in the domain $[a, b]$ in the time interval Δt is found by summing up all the $f(x,t)\varrho\Delta t\Delta x$ contributions, from $x = a$ to $x = b$, as $\Delta x \to 0$. This sum is just the integral of $\varrho f\Delta t$, or

$$\int_a^b \varrho f\Delta t\,dx.$$

When $f > 0$ we have mass injection, whereas $f < 0$ means mass extraction. We need to add the f–integral to the balance in (7.11), i.e., the sum of the mass injection and the net flow into Ω must equal the net increase of the mass of the substance inside Ω:

$$\varrho q(a)\Delta t - \varrho q(b)\Delta t + \int_a^b \varrho f\Delta t\,dx = \int_a^b \varrho\Delta c\,dx. \tag{7.12}$$

An increase Δc in the concentration in the time interval Δt can be expressed by the time-rate-of-change of c, multiplied by Δt:

$$\Delta c \approx \frac{\partial c}{\partial t}\Delta t. \tag{7.13}$$

Alternatively, one can arrive at the same result using a Taylor series,

$$c(x,t+\Delta t) = c(x,t) + \frac{\partial c}{\partial t}\Delta t + \mathcal{O}(\Delta t^2),$$

which, when neglecting higher-order terms in Δt, gives (7.13), since $c(x,t+\Delta t) - c(x,t) = \Delta c$.

We can then, after dividing by Δt, write the mass conservation equation in the form

$$\varrho(q(a,t) - q(b,t)) + \int_a^b \varrho f \, dx = \int_a^b \varrho \frac{\partial c}{\partial t} \, dx. \tag{7.14}$$

Assuming that ϱ is constant, the terms on the left-hand side can be transformed to an integral over Ω by using integration by parts. We have

$$\int_a^b \varrho \frac{\partial q}{\partial x} \, dx = - \int_a^b q \frac{\partial \varrho}{\partial x} \, dx + \varrho [q]_a^b,$$

that is,

$$\varrho(q(a,t) - q(b,t)) = - \int_a^b \varrho \frac{\partial q}{\partial x} \, dx.$$

Collecting the two integrals, we can write the mass conservation principle in the form

$$\int_a^b \varrho \left[\frac{\partial c}{\partial t} + \frac{\partial q}{\partial x} - f \right] dx = 0.$$

Since this equation must be valid for any arbitrarily chosen interval $[a, b]$, the integrand must be zero. The rigorous proof of this assertion is mathematically complicated, but a rough justification goes as follows. Think of the integrand being non-zero in a small interval and then choose $[a, b]$ to be this interval to obtain a non-zero (i.e., incorrect) value of the integral; the integrand must vanish.

Using the fact that the integrand must vanish gives us the PDE for mass conservation:

$$\frac{\partial c}{\partial t} + \frac{\partial q}{\partial x} - f = 0. \tag{7.15}$$

There are two unknowns here, c and q, and one equation. Thus, we need an additional equation relating the concentration c to the velocity q. This equation is called Fick's law:

$$q = -k \frac{\partial c}{\partial x}. \tag{7.16}$$

Fick's law states that the velocity of the substance depends on variations in the concentration, and that the flow is directed where c decreases, i.e., the substance flows from regions of high concentration to regions of low concentration. This sounds reasonable. The parameter k depends on the substance and the medium in which the substance diffuses and must be measured in physical experiments. In most applications of the diffusive transport of a substance in a fluid, k can be regarded as a constant. Thinking of ink in water or sugar in coffee, it seems reasonable to assume that the diffusive transport ability is the same throughout the entire medium (water or coffee), i.e., $k = \text{const}$.

By inserting Fick's law (7.16) in the mass conservation equation (7.15), we can eliminate q and get a PDE with only one unknown function, c:

$$\frac{\partial c}{\partial t} = k\frac{\partial^2 c}{\partial x^2} + f(x,t).\tag{7.17}$$

Deriving a Three-Dimensional PDE

In three space dimensions the derivation is very similar, but we need to make use of some results from vector calculus and surface integrals. Just a brief sketch is presented here. We consider an arbitrarily chosen volume V and state that conservation of mass inside V must imply that the net flow of mass into V equals the increase in mass inside V. Let q be the velocity of the substance. The net flow of mass into V in a time interval Δt can be expressed as a surface integral of the normal component of q (the tangential component does not bring the substance into V!):

$$-\int_{\partial V} \varrho q \cdot n\Delta t\, dS.$$

The symbol ∂V denotes the surface of V, and n is an outward unit normal vector to ∂V.

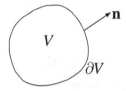

The minus sign originates from the fact that we seek the net *inflow*, and the fact that n points *outward*, i.e., $q \cdot n$ represents the outflow; so to get the inflow, we must change the sign.

The inflow is to be balanced by the increase in concentration times the density and the total injected mass in V, exactly as in the one-dimensional case:

$$-\int_{\partial V} \varrho q \cdot n\Delta t\, dS + \int_V \varrho f\Delta t\, dV = \int_V \varrho\frac{\partial c}{\partial t}\Delta t\, dV.$$

The surface integral can be transformed via the divergence theorem[7] to a volume integral:

$$\int_{\partial V} \varrho q \cdot n\Delta t\, dS = \int_V \nabla \cdot (\varrho q)\Delta t\, dV.$$

Collecting the two volume integrals, assuming constant ϱ, dividing by $\varrho\Delta t$, and requiring that the integrand must vanish, we arrive at the equation

---

[7] Also called Gauss' theorem or Gauss' divergence theorem, cf. your favorite book on surface integrals.

$$\frac{\partial c}{\partial t} + \nabla \cdot \boldsymbol{q} - f = 0 \,. \tag{7.18}$$

We have here assumed that ϱ is constant.

Fick's law reads, in the three dimensional case,

$$\boldsymbol{q} = -k\nabla c \,, \tag{7.19}$$

The physical principle is that the substance moves from regions of high concentration to regions of low concentration, the direction being determined by the greatest decrease in c, which is in the direction of the negative gradient $(-\nabla c)$. Inserting (7.19) in (7.18) eliminates $\boldsymbol{q}$ and gives the three-dimensional diffusion equation for c:

$$\frac{\partial c}{\partial t} = k\nabla^2 c + f(x, y, z, t) \,. \tag{7.20}$$

Here we have used $\nabla \cdot \nabla c = \nabla^2 c$:

$$\left(\frac{\partial}{\partial x}, \frac{\partial}{\partial y}, \frac{\partial}{\partial z} \right) \cdot \left(\frac{\partial c}{\partial x}, \frac{\partial c}{\partial y}, \frac{\partial c}{\partial z} \right) = \frac{\partial^2 c}{\partial x^2} + \frac{\partial^2 c}{\partial y^2} + \frac{\partial^2 c}{\partial z^2} \equiv \nabla^2 c \,.$$

Boundary Conditions

The common boundary conditions are typically

- Prescribed concentration at parts of the boundary,
- Impermeable boundaries, i.e., no normal flow,

$$q = 0 \quad \text{or by Fick's law} \quad \frac{\partial c}{\partial x} = 0,$$

in one-dimensional problems and

$$\boldsymbol{q} \cdot \boldsymbol{n} = 0 \quad \text{or by Fick's law} \quad \frac{\partial c}{\partial n} = 0 \quad \left(\frac{\partial c}{\partial n} \equiv \nabla c \cdot \boldsymbol{n} \right)$$

in three-dimensional problems, or
- Prescribed inflow Q at a boundary,

$$\boldsymbol{q} \cdot \boldsymbol{n} = Q \quad \text{or, by Fick's law,} \quad -k\frac{\partial c}{\partial n} = Q \,.$$

More information about such boundary conditions is given at the end of Sect. 7.3.2. The initial condition for diffusive transport is simply a specification of the concentration at initial time $(t = 0)$: $c(x, 0) = I(x)$, where $I(x)$ is a known function.

7.3.2   Heat Conduction

Energy Balance

Derivation of the diffusion equation for heat conduction is much more complicated, from a physical point of view, than the derivation of the same equation for the diffusive transport of a substance. This is due to the fact that the diffusion equation for heat conduction is based on the *First Law of Thermodynamics* combined with Fourier's law, which is closely connected to the *Second Law of Thermodynamics*, plus other relations from thermodynamics. These models and equations look more complicated than the simple mass conservation principle used in diffusive transport.

The First Law of Thermodynamics expresses an energy balance (some call it the conservation of energy): The increase in total energy of a system equals the work on the system plus the supplied heat. In a simplified version, adapted to the present heat conduction application, one can state that the *increase in internal energy of a system equals the supplied heat*. We will use this principle to derive a PDE, like (7.1), governing temperature distributions. The derivation is more compact than in Sect. 7.3.1. Therefore, you should study Sect. 7.3.1 carefully before proceeding, since many of the ideas and mathematical details are the same, but exposed in a different physical context.

Deriving a One-Dimensional PDE

Again we consider a one-dimensional heat conduction application and apply our refined version of energy balance to an arbitrarily chosen interval $\Omega = (a, b)$. Physically, we have a three-dimensional problem, like heat conduction in an insulated tube, such that we can assume that all quantities depend on x only (if the x axis coincides with the center line of the tube). Let $e(x, t)$ be the internal energy per unit mass, let ϱ be the mass density, and let $q(x, t)$ be the flow of heat (defined per unit time). During a time interval Δt, the increase in total energy in Ω is

$$\int_a^b \varrho \frac{\partial e}{\partial t} \Delta t \, dx \, .$$

The supplied heat consists in our case of heat conducted from the surroundings plus heat generated in the medium, e.g., by chemical reactions or radioactivity. The heat conducted from the surroundings is represented by the flow of heat, q. The net inflow of heat into $[a, b]$ in a time interval Δt is

$$\Delta t (q(a, , t) - q(b, t)) \, .$$

The generated heat per unit time and unit volume is supposed to be represented by a function $f(x, t)$, such that the total amount of heat generated in Ω in the time interval Δt is simply $\int_a^b \Delta t f \, dx$. We now have the following mathematical expression for our basic energy balance:

$$\Delta t (q(a,t) - q(b,t)) + \int_a^b \Delta t f \, dx - \int_a^b \varrho \frac{\partial e}{\partial t} \Delta t \, dx. \tag{7.21}$$

The *mathematical* similarity with the mass conservation equation (7.14) is striking; only a factor ϱ differs due to a slightly different definition of q. Using integration by parts "backward" on the left-hand side transforms this difference to an integral over $[a,b]$:

$$-\int_a^b \frac{\partial q}{\partial x} dx + \int_a^b \Delta t f dx = \int_a^b \varrho \frac{\partial e}{\partial t} \Delta t \, dx.$$

Collecting the integrals yields

$$\int_a^b \left(\varrho \frac{\partial e}{\partial t} + \frac{\partial q}{\partial x} - f \right) dx = 0.$$

For this equation to hold for an arbitrary interval $[a,b]$, the integrand must vanish, a requirement that leads to the PDE

$$\varrho \frac{\partial e}{\partial t} + \frac{\partial q}{\partial x} - f = 0. \tag{7.22}$$

We have two unknown functions, $e(x,t)$ and $q(x,t)$, and thus we need an additional equation. Observe also that the *temperature*, which is the quantity we want to compute, does not yet enter our model. However, we will relate both e and q to the temperature T.

In thermodynamics there is a class of relations called *equations of state*. One such relation reads $e = e(T,V)$, relating internal energy e to temperature T and density $\varrho = 1/V$ (V is referred to as volume). By the chain rule, we have

$$\frac{\partial e}{\partial t} = \frac{\partial e}{\partial T}\bigg|_V \frac{\partial T}{\partial t} + \frac{\partial e}{\partial V}\bigg|_T \frac{\partial V}{\partial t}.$$

The first coefficient $(\partial e / \partial T)_V$ is called the *specific heat capacity at constant volume*, denoted by c_v,

$$c_v \equiv \frac{\partial e}{\partial T}\bigg|_V.$$

The heat capacity will in general vary with T and V ($= 1/\varrho$), but taking it as a constant is reasonable in many applications. We will do so here. One can find tables in the literature with the constant values of c_v for many substances.

The other term, $(\partial e / \partial V)_T$, can, by the second law of thermodynamics, be shown to be proportional to the derivative of T with respect to V (or ϱ). The term models temperature effects due to compressibility and volume expansion. These effects are often small and can be neglected. We will therefore set $(\partial e / \partial V)_T = 0$.

In (7.22) we can now replace the appearance of e by T,

$$\frac{\partial e}{\partial t} = c_v \frac{\partial T}{\partial t},$$

but we still have two unknowns, T and q, and only one equation:

$$\varrho c_v \frac{\partial T}{\partial t} + \frac{\partial q}{\partial x} - f = 0. \tag{7.23}$$

The solution to this problem is to use Fourier's law of heat conduction. This law says that the flow of heat is from hot to cold regions, in the direction of the greatest decrease of T, i.e., along the negative gradient of T. Mathematically we can express Fourier's law as

$$q = -k \frac{\partial T}{\partial x}. \tag{7.24}$$

The mathematical similarity with Fick's law should be noticed. From a physics point of view, Fourier's law, with the transport of heat from hot to cold areas, is closely connected to The Second Law of Thermodynamics.

The coefficient k reflects the medium's ability to conduct heat. In many applications, one desires to study heat conduction from one medium to another. The domain will then consist of several different materials, each material having a specific value of k. This means that k is a function of x, or, more specifically, a piecewise constant function of x. The example of cooling a can of beer in a refrigerator involves many materials: air, beer, aluminum (in the can), other goods, their packages, and so forth. Besides having different k values, the materials also have different densities (ϱ) and heat capacities (c_v). Hence, ϱ and c_v will also typically be piecewise constant functions.

Inserting (7.24) in (7.23) eliminates q and gives us a PDE governing the temperature T:

$$\varrho(x) c_v(x) \frac{\partial T}{\partial t} = \frac{\partial}{\partial x} \left(k(x) \frac{\partial T}{\partial x} \right) + f(x, t). \tag{7.25}$$

Contrary to the diffusive transport PDE (7.17), where a constant k is a reasonable assumption, we have in the heat conduction PDE (7.25) *variable coefficients* ϱ, c_v, and k. These are constant if we consider heat conduction in a homogeneous medium.

We note that there are other ways of replacing $\partial e/\partial t$ by a temperature expression. One common model ends up with a slightly different heat capacity, called heat capacity at constant pressure. When looking for values of c_v in the literature, one should know that there is a related type of heat capacity.

Physical Interpretation of the Parameters

Let us try to explain what the parameters ϱ, c_v, and k mean physically. First we discuss the effect of changing ϱ and c_v in a heat problem. Assume that we supply heat to a system by the function f (think of an oven in a room or a heat generation device embedded in a solid material). If we neglect the spread of heat due to conduction, we can omit the term with spatial derivatives, such that the PDE reduces to

$$\varrho c_v \frac{\partial T}{\partial t} = f.$$

With a constant heat supply in time and space, $f = C = \text{const}$, and constant ϱc_v, we obtain

$$T = \frac{C}{\varrho c_v} t + \text{integration constant}.$$

That is, the density and heat capacity of the material determine how fast the temperature rises due to the heat supply. When the sun shines and thereby heats up water and land, the land is heated much more quickly than the water, because the heat capacity is much smaller for land compared to water. At night, land loses heat much faster than water, for the same reason. The air above land gets warmer than the air above the ocean during the day. Since warm air flows upward, wind from the ocean toward land often arises in the afternoon. A wind in the opposite direction is frequently experienced in the morning during periods of good weather.

The parameter k is connected to the flow of heat. If k is large, heat is efficiently transported *through* the material. Think of the application of bringing two metal pieces, at different temperatures, into contact with each other, as depicted in Figs. 7.4 and 7.5. A large k in the two materials makes heat flow quickly, and the temperature difference is rapidly smoothed out. The smoothing process runs more slowly when k is small. In Sect. 7.3.5 we argue that a characteristic time for changes in the temperature behaves like $\varrho c_v / k$. A large k or small c_v yields quick time-dependent responses in temperature. While c_v is related to temperature changes in time at a point in space, k is related to heat transport in space.

Deriving a Three-Dimensional PDE

The derivation of a three-dimensional version of (7.25) follows the same mathematical steps as the derivation of the three-dimensional PDE for diffusive transport. The energy balance in an arbitrarily chosen volume V can be written

$$-\int_{\partial V} \boldsymbol{q} \cdot \boldsymbol{n} \Delta t \, dS + \int_V f \Delta t \, dV = \int_V \varrho \frac{\partial e}{\partial t} \Delta t \, dV.$$

Here, $\boldsymbol{q}$ is the heat flow vector field, e and f are internal energy and heat generation, respectively, while the rest of the symbols are as defined in Sect. 7.3.1. The integral on the left-hand side is transformed, using the divergence theorem, to a volume integral $\int_V \nabla \cdot \boldsymbol{q} \, dV$. Writing the equation as one integral that equals zero reveals that the integrand must vanish, resulting in the PDE

$$\varrho \frac{\partial e}{\partial t} + \nabla \cdot \boldsymbol{q} - f = 0 .$$

The thermodynamic relation used to introduce T in the first term is as explained in the one-dimensional case. In addition, we need a three-dimensional version of Fourier's law:

$$\boldsymbol{q} = -k\nabla T .$$

Performing the same substitutions and eliminations as in the one-dimensional case, we get the three-dimensional PDE for $T(x, y, z, t)$:

$$\varrho c_v \frac{\partial T}{\partial t} = \nabla \cdot (k\nabla T) + f . \tag{7.26}$$

The term $\nabla \cdot (k\nabla T)$ arises from inserting $\boldsymbol{q} = -k\nabla T$ in $\nabla \cdot \boldsymbol{q}$. Observe that we do not assume that k is constant here. If k is constant, we can write $\nabla \cdot (k\nabla T) = k\nabla \cdot \nabla T = k\nabla^2 T$. Equation (7.26) then takes a form similar to that of (7.4):

$$\frac{\partial T}{\partial t} = \lambda \nabla^2 T + \hat{f}, \tag{7.27}$$

where $\hat{f} = f/(\varrho c_v)$ and $\lambda = k/(\varrho c_v)$, since we have divided by ϱc_v.

Boundary Conditions

There are three common types of boundary conditions in heat transfer applications. The simplest condition is to prescribe T at a part of the boundary. Such a condition requires some external device at the boundary to control the temperature. Another common condition models insulated boundaries, i.e., boundaries effectively covered by some isolating material such that the heat cannot escape. This means that there is no heat flow out of the boundary: $\boldsymbol{q} \cdot \boldsymbol{n} = 0$, which, by Fourier's law, becomes $-k\nabla T \cdot \boldsymbol{n} = 0$, often written as $-k\frac{\partial T}{\partial n} = 0$. For a one-dimensional problem posed on $[a, b]$, the outward unit normal vector points in the positive x-direction on $x = a$ and in negative x-direction on $x = b$. In both cases, $\partial T/\partial x = 0$ is the relevant condition.

A third boundary condition, modeling heat exchange with the surroundings, is often used:

$$-\boldsymbol{q} \cdot \boldsymbol{n} = h_T (T - T_s),$$

or, equivalently,

$$- k\frac{\partial T}{\partial n} = h_T (T - T_s) . \tag{7.28}$$

This law, commonly referred to as *Newton's law of cooling*, states that the net heat flow out of the boundary is proportional to the difference between the temperature in the body (T) and the temperature in the surroundings (T_s). The constant of proportionality is referred to as the *heat transfer coefficient*, and it often requires quite sophisticated modeling to determine its value. When applying (7.28) in one-dimensional problems, one must remember what $\partial T / \partial n$ is, i.e., $\partial T / \partial x$ or $-\partial T / \partial x$, as explained above.

Some names are commonly associated with the three boundary conditions mentioned here:

- Prescribed T is called a *Dirichlet*[8] *condition*
- Prescribed heat flow, $-k \frac{\partial T}{\partial n}$, is called a *Neumann*[9] *condition*
- The cooling law (7.28) is called a *Robin*[10] *condition*

Energy transfer by electromagnetic radiation can also heat bodies. Mathematically, this is often modeled by a nonlinear boundary condition containing the temperature to the fourth or fifth power. This subject is not discussed further here.

At the initial time, $t = 0$, the complete temperature field must be prescribed: $T(x, 0) = I(x)$, where $I(x)$ is a specified function.

7.3.3 Viscous Fluid Flow

Fluid flows are frequently dominated by convection, but one can occasionally encounter flows displaying solely diffusive effects. This happens when the geometry of the fluid has certain symmetry properties. The following example or derivation will give an outline of what we mean.

We assume that the fluid is incompressible and homogeneous, which are relevant assumptions for a range of fluids, including water, air (at flow speeds less than about 350 km/h), and oil. The flow of a homogeneous, incompressible fluid is governed by the Navier[11]–Stokes[12] equations. These can be written in the form

$$\nabla \cdot \boldsymbol{v} = 0, \tag{7.29}$$

$$\varrho \left(\frac{\partial \boldsymbol{v}}{\partial t} + (\boldsymbol{v} \cdot \nabla) \boldsymbol{v} \right) = -\nabla p + \mu \nabla^2 \boldsymbol{v} + \varrho \boldsymbol{g} . \tag{7.30}$$

---

[8] Johann Peter Gustav Lejeune Dirichlet, 1805–1859, was a German mathematician. His contributions to mathematics are remarkable, including work on Fourier series and the mathematical theory of heat. On Gauss's death in 1855, Dirichlet was appointed to fill Gauss's vacant chair at the University of Göttingen.

[9] Carl Gottfried Neumann, 1832–1925, was a German mathematician known to be one of the initiators of the theory of integral equations, and is famous for the very important boundary condition that bears his name.

[10] Victor Gustave Robin, 1855–1897, was a French mathematician who worked in particular in the field of thermodynamics. Today he is mostly known for a widely used boundary condition in heat transfer problems.

[11] Claude-Louis Navier, 1785–1836, was a French engineer and physicist who made important contributions to mechanics, in particular by formulating the PDE¡s that govern fluids and solids.

[12] Sir George Gabriel Stokes, 1819–1903, was a British mathematician and physicist who performed important research in the fields of fluid dynamics, optics, and mathematical physics.

Here, v is the velocity of the fluid, p is the pressure, ϱ is the density, μ is the viscosity, and g denotes the acceleration of gravity (i.e., gravity forces). The velocity and the pressure are the primary unknowns in these PDEs; all the other quantities are assumed known. The Navier–Stokes equations are in general difficult to solve, but in some cases they reduce to simpler equations. One type of simplification ends up with the diffusion equation.

Consider the flow of a fluid between two straight, parallel, infinite plates. The flow can be driven by a pressure gradient, by gravity, or by moving the plates. Infinite plates do not occur in nature or in technological devices, of course. Nevertheless, sometimes one encounters two almost parallel surfaces with a small gap filled with a fluid. Surfaces in machinery constitute an example. Although these surfaces are curved, the gap between them can be so small that a model based on assuming the surfaces to be infinite, parallel, flat plates can be very good. Figure 7.9 depicts the type of problem we are addressing.

It is reasonable to assume that the flow between two straight, parallel plates is parallel to the plates. To express our assumptions with mathematics, we introduce the plates as the planes $x = 0$ and $x = H$, and direct the y axis such that it is aligned with the flow. The velocity field is now of the form $v(x, y, z, t) = v(x, y, z, t)j$. The fact that the velocity has this uni-directional form leads to a substantial simplification of the Navier–Stokes equations. The mathematical details and physical arguments of the simplification can be somewhat difficult to grasp for a novice modeler, but that should be of less concern here. Our message is that diffusion occurs in many different physical contexts. The derivation below is therefore compact such that you will not lose track and miss the main result, i.e., yet another diffusion equation.

It is reasonable to assume that there is no variation with z in this problem (infinite planes and no flow directed in the z direction); thus $v = v(x, y, t)$. Equation (7.30), $\nabla \cdot v = 0$, simplifies to $\partial v / \partial y = 0$, leaving v as a function of x and t only. Inserting $v = v(x, t)j$ in (7.30) requires some algebra. The result is that the term $(v \cdot \nabla)v$ vanishes, the time derivative term becomes $j\varrho \partial v / \partial t$, and $\mu \nabla^2 v$ is reduced to $j\mu \nabla^2 v$. Equation (7.30) is a *vector* equation, giving us three scalar equations:

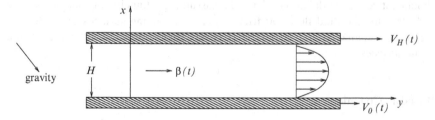

Fig. 7.9 Sketch of viscous fluid flow between two flat plates

$$0 = -\frac{\partial p}{\partial x} + \varrho g_x, \tag{7.31}$$

$$\varrho \frac{\partial v}{\partial t} = -\frac{\partial p}{\partial y} + \mu \frac{\partial^2 v}{\partial x^2} + \varrho g_y, \tag{7.32}$$

$$0 = -\frac{\partial p}{\partial z} + \varrho g_z. \tag{7.33}$$

The gravity vector g is decomposed as $\boldsymbol{g} = g_x \boldsymbol{i} + g_y \boldsymbol{j} + g_z \boldsymbol{k}$. Component (7.32) is of main interest here. Differentiation of (7.32) with respect to y shows that $\partial^2 p/\partial y^2 = 0$, so p can only depend on y in a linear way. The two other equations, (7.31) and (7.33), imply that p is linear in x and z too. Thus, $\partial p/\partial y$ is either a constant or a function of time only. The latter case allows a pulsating pressure and thereby a time-dependent velocity field too. Let us denote $-\partial p/\partial y$ as $\beta(t)$. Since ϱg_y is a constant, we can add this constant to $\beta(t)$ and name the new function $C(t)$. Physically, (7.32) expresses that the flow is driven by a pressure gradient and gravity (boundary conditions can in addition express that the flow can be driven by moving plates, i.e., moving boundaries). The $C(t)$ function is hence the combined effect of the pressure of gravity that drives the flow. Equation (7.32) is now a standard diffusion equation:

$$\frac{\partial v}{\partial t} = v \frac{\partial^2 v}{\partial x^2} + f(t), \tag{7.34}$$

where we have divided by ϱ and introduced a new function $f(t) = C(t)/\varrho$. The parameter v equals μ/ϱ and is actually a common viscosity coefficient easily found in the fluid dynamics literature.

What can the diffusion model (7.34) be used for in a fluid flow context? Normally, (7.34) is used to study friction between lubricated surfaces. The fluid of interest is then some kind of greasy oil. Engineers designing machinery may want to compute the relation between the friction force (from the fluid) on the surfaces, since this is relevant for the required applied force in the machinery. The friction force is calculated by evaluating $\mu \partial v/\partial x$ at the boundaries, i.e., the plates $x=0$ and $x=H$. If the fluid is also driven by pressure differences (some kind of pumping, for instance) and perhaps also gravity, these effects are trivially incorporated in f. One should notice that model (7.34), derived by assuming infinitely long flat plates, may well describe the fluid flow and friction between rotating surfaces in machinery, e.g. in a car engine, as long as the gap H is very small compared with the curvature of the surfaces.

Boundary Conditions

The boundary conditions are simple, because viscous fluids stick to walls: The velocities $v(0,t)$ and $v(H,t)$ must equal the velocity of the planes $x = 0$ and $x = H$, respectively.

Flow in Tubes

Viscous fluid flow in a straight tube with a circular cross section can be modeled analogously to the flow between two flat plates. The key idea is to introduce cylindrical coordinates (y, r, θ), where the y axis coincides with the tube axis, i.e., the flow direction. The r and θ coordinates represent polar coordinates in the circular cross section of the tube. The velocity field can then be written as $v = v(y, r, \theta, t)\boldsymbol{j}$. Due to symmetry, we do not expect v to vary with θ. Inserting $v = v(y, r, t)\boldsymbol{j}$ in (7.30) yields $\partial v/\partial y = 0$, i.e., $v = v(r, t)\boldsymbol{j}$. With this v (7.30) leads to a slight modifications of the Cartesian counterpart (7.31)–(7.33):

$$0 = -\frac{\partial p}{\partial r}, \tag{7.35}$$

$$\varrho \frac{\partial v}{\partial t} = -\frac{\partial p}{\partial y} + \mu \frac{1}{r} \frac{\partial}{\partial r} \left(r \frac{\partial v}{\partial r} \right), \tag{7.36}$$

$$0 = -\frac{1}{\theta} \frac{\partial p}{\partial \theta}. \tag{7.37}$$

The details of the derivation of (7.35)–(7.37) requires familiarity with vector calculus in cylindrical coordinates. We have omitted gravity effects, since these involve somewhat complicated calculations in cylindrical coordinates without adding any interesting aspects of the PDE we want to derive.

Again, differentiation of (7.36) with respect to y reveals that $-\partial p/\partial y$ cannot depend on y. Equations (7.35) and (7.37) imply that p is independent of r and θ. Therefore p can only depend on time (in a general way) and be linear in y. We denote $-\partial p/\partial y$ as $C(t)$, assumed to be a prescribed function, as in the flow between flat plates. The PDE governing $v(r, t)$ then follows from (7.36):

$$\varrho \frac{\partial v}{\partial t} = +\mu \frac{1}{r} \frac{\partial}{\partial r} \left(r \frac{\partial v}{\partial r} \right) + C(t). \tag{7.38}$$

We mention this model since flow in a tube or pipeline is quite common and the model is so closely related to flow between plates. In the fluid mechanics literature, (7.34) and (7.38) are known as Couette[13] and Poiseuille[14] flows, respectively.

---

[13] Maurice Marie Alfred Couette, 1858–1943, was a French physicist known for his studies in fluid mechanics.

[14] Jean Louis Marie Poiseuille, 1797–1869, was a French physician and physiologist. He was particularly interested in the flow of human blood in narrow tubes and performed physical experiments and mathematical formulations of flow in pipelines.

7.3.4 Summarizing the Models

We have seen how the diffusion equation arises in three different physical problems. The PDEs, in one-dimensional form, with appropriate boundary conditions are reviewed here for easy reference. The initial condition is always that the primary unknown $u(x,t)$ must be known at $t = 0$. The governing PDEs are to be solved in an interval (a,b) for time $t > 0$. One boundary condition must be applied at $x = a$ and one at $x = b$. Below, we have used a slightly different notation than we did when deriving the different diffusion equations, because now we want to use a common set of symbols for the various physical applications (e.g., the unknown from now on reads $u(x,t)$). This common set of symbols emphasizes that the mathematics are the same, a fact that is advantageous to explore when developing algorithms and implementations; the code and its methodology are then obviously applicable in different physical contexts.

We remark that for all the models below we need to prescribe an initial condition. This condition takes the form

$$u(x,0) = I(x)$$

for all three physical applications, where $I(x)$ is a known function. Physically it means that the concentration, the temperature, or the velocity must be known everywhere in space when the process starts.

Diffusive Transport

Transport by molecular diffusion is governed by the PDE

$$\frac{\partial u}{\partial t} = k\frac{\partial^2 u}{\partial x^2} + f(x,t).$$

(7.39)

The parameters are

- $u(x,t)$: the concentration of a substance in a fluid
- k: the diffusive transport coefficient
- $f(x,t)$: the external supply of the substance

Common boundary conditions for diffusive transport are as follows:

(a) Controlled concentration,

$$u = U_0(t), \quad U_0(t) \text{ is prescribed.}$$

(7.40)

(b) Impermeable boundary for the substance ("wall"), and

$$\frac{\partial u}{\partial n} = 0.$$

(7.41)

(c) Controlled mass (in-)flow of the substance,

$$-k\frac{\partial u}{\partial n} = Q(t), \quad Q(t) \text{ is prescribed.} \tag{7.42}$$

Note: $\frac{\partial u}{\partial n}$ is the derivative in the outward normal direction to the boundary, i.e., $\frac{\partial u}{\partial n} = -\frac{\partial u}{\partial x}$ at $x = a$ and $\frac{\partial u}{\partial n} = \frac{\partial u}{\partial x}$ at $x = b$.

Heat Conduction

The governing PDE for heat conduction in a material at rest reads

$$c_v \varrho \frac{\partial u}{\partial t} = \frac{\partial}{\partial x}\left(k\frac{\partial u}{\partial x}\right) + f(x,t). \tag{7.43}$$

The parameters are the following:

- $u(x,t)$, the temperature
- $c_v(x)$, the heat capacity, at constant volume, of the material
- $\varrho(x)$m the density of the material
- $k(x)$, the heat conduction coefficient of the material
- $f(x,t)$, the supply of heat per unit time and volume

The set of relevant boundary conditions in heat conduction problems is listed next.

(a) Controlled temperature:

$$u = U_0(t), \quad U_0(t) \text{ is prescribed.} \tag{7.44}$$

(b) Insulated boundary (no heat flow):

$$-k\frac{\partial u}{\partial n} = 0. \tag{7.45}$$

(c) Controlled heat flow:

$$\frac{\partial u}{\partial n} = Q(t), \quad Q(t) \text{ is prescribed.} \tag{7.46}$$

(d) Cooling law:

$$-k\frac{\partial u}{\partial n} = h_T(u - u_s), \quad h_T, u_s(t) \text{ are prescribed,} \tag{7.47}$$

where h_T is a heat transfer coefficient, modeling the heat exchange with the surrounding medium, which has temperature $u_s(t)$.

Viscous Thin-Film Flow

This diffusion problem is a special case of the incompressible Navier–Stokes equations, where the fluid flows between two plates, described by the equations $x = a$ and $x = b$. The simplified Navier–Stokes equations take the form

$$\frac{\partial u}{\partial t} = k \frac{\partial^2 u}{\partial x^2} + f . \tag{7.48}$$

The parameters are as follows:

(a) $u(x, t)$, the velocity directed along the boundary plates[15]
(b) k, the fluid viscosity
(c) $f(t)$, the effects of pressure gradient and gravity

The relevant boundary condition is[16]

(a) Controlled plate velocities:

$$u = U_0(t), \quad U_0 \text{ is prescribed.} \tag{7.49}$$

The flow can be driven by the plates (boundary conditions), by a pressure gradient (source term) and/or gravity (source term).

Spherical Symmetry

For three-dimensional diffusion problems with spherical symmetry and a constant diffusion coefficient k, the governing PDE is

$$\frac{\partial u}{\partial t} = k \frac{\partial^2 u}{\partial x^2} + x f(x), \tag{7.50}$$

where x denotes the radial coordinate, and the function of physical significance is $v(x, t) = r^{-1} u(x, t)$. The x domain is now $[a, b]$, with a possibility of $a = 0$. At the boundary we normally have conditions on v, either v known or $-k \partial v / \partial x$ (radial flux) known. In the former case, u is also known by dividing the boundary values by x. The latter case is somewhat more involved, since we can have

$$-k \frac{\partial v}{\partial x} = q(x),$$

---

[15] The velocity direction is perpendicular to the x coordinate.
[16] We could easily model *free surface* thin-film flow by the boundary condition $\partial u / \partial x = 0$, but the applications are somewhat limited, since instabilities in the form of waves (and hence a three-dimensional problem) often develops at free thin-film surfaces. The wavy thin-film motion of rain on the front window of a car is an example.

which is transformed to

$$-k\frac{\partial u}{\partial x} = xq(x) - \frac{k}{x}v. \tag{7.51}$$

This condition is similar to a cooling law in heat conduction. The bottom line is that programs for the standard one-dimensional diffusion equation can also handle a class of spherically symmetric diffusion problems.

7.3.5 Scaling

Scaling is a very useful but difficult topic in mathematical modeling. Readers who are eager to see how a diffusion problem is solved on a computer can jump to Sect. 7.4. There we work with a scaled problem, but knowing the details of scaling is not a prerequisite for Sect. 7.4; one can just imagine that the input data in the problem have been given specific values (typically values of unity). The forthcoming material on scaling gives a gentle introduction to the scaling of PDEs and must be viewed as a collection of a few examples, and not a complete exposition of scaling as a mathematical technique. In a sense, scaling is, mathematically, a very simple mechanical procedure, but applying scaling correctly to a problem is often very demanding. The reader is hereby warned about the inherent difficulties in the material on the forthcoming pages.

A Heat Conduction Model Problem

Let us look at a heat conduction problem without heat sources and with the temperature controlled at constant values at the boundaries:

$$\varrho c_v \frac{\partial u}{\partial t} = k\frac{\partial^2 u}{\partial x^2}, \quad x \in (a,b), \ t > 0, \tag{7.52}$$

$$u(a,t) = U_a, \quad t > 0, \tag{7.53}$$

$$u(b,t) = U_b, \quad t > 0, \tag{7.54}$$

$$u(x,0) = I(x), \quad x \in [a,b]. \tag{7.55}$$

The heat conduction takes place in a homogeneous material such that ϱ, c_v, and k are constants. To be specific, we assume that $I(x)$ is a step function (relevant for the introductory example of bringing together two metal pieces at different temperatures):

$$I(x) = \begin{cases} U_a, \ a \leq x < (b-a)/2, \\ U_b, \ (b-a)/2 \leq x \leq b. \end{cases} \tag{7.56}$$

If you wonder how the solution $u(x,t_\ell)$ behaves, you can take a look at Fig. 7.10 on page 313. Here we have plotted $u(x,t_\ell)$ at some time points t_ℓ. You should notice how the initial jump in the temperature (cf. $I(x)$) is smoothed in time and how the solution stabilizes at a straight line, going from one boundary value to the other.

Motivation for Scaling

The solution $u(x, t)$ of (7.52)–(7.55) depends on the physical parameters ϱ, c_v, k, U_a, U_b, as well as the geometry parameters a and b. We can write

$$u(x, t; \varrho, c_v, k, U_a, U_b, a, b).$$

Investigations of u based on numerical experiments require seven parameters to be varied. Say we try three values for each parameter: low, medium, and high. This implies $3^7 = 2{,}187$ numerical experiments. Varying each parameter over five values results in a demand of 78,125 experiments to cover all combinations of the input values.

However, common sense tells us that not all of these experiments will yield different results. It must be the length of the domain, $b - a$, which has significance, and not the individual a and b values. Regarding the boundary conditions, it must be the difference $U_a - U_b$ that influences u. Dividing the PDE by ϱc_v shows that the PDE remains the same as long as the fraction between k and ϱc_v remains the same. Therefore, the number of independent parameters in the mathematical model is obviously less than seven.

There is a systematic procedure, called *scaling*, for reducing the number of parameters in a mathematical model. Scaling is both an art and a cookbook procedure, and here we shall only give a glimpse of the recipe for carrying out the scaling technically. The result will be appealing; we can get full insight into the dependence of u on the seven input parameters by performing just a single numerical computation of u! Sounds like magic? Not really: Scaling is comprised technically of calculations with fractions and the chain rule of differentiation.

The Mechanics of Scaling

The idea of scaling is to replace a variable q, which has a physical dimension, by a dimensionless variable $\bar{q}$. The aim is to have the maximum of $|\bar{q}|$ equal to unity or be of order unity. If q_r is a characteristic reference value of q and q_c is a characteristic magnitude of $q - q_r$, a common scaling is

$$\bar{q} = \frac{q - q_r}{q_c}.$$

We observe that if q is measured in, say, seconds, q_r and q_c also have dimensions seconds, and this dimension cancels out in the fraction, i.e., $\bar{q}$ is dimensionless.

As an example, consider the function

$$f(x) = A + B \sin(\omega x).$$

Since f lies between $A + B$ and $A - B$ (the absolute value of the sine function is at most one), a characteristic reference value f_r is the average A. The characteristic magnitude of $f - A$ is B so we can introduce the scaling

$$\bar{f}(x) = \frac{f(x) - A}{B} = \sin \omega x.$$

We clearly see that $\max_x |\bar{f}(x)| = 1$, as desired. In $\bar{f}(x)$ there is only one parameter left, ω, i.e., A and B are in a sense "parameterized away". No information is lost, since we can easily construct f on the basis of $\bar{f}$: $f(x) = A + B\bar{f}(x)$. In this example the reference value and the characteristic magnitude were obvious. Finding the characteristic magnitude is occasionally quite demanding in PDE problems, and it requires quite a bit of physical insight and approximate mathematical analysis of the problem.

Let us apply a scaling procedure to (7.52)–(7.55). Our aim to is scale x, t, u, and I such that these variables have a magnitude between zero and order unity. The x parameter varies between a and b, so

$$\bar{x} = \frac{x - a}{b - a},$$

is a scaling where the scaled parameter $\bar{x}$ varies between 0 and 1, as desired. Since the initial time is 0, the reference value is 0. The characteristic magnitude of time, call it t_c, is more problematic. This t_c can be the time it takes to experience significant changes in u. More analysis is needed to find candidate values for t_c; thus we just scale t as

$$\bar{t} = \frac{t}{t_c},$$

and remember that t_c must be determined later in the procedure. The function $I(x)$ is easy to scale since it has only two values. We can take U_a as a reference value and $U_b - U_a$ as a characteristic magnitude:

$$\bar{I}(\bar{x}) = \frac{I(\bar{x}L) - U_a}{U_b - U_a}.$$

Finally we need to scale u. Initially we could use the same scaling as for I, but, as time increases, the PDE governs the magnitude of u. Without any idea of how the solution of the PDE behaves, it is difficult to determine an appropriate scaling. Choosing a scaling based on the initial data,

$$\bar{u} = \frac{u - U_a}{U_b - U_a},$$

is fortunately a good choice in the present example, since it can be shown mathematically that u in (7.52)–(7.55) is bounded by the initial condition and the boundary values.

We are now ready to replace the physical variables x, t, u, and I, i.e. all independent and dependent variables plus all function expressions, by their dimensionless equivalents. From

$$\bar{x} = \frac{x-a}{b-a}, \quad \bar{t} = \frac{t}{t_c}, \quad \bar{I} = \frac{I-U_a}{U_b-U_a}, \quad \bar{u} = \frac{u-U_a}{U_b-U_a},$$

we get

$$x = a + (b-a)\bar{x}, \quad t = t_c\bar{t}, \quad I = U_a + (U_b - U_a)\bar{I}, \quad u = U_a + (U_b - U_a)\bar{u},$$

which we insert into (7.52)–(7.55). Noting that

$$\frac{\partial u}{\partial t} = \frac{\partial \bar{t}}{\partial t}\frac{\partial}{\partial \bar{t}}(U_a + (U_b - U_a)\bar{u}) = \frac{1}{t_c}(U_b - U_a)\frac{\partial \bar{u}}{\partial \bar{t}},$$

with a similar development for the $\partial u/\partial x$ expression, we arrive at

$$\varrho c_v \frac{U_b - U_a}{t_c}\frac{\partial \bar{u}}{\partial \bar{t}} = k\frac{U_b - U_a}{(b-a)^2}\frac{\partial^2 \bar{u}}{\partial \bar{x}^2}, \quad \bar{x} \in (0,1), \bar{t} > 0, \tag{7.57}$$

$$\bar{u}(0,\bar{t}) = 0, \quad \bar{t} > 0, \tag{7.58}$$

$$\bar{u}(1,\bar{t}) = 1, \quad \bar{t} > 0, \tag{7.59}$$

$$\bar{u}(\bar{x},0) = \begin{cases} 0, 0 \leq \bar{x} \leq \frac{1}{2}, \\ 1, \frac{1}{2} < \bar{x} \leq 1. \end{cases} \tag{7.60}$$

The PDE (7.57) can be written in the form

$$\frac{\partial \bar{u}}{\partial \bar{t}} = \delta\frac{\partial^2 \bar{u}}{\partial \bar{x}^2}, \tag{7.61}$$

with δ being a dimensionless number,

$$\delta = \frac{k t_c}{\varrho c_v (b-a)^2}.$$

We have not yet determined t_c, which is the task of the next paragraph.

Finding a Non-Trivial Scale

The goal of scaling is to have the maximum value of all independent and dependent variables in the PDE problem of order unity. One can argue that if $\bar{u}$ is of order unity, the coordinates are of order unity, and $\bar{u}$ is sufficiently smooth, we expect the derivatives of $\bar{u}$ to also be of order unity. Looking at (7.61), we should then have δ of order unity (since both derivative terms are of order unity). Choosing $\delta = 1$ implies

$$t_c = \frac{1}{k}\varrho c_v (b - a)^2,$$

which determines the time scale. Newcomers to scaling may find these arguments a bit too vague. A more quantitative reasoning is based on a typical solution of (7.52). Inserting

$$u(x,t) = e^{-vt} \sin\left(\pi \frac{x - a}{b - a}\right)$$

in (7.52) reveals that this is a solution for the special case $U_a = U_b = 0$ and $I(x) = \sin\left(\pi \frac{x-a}{b-a}\right)$, provided that

$$v = \frac{k\pi^2}{\varrho c_v (b - a)^2}.$$

The boundary and initial values are not important here; what is important is the temporal evolution of u, described by the factor e^{-vt}. The solution is exponentially damped. A common choice of the characteristic time t_c is the time that reduces the initial (unit) amplitude by a factor of $1/e$ (the e-folding time[17]):

$$e^{-vt_c} = e^{-1} \quad \Rightarrow \quad t_c = \frac{1}{v} = \frac{\varrho c_v (b - a)^2}{k\pi^2}.$$

This means that significant changes (here reduction of the amplitude by a factor of $e^{-1} \approx 0.37$) in u take place during a time interval of length t_c. For esthetic reasons, many prefer to remove the factor π^2 from the expression for t_c and write

$$t_c = \frac{1}{k}\varrho c_v (b - a)^2. \tag{7.62}$$

The consequence is that we just use a different amplitude reduction factor to derive t_c. There are many different solutions of (7.52), with different (typically shorter) time scales, so there is no single correct value for t_c. The main point is that t_c is not orders of magnitude away from the dominating time scales of the problem.

You should observe that two different ways of reasoning led to the same time scale t_c, thus providing some confidence in the result. Numerical experiments or available analytical solutions of the problem can be used to test if our choice of t_c is appropriate. In general, the dominating time scales depend on the PDE, the initial conditions, and the boundary conditions. Hopefully we have managed to explain

---

[17] The specific choice $1/e$ as a suitable reduction factor is mostly inspired by esthetics. For a function e^{-vt} we get v^{-1} as the characteristic time. A reduction factor of 10^{-5} gives a characteristic time $v^{-1} 5 \ln 10$.

that finding the right scales is potentially very difficult, and that there are even some non-trivial points in this simple introduction to diffusion PDEs.

Summary of the Scaled Problem

We can now summarize the equations in the scaled diffusion problem:

$$\frac{\partial \bar{u}}{\partial \bar{t}} = \frac{\partial^2 \bar{u}}{\partial \bar{x}^2}, \quad \bar{x} \in (0,1), \; \bar{t} > 0, \tag{7.63}$$

$$\bar{u}(0, \bar{t}) = 0, \quad \bar{t} > 0, \tag{7.64}$$

$$\bar{u}(1, \bar{t}) = 1, \quad \bar{t} > 0, \tag{7.65}$$

$$\bar{u}(\bar{x}, 0) = \begin{cases} 0, 0 \le \bar{x} \le \frac{1}{2}, \\ 1, \frac{1}{2} < \bar{x} \le 1. \end{cases} \tag{7.66}$$

The most striking feature of (7.63)–(7.66) is that *there are no physical parameters*, no density, heat capacity, and so on. This means that $\bar{u}(\bar{x}, \bar{t})$ is in (7.63)–(7.66) independent of ϱ, c_v, k, U_a, U_b, a, and b! More importantly, we can solve (7.63)–(7.66) just once, and all information about the solution's dependence on the input data such as k, ϱ, and so on, must be inherited in $\bar{u}(\bar{x}, \bar{t})$ and the scaling. Suppose we produce some curves $\bar{u}(\bar{x}, \bar{t}_\ell)$ for some time values $\bar{t}_\ell$, say $\bar{t}_1 = 0.1$, $\bar{t}_2 = 0.25$, and $\bar{t}_3 = 1$, as depicted in Fig. 7.10. The physical temperatures $u(x, t)$ are related to the $\bar{u}(\bar{x}, \bar{t}_\ell)$ curves through the scaling; i.e., the real temperatures are

$$u(x,t) = U_a + (U_b - U_a)\bar{u}\left(\frac{x-a}{b-a}, \frac{tk}{\varrho c_v(b-a)^2}\right). \tag{7.67}$$

This relation tells us everything about the influence of the original input data ϱ, c_v, k, U_a, U_b, a, and b on the temperature. In other words, the 2,187 or 78,125 numerical experiments we mentioned are reduced to the need for a single experiment! This example should illustrate the usefulness of scaling.

Unfortunately, scaling away all the physical parameters is only possible in the very simplest PDE problems, but frequently scaling helps reduce the number of input data in the model, and hence the amount of numerical experiments.

It is tedious to write all the bars if we want to work further with the scaled problem. Therefore, simply dropping the bars is a common habit. We then write the scaled PDE problem as

$$\frac{\partial u}{\partial t} = \frac{\partial^2 u}{\partial x^2}, \quad x \in (0,1), \; t > 0, \tag{7.68}$$

$$u(0, t) = 0, \quad t > 0, \tag{7.69}$$

$$u(1, t) = 1, \quad t > 0, \tag{7.70}$$

$$u(x, 0) = \begin{cases} 0, 0 \le x \le \bar{c}, \\ 1, \bar{c} < x \le 1. \end{cases} \tag{7.71}$$

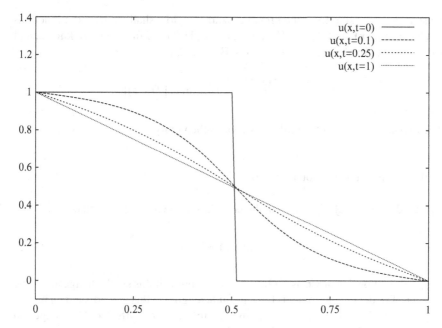

Fig. 7.10 Solution $u(x, t)$ of (7.63)–(7.66)

When you meet a problem such as (7.68)–(7.71), where many (in this case all) parameters have unit values, you should immediately recognize the problem as a scaled problem. A common misinterpretation is to think that many physical parameters are set to the value one, but this can be completely unphysical. If you desire to work with unit values, you should always scale the underlying physical problem to arrive at the correct model, where "physical parameters" are removed. The information provided here may be sufficient for scaling diffusion problems, but in other PDE problems you probably have to gain more knowledge and experience about the art of scaling in general in order to construct sound scalings. The technique of scaling has been particularly important in applied mathematics, especially in perturbation and similarity methods, so there is a vast literature on scaling. Unfortunately, this literature does not discuss scaling in a modern scientific computing context, i.e., with the aim of reducing the amount of numerical experimentation to gain insight into a model.

Extension: Variable Coefficients

A PDE with variable coefficients, e.g.,

$$\varrho(x)c_v(x)\frac{\partial u}{\partial t} = \frac{\partial}{\partial x}\left(k(x)\frac{\partial u}{\partial x}\right) + f(x,t),$$

presents an extension from the previous example: How shall we treat the functions $\varrho(x)$, $c_v(x)$, $k(x)$, and $f(x,t)$? The idea is simple: These functions are known, and we scale them by their maximum values. For example,

$$\bar{f}(x,t) = \frac{f(x,t)}{f_c}, \quad f_c = \max_{x,t} |f(x,t)| .$$

The maximum value of the scaled function is then unity.

Scaling with Neumann Boundary Condition

Let us address (7.52)–(7.55) when the boundary value $u = U_b$ is replaced by

$$- k \frac{\partial}{\partial x} u(b,t) = -Q_b, . \tag{7.72}$$

i.e., the heat flow is known as $-Q_b$ on $x = b$. How will this small change affect the scaling? The detailed arguments become quite lengthy in this case too, but an important result is that all scalings, except the one for u, are not affected. Our suggested scaling of u in (7.52)–(7.55) reads

$$\bar{u} = \frac{u - U_a}{U_b - U_a} . \tag{7.73}$$

The question now is whether $U_b - U_a$ is a characteristic magnitude of $u - U_a$. The answer depends on the value of the input data in the problem.

Suppose we apply (7.73). The scaled PDE problem becomes (with bars dropped)

$$\frac{\partial u}{\partial t} = \frac{\partial^2 u}{\partial x^2}, \quad x \in (0,1), \ t > 0, \tag{7.74}$$

$$u(0,t) = 0, \quad t > 0, \tag{7.75}$$

$$\frac{\partial}{\partial x} u(1,t) = \beta, \quad t > 0, \tag{7.76}$$

$$u(x,0) = \begin{cases} 0, \ 0 \le x \le \bar{c}, \\ 1, \ \bar{c} < x \le 1, \end{cases} \tag{7.77}$$

where

$$\beta = \frac{Q_b(b-a)}{k(U_b - U_a)} \tag{7.78}$$

is a dimensionless number. If β is too far from unity, u has a large derivative at the boundary, suggesting that u itself may have a magnitude much larger than unity. In that case it would be more natural to base the magnitude of u on Q_b. Estimating u as a straight line based on the knowledge that u is U_a at $x = a$ and u has a derivative $-Q_b/k$ at $x = b$ (yes, we use the same symbols now for scaled and unscaled quantities, but the context should imply the right distinction), we can find that

$$u|_{x=b} \approx U_a + \frac{Q_b}{k}(b-a),$$

resulting in a characteristic magnitude $u|_{x=b} - U_a$ and a scaling

$$\bar{u} = \frac{k(u - U_a)}{Q_b(b-a)}. \tag{7.79}$$

The new characteristic magnitude cancels out in the PDE and does not influence our arguments for the choice of t_c. The boundary condition at $x = b$ gets a simpler form with this new scaling of u:

$$\frac{\partial}{\partial \bar{x}}\bar{u}(1, \bar{t}) = 1. \tag{7.80}$$

The initial condition is also affected:

$$\bar{u}(\bar{x}, 0) = \begin{cases} 0, & 0 \leq \bar{x} \leq \bar{c}, \\ \beta^{-1}, & \bar{c} < \bar{x} \leq 1. \end{cases} \tag{7.81}$$

We see that for both scalings, there is one physical parameter in the problem, β, and it enters the PDE problem in different ways (in the boundary condition or in the initial condition), depending on the choice of scale for u.

If β is of order unity, there is no significant difference between the two scalings. For $\beta \gg 1$, i.e., $Q_b(b-a)$ is much larger than $k(U_b - U_a)$, the scaling (7.79) is advantageous, whereas for $\beta \ll 1$, (7.73) is the best choice. Project 7.7.4 on page 356 asks you to perform detailed calculations to understand the difference between the two suggested scalings.

This discussion shows that the *scales depend on the regimes of the physical parameters* in the problem. In this example there is no unified scaling that can ensure values of order unity. This is also the limitation of scaling. In complicated problems, there can be a need for different scalings for different physical regimes.

When performing numerical experiments in the scaled problem with a derivative boundary condition, we need to vary one physical parameter, i.e., we compute $\bar{u}(\bar{x}, \bar{t}; \beta)$ for various values of the β parameter. The physical solution is obtained by the "inverse scaling", i.e., expressing $u(x, t)$ in terms of $\bar{u}(\bar{x}, \bar{t}; \beta)$ using the various physical input data in the problem.

Remark

Before writing a computer program, one has to decide whether to use the scaled form or use the variables with physical dimensions. One solution is to implement the original form of the problem, with dimensions, but set the parameters in the experiments such that one simulates a scaled form.

7.4   Explicit Numerical Methods

Our purpose now is to solve a one-dimensional diffusion equation, such as (7.1), by means of a computer. To this end, we need to look at a mathematical problem consisting of the diffusion PDE *and* boundary and initial conditions. Our goal is to solve the (scaled) problem

$$\varrho(x)c_v(x)\frac{\partial u}{\partial t} = \frac{\partial}{\partial x}\left(k(x)\frac{\partial u}{\partial x}\right) + f(x,t), \quad x \in (0,1),\ t > 0, \quad (7.82)$$

$$u(0,t) = D_0(t), \quad t > 0, \quad (7.83)$$

$$\frac{\partial}{\partial x}u(1,t) = N_1(t), \quad t > 0, \quad (7.84)$$

$$u(x,0) = I(x), \quad x \in [0,1]. \quad (7.85)$$

We refer to (7.82)–(7.85) as the *continuous* problem. The PDE (7.82) is fulfilled at all the *infinite* number of continuously distributed points, $0 < x < 1$ and $t > 0$. Moreover, u is defined at the same set of infinite points. Solving the problem (7.82)–(7.85) on a computer requires us to construct an algorithm with a *finite* number of steps for computing a *finite* number of parameters that describe u. This algorithm must solve a *discrete* version of (7.82)–(7.85), and the process of constructing such a discrete problem is called *discretization*. The *finite difference method* is one way of discretizing problems involving PDEs.

7.4.1   The Basics of Finite Difference Discretizations

Applying the finite difference method to the problem (7.82)–(7.85) implies

(a) Constructing a *grid*, with a finite number of points in (x,t) space, see Fig. 7.11
(b) Requiring the PDE (7.82) to be satisfied at each point in the grid
(c) Replacing the derivatives by finite difference approximations
(d) Calculating (an approximation of) u at the grid points only

The finite difference discretization is characterized by these four steps. Requiring the PDE to be satisfied at a number of discrete grid points, instead of a continuously distributed set of (x,t), represents an approximation. Replacing derivatives by finite difference approximations is another source of error.

We will start with a version of the PDE (7.82) where $\varrho(x)$, $c_v(x)$, and $k(x)$ are constant. These parameters are then "scaled away" such that the PDE reads

$$\frac{\partial u}{\partial t} = \frac{\partial^2 u}{\partial x^2} + f(x,t), \quad x \in (0,1), \ t > 0. \tag{7.86}$$

Discrete Functions on a Grid

Look at the grid in Fig. 7.11. We have drawn grid lines, $x = x_i$ and $t = t_\ell$. It is convenient to operate with a subscript or index i to count the grid lines $x = \mathrm{const}$. Similarly, we introduce the subscript or index ℓ to count the grid lines $t = \mathrm{const}$. The spacing between the lines $x = \mathrm{const}$ is denoted by Δx. Not surprisingly, Δt is the symbol representing the distance between two neighboring lines $t = \mathrm{const}$. In Fig. 7.11 you can see the lines $x = x_0 = 0$, $x = x_1 = \Delta x$, $x = x_2 = 2\Delta x$, ..., $x = x_9 = 9\Delta x = 1$ and $t = t_0 = 0$, $t = t_1 = \Delta t$, $t = t_2 = 2\Delta t$, $t = t_3 = 3\Delta t$.

In the general case, we can have n grid lines $x = x_i = i\Delta x$, $i = 0, \ldots, n$. Of course, $x_0 = 0$ and $x_n = 1$. The corresponding grid spacing Δx between the n intervals, made up by the grid lines, is then

$$\Delta x = \frac{1}{n}.$$

We also assume that we want to compute the solution at $m + 1$ time levels, i.e., for $t = t_\ell = \ell \Delta t$. We count i from 0 and ℓ from 0; if you find this unnatural you can introduce your own counting, the only requirement is that you be consistent (which is hard in practice).

A general grid point is denoted by (x_i, t_ℓ). The value of an arbitrary function $Q(x,t)$ at the grid point (x_i, t_ℓ) is written as Q_i^ℓ, i.e.,

$$Q(x_i, t_\ell) = Q_i^\ell, \quad i = 0, \ldots, n, \ \ell = 0, \ldots, m.$$

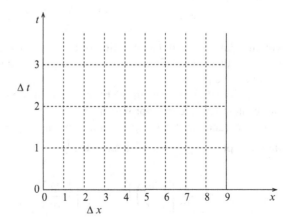

Fig. 7.11 A computational grid in the $x - -t$ plane. The grid points are located at the points of intersection of the *dashed lines*

In our problem we have the functions $u(x,t)$ and $f(x,t)$, which at the grid points have the values u_i^ℓ and f_i^ℓ, respectively, in our more compact notation.

The purpose of a finite difference method is to compute the values u_i^ℓ for $i = 0, \ldots, n$ and $\ell = 0, \ldots, m$.

Fortunately, you now have an idea of what a grid in $x - -t$ space is. The second item in the list of characteristics of the finite difference method is to sample the PDE at these grid points. To this end, we write the PDE (7.86) as

$$\frac{\partial}{\partial t} u(x_i, t_\ell) = \frac{\partial^2}{\partial x^2} u(x_i, t_\ell) + f(x_i, t_\ell), \tag{7.87}$$

where i and ℓ denote points in the computational $x - -t$ grid.

Finite Difference Approximations

The third step in the list of characteristics of the finite difference method consists of replacing derivatives by finite differences. Since there are many possible finite difference approximations for a derivative, this step has no unique recipe. We will try different approximations of the time derivative later, but for now we use the specific choices

$$\frac{\partial}{\partial t} u(x_i, t_\ell) \approx \frac{u_i^{\ell+1} - u_i^\ell}{\Delta t}, \tag{7.88}$$

$$\frac{\partial^2}{\partial x^2} u(x_i, t_\ell) \approx \frac{u_{i-1}^\ell - 2u_i^\ell + u_{i+1}^\ell}{\Delta x^2}. \tag{7.89}$$

The finite difference approximation (7.88) is a one-sided difference, typically like the fraction

$$\frac{g(t + h) - g(t)}{h}$$

you may have seen in the definition of the derivative $g'(t)$ in introductory calculus books (in that case the limit $h \to 0$ is a central point; now we have a finite h, or Δt as we denote it here).

The finite difference on the right-hand side of (7.89) is constructed by combining two centered difference approximations. We first approximate the "outer" derivative at $x = x_i$ (and $t = t_\ell$), using a fictitious point $x_{i+\frac{1}{2}} = x_i + \frac{1}{2}\Delta x$ to the right and a fictitious point $x_{i-\frac{1}{2}} = x_i - \frac{1}{2}\Delta x$ to the left:

$$\frac{\partial}{\partial x}\left[\left(\frac{\partial u}{\partial x}\right)\right]_i^\ell \approx \frac{1}{\Delta x}\left[\left[\frac{\partial u}{\partial x}\right]_{i+\frac{1}{2}}^\ell - \left[\frac{\partial u}{\partial x}\right]_{i-\frac{1}{2}}^\ell\right].$$

The first-order derivative at $x_{i+\frac{1}{2}}$ can be approximated by a centered difference using the point x_{i+1} to the right and the point x_i to the left:

$$\left[\frac{\partial u}{\partial x}\right]_{i+\frac{1}{2}}^{\ell} \approx \frac{u_{i+1}^{\ell} - u_i^{\ell}}{\Delta x}.$$

Similarly, the first-order derivative at $x_{i-\frac{1}{2}}$ can be approximated by a centered difference using the point x_i to the right and the point x_{i-1} to the left:

$$\left[\frac{\partial u}{\partial x}\right]_{i-\frac{1}{2}}^{\ell} \approx \frac{u_i^{\ell} - u_{i-1}^{\ell}}{\Delta x}.$$

Combining the differences gives (7.89), a calculation that you should verify. The derivation of (7.89) shows that we used only centered difference approximations. Therefore, (7.89) is also a centered difference. Centered differences are known to be more accurate than one-sided differences. You may thus wonder why we do not use a centered difference for $\partial u/\partial t$. The reason is because our one-sided difference gives a computational algorithm that is much simpler to implement than if we use a centered difference in time.

The Finite Difference Scheme

Inserting the difference approximations (7.88) and (7.89) in (7.87) results in

$$\frac{u_i^{\ell+1} - u_i^{\ell}}{\Delta t} = \frac{u_{i-1}^{\ell} - 2u_i^{\ell} + u_{i+1}^{\ell}}{\Delta x^2} + f_i^{\ell}. \tag{7.90}$$

This is our discrete version of the PDE (7.87). Alternative versions will be derived later.

   The computational algorithm consists of computing u along the $t = t_{\ell}$ lines, one line at a time. That is, we compute u_i^{ℓ} values, for $i = 0, \ldots, n$, first for $t = 0$, which is trivial since u is known from the initial condition (7.85) at $t = 0$, then for $t = t_1 = \Delta t$, then for $t = t_2 = 2\Delta t$, and so forth. Suppose that the u_i^{ℓ} values at time level $t = t_{\ell}$ are known for all the spatial points ($i = 0, \ldots, n$). We can then use (7.90) to find values at the next time level $t = t_{\ell+1}$. We solve (7.90) with respect to $u_i^{\ell+1}$, yielding a simple formula for the solution at the new time level:

$$u_i^{\ell+1} = u_i^{\ell} + \frac{\Delta t}{\Delta x^2}\left(u_{i-1}^{\ell} - 2u_i^{\ell} + u_{i+1}^{\ell}\right) + \Delta t f_i^{\ell}. \tag{7.91}$$

We refer to (7.90) or (7.91) as a *finite difference scheme* for the PDE (7.86). At a general point $(x_i, t_{\ell+1})$ we can compute the function value $u_i^{\ell+1}$ from the known values $u_{i-1}^{\ell}, u_i^{\ell}$, and u_{i+1}^{ℓ}. This procedure is graphically illustrated in Fig. 7.12. Many refer

Fig. 7.12 Illustration of the updating formula (7.91); u_4^3 is computed from u_3^2, u_4^2, and u_5^2

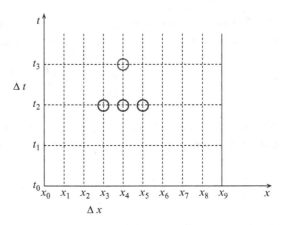

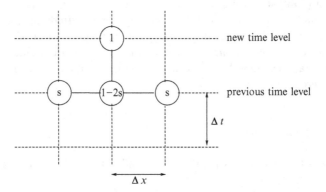

new time level

previous time level

Fig. 7.13 Illustration of the computational molecule corresponding to the finite difference scheme (7.91). The weight s is equal to $\Delta t / \Delta x^2$

to the circles in Fig. 7.12 as a *computational molecule*, especially if we write the weights inside the circles and connect the circles by lines, as depicted in Fig. 7.13.

Because new $u_i^{\ell+1}$ values can be computed explicitly from a simple formula, where all the quantities are known, we refer to the finite difference scheme as *explicit*. The scheme is often referred to as the explicit Euler scheme or the forward Euler scheme (the term "forward" relates to the forward difference in time). The opposite type of schemes, the so-called *implicit* schemes, typically couple all the new $u_i^{\ell+1}$ values for $i = 0, \ldots, n$ in a linear system. This means that we need to solve a linear system to find the u values at a new time level. Implicit schemes hence imply a more complicated computational procedure, but implicit schemes have some numerical advantages over explicit schemes. Section 7.5 is devoted to implicit schemes for diffusion problems.

7.4.2 Incorporating Dirichlet Boundary Conditions

The observant reader will notice that it is difficult to use (7.91) for computing new values $u_1^{\ell+1}$ and $u_n^{\ell+1}$ at the boundary, because (7.91) for $i = 1$ and $i = n$ involves values u_{-1}^{ℓ} and u_{n+1}^{ℓ} *outside* the grid. The solution is to use the boundary conditions (7.83) and (7.84) for $i = 0$ and $i = n$. These must be combined with scheme (7.91).

Suppose we have the boundary conditions $u(0, t) = 0$ and $u(1, t) = 0$, previously categorized as Dirichlet conditions. We can then use (7.91) to update $u_i^{\ell+1}$ at all *internal* grid points, $i = 1, \ldots, n-1$. For $i = 1$, (7.91) involves the known values u_0^{ℓ}, u_1^{ℓ}, and u_2^{ℓ}, and u_0^{ℓ} is zero from the boundary condition. Similarly, for $i = n - 1$, (7.91) involves the known values u_{n-2}^{ℓ}, u_{n-1}^{ℓ}, and the boundary condition $u_n^{\ell} = 0$.

The computational procedure can be summarized as we have done in Algorithm 7.1.

Let us generalize the simple boundary conditions $u(0, t) = u(1, t) = 0$ a bit to the choices

$$u(0, t) = D_0(t), \quad u(1, t) = D_1(t), \tag{7.92}$$

where $D_0(t)$ and $D_1(t)$ are prescribed functions. The computational procedure is hardly affected by this slight generalization. We use scheme (7.91) for the internal points $i = 2, \ldots, n - 1$, and at the boundary points we just insert the correct values:

$$u_0^{\ell+1} = D_0(t_{\ell+1}), \quad u_n^{\ell+1} = D_1(t_{\ell+1}).$$

Algorithm 7.2 incorporates the non-homogeneous Dirichlet conditions (7.92).

7.4.3 Incorporating Neumann Boundary Conditions

Implementing Neumann boundary conditions, such as (7.84),

$$\frac{\partial}{\partial x} u(1, t) = N_1(t),$$

Algorithm 7.1

Diffusion Equation with $u = 0$ at the Boundary.

$\alpha = \Delta t / \Delta x^2$

SET INITIAL CONDITIONS:

$u_i^0 = I(x_i), \quad$ for $i = 0, \ldots, n$

for $\ell = 0, 1, \ldots, m - 1$

    UPDATE ALL INNER POINTS:

    for $i = 1, \ldots, n - 1$

        $u_i^{\ell+1} = u_i^{\ell} + \alpha \left(u_{i-1}^{\ell} - 2u_i^{\ell} + u_{i+1}^{\ell} \right) + \Delta t f_i^{\ell}$

    INSERT BOUNDARY CONDITIONS:

    $u_0^{\ell+1} = 0, \quad u_n^{\ell+1} = 0$

Algorithm 7.2

Diffusion Equation with $u(0, t) = D_0(t)$ and $u(1, t) = D_1(t)$.

$\alpha = \Delta t / \Delta x^2$

SET INITIAL CONDITION:

$u_i^0 = I(x_i)$,   for $i = 0, \dots, n$

$t_0 = 0$

for $\ell = 0, 1, \dots, m - 1$

    $t_{\ell+1} = t_\ell + \Delta t$

    UPDATE ALL INNER POINTS:

    for $i = 1, \dots, n - 1$

        $u_i^{\ell+1} = u_i^\ell + \alpha \left(u^\ell - 2u_i^\ell + u_{i+1}^\ell \right) + \Delta t f_i^\ell$

    INSERT BOUNDARY CONDITIONS:

    $u_0^{\ell+1} = D_0(t_{\ell+1}), \;\; u_n^{\ell+1} = D_1(t_{\ell+1})$

---

where N_1 is known, is more difficult than implementing the conditions where u is known (Dirichlet conditions). The recipe goes as follows. First, we discretize the derivative at the boundary, i.e., replace $\partial u / \partial x$ by a finite difference. Since we use a centered spatial difference in the PDE, we should use a centered difference at the boundary too:

$$\frac{\partial}{\partial x} u(1, t_\ell) \approx \frac{u_{n+1}^\ell - u_{n-1}^\ell}{2\Delta x}. \tag{7.93}$$

The discrete version of the boundary condition then reads

$$\frac{u_{n+1}^\ell - u_{n-1}^\ell}{2\Delta x} = N_1(t_\ell). \tag{7.94}$$

Unfortunately, this formula involves a u value outside the grid, u_{n+1}^ℓ. However, if we use scheme (7.91) at the boundary point $x = x_n$,

$$u_n^{\ell+1} = u_n^\ell + \frac{\Delta t}{\Delta x^2} \left(u_{n-1}^\ell - 2u_n^\ell + u_{n+1}^\ell \right) + \Delta t f_n^\ell, \tag{7.95}$$

we have two equations, both involving the fictitious value u_{n+1}^ℓ. This enables us to eliminate this value. Solving (7.94) with respect to u_{n+1}^ℓ,

$$u_{n+1}^\ell = u_{n-1}^\ell + 2N_1^\ell \Delta x. $$

Inserting this expression in (7.95) yields a modified scheme at the boundary:

$$u_n^{\ell+1} = u_n^\ell + 2\frac{\Delta t}{\Delta x^2} \left(u_{n-1}^\ell - u_n^\ell + N_1^\ell \Delta x \right) + \Delta t f_n^\ell. \tag{7.96}$$

Algorithm 7.3

Diffusion Equation with $u(0,t) = D_0(t)$ and $u'(1,t) = N_1(t)$.

$\alpha = \Delta t / \Delta x^2$

SET INITIAL CONDITIONS:

$u_i^0 = I(x_i)$, for $i = 0, \ldots, n$

$t_0 = 0$

for $\ell = 0, 1, \ldots, m - 1$

$\quad t_{\ell+1} = t_\ell + \Delta t$

$\quad$ UPDATE ALL INNER POINTS:

$\quad$ for $i = 1, \ldots, n - 1$

$\qquad u_i^{\ell+1} = u_i^\ell + \alpha \left(u_{i-1}^\ell - 2u_i^\ell + u_{i+1}^\ell \right) + \Delta t f_i^\ell$

$\quad$ INSERT DIRICHLET BOUNDARY CONDITION:

$\quad u_0^{\ell+1} = D_0(t_{\ell+1})$

$\quad$ INCORPORATE NEUMANN BOUNDARY CONDITION:

$\quad u_n^{\ell+1} = u_n^\ell + 2\alpha \left(u_{n-1}^\ell - u_n^\ell + N_1^\ell \Delta x \right) + \Delta t f_n^\ell$

We are now in a position to formulate the complete computational algorithm for our model problem consisting of the simplified PDE (7.86) and the boundary and initial conditions (7.83)–(7.85), see Algorithm 7.3.

Remark

Thoughtful readers with a lazy attitude may have wondered about a much simpler way to incorporate the Neumann boundary condition in (7.96). Instead of using a centered difference, as in (7.94), we could use a one-sided difference,

$$\frac{\partial}{\partial x} u(1, t_\ell) \approx \frac{u_n^\ell - u_{n-1}^\ell}{\Delta x},$$

and hence

$$\frac{u_n^\ell - u_{n-1}^\ell}{\Delta x} = N_1^\ell,$$

which, when solved with respect to u_n^ℓ, yields

$$u_n^\ell = u_{n-1}^\ell + N_1^\ell \Delta x.$$

This updating formula of the boundary value does not involve any fictitious value (u_{n+1}^ℓ) outside the boundary, and there is hence no need to combine the discrete boundary condition with the discrete PDE evaluated at the boundary. Nevertheless, the one-sided difference is less accurate than the centered difference approximation we use in the PDE and may therefore decrease the accuracy of the whole computation.

7.4.4 How to Verify a Computer Implementation

After having created a program, we need to check that the program is correct. This
is a very demanding task. Ideally, we would like to prove correctness, but this turns
out to be extremely difficult except in very special cases. Instead we have to provide
sufficient evidence that the program works correctly. This means that we establish a
series of test cases and show, for each case, that the results are reasonable. But what
is "reasonable"? The difficulty with this procedure is that our solution procedure
is only approximate, so if we compare numerical results with an exact analytical
solution to the diffusion equation, there will always be differences, i.e., errors. The
question is therefore whether the observed errors are due to approximations or to a
combination of approximations and programming errors. In the forthcoming examples, we will take a few steps toward providing evidence that a program correctly
implements a solution algorithm for the diffusion equation.

A Trivial Test Case

As a first test of a program, we recommend constructing a case where the solution
is constant in space and time. The reason for this is that many coding errors in
a program can be uncovered by such a simple test. However, this simple test can
also hide many programming errors; so to provide more evidence that the program
works, we need to proceed with more complicated solutions.

To obtain a solution $u(x, t) = C$, where C is a constant, we specify $f(x, t) = 0$,
$I(x) = C$, $D_0(t) = C$, and $D_1(t) = C$. Inserting $u = C$ in the diffusion equation
and the initial and boundary conditions shows that this is indeed a valid solution.
The constant C should not be chosen as 0 or 1, since these values are less likely to
uncover coding errors – forgetting a factor or term can still give a correct u value of
0 or 1. We therefore set C to some "arbitrary value", say 0.8.

Let us compute u_i^ℓ values in this case by hand, following Algorithm 7.2. From
the initial condition we have $u_i^0 = C$ for $i = 0, \ldots, n$. The first pass in the for loop
has $\ell = 0$, and we compute u_i^1 from

$$u_i^1 = C + \alpha(C - 2C + C) + 0 = C$$

for all inner points ($i = 1, \ldots, n - 1$). At the boundary we set

$$u_0^1 = C, \quad u_n^1 = C .$$

Repeating the procedure for the next pass in the loop, we find that $u_i^2 = C$ for
$i = 0, \ldots, n$, and, in fact, $u_i^\ell = C$ for all ℓ and i values.

Comparing with an Exact Solution

Exact solutions to PDEs are, in general, hard to find. However, for the diffusion equation some typical solutions are well known. Inserting

$$u(x,t) = e^{-\pi^2 t} \sin(\pi x) \tag{7.97}$$

in the governing equation shows that this is a solution if $f = 0$. We can then adjust I, D_0, and D_1 such that (7.97) is the solution, i.e.,

$$I(x) = \sin(\pi x), \quad D_0(t) = 0, \quad D_1(t) = 0.$$

Running a program that implements Algorithm 7.2 should result in a numerical solution that is close to the exact solution (7.97), but how close? We know that our numerical approximations used to derive Algorithm 7.2 contain errors, so how can we distinguish between a numerical approximation error and a programming error? This is a really difficult question, which indicates how hard it can be to check if simulation programs are free of programming errors. A good approach to answering the question goes as follows.

One can mathematically show that the error E implied by the scheme in Algorithm 7.2 is of the form

$$E = C_1 \Delta x^2 + C_2 \Delta t . \tag{7.98}$$

For a given α value we can replace Δt by $\alpha \Delta x^2$:

$$E = C_3 \Delta x^2, \quad C_3 = C_1 + C_2 \alpha . \tag{7.99}$$

This means that if we halve the grid spacing Δx, the error should be reduced by a factor of four. This is something we can test to provide evidence that the program is working correctly. It appears that this is one of the best test methods we can apply.

Let $u(x,t)$ be the exact solution and $\tilde{u}$ the corresponding numerical solution. The error can be defined as

$$E = \sqrt{\int_0^1 (u - \tilde{u})^2 dx} . \tag{7.100}$$

Since $\tilde{u}$ is only known at the grid points, we approximate the integral by the trapezoidal rule:

$$E \approx \left(\Delta x \left(\frac{1}{2}(u(0, t_\ell) - u_0^\ell)^2 + \frac{1}{2}(u(1, t_\ell) - u_n^\ell)^2 + \sum_{i=1}^{n-1}(u(x_i, t_\ell) - u_i^\ell)^2 \right) \right)^{1/2} . \tag{7.101}$$

With the exact solution (7.97), we can compute E corresponding to four grid resolutions: $\Delta x = 0.1, 0.05, 0.025, 0.0125$. Since $E/\Delta x^2$ should be a constant (called

C_3 above), we can compute the ratio $E/\Delta x^2$ for the four difference cases. The ratio becomes 0.0633159, 0.0647839, 0.0655636, and 0.0659656. Since these numbers are almost equal, we conclude that $E/\Delta x^2$ is constant, as it should be. This provides solid evidence that the program works correctly when the boundary conditions are $u = 0$ and there is no source term. Finding a solution when the boundary conditions and the source term are non-trivial can be done as described next.

The Method of Manufactured Solutions

We can, in fact, always find an exact solution to a PDE problem. The procedure, often called the method of manufactured solutions, goes as follows. Pick any function $v(x,t)$ as the exact solution. Insert v in the PDE; it does not fit, but adding a source term will make it fit. As an example, take

$$v(x,t) = t^2 + 4x^3.$$

Inserting this function in the PDE

$$\frac{\partial u}{\partial t} = \frac{\partial^2 u}{\partial x^2} + f(x,t),$$

shows that $v(x,t)$ is a solution if we choose

$$f(x,t) = 2t - 24x.$$

Then we set $I(x) = v(x,0)$, $D_0(t) = v(0,t)$, and $D_1(t) = v(1,t)$. To test a program, we set I, D_0, and D_1 as just listed and compute E for a sequence of Δx values. An approximately constant ratio $E/\Delta x^2$ (for the different E and Δx values) provides evidence that the program works.

A nice manufactured solution to use in the debugging phase is a linear function of space and time. Such a function will very often be (almost) exactly reproduced by numerical schemes for PDEs, regardless of the values of the grid resolution. That is, with such a solution we can control that E is sufficiently close to zero (to machine precision) also on very coarse grids. In the present problem we can choose $v(x,t) = t/2 + 3x$ as a manufactured solution. The corresponding source term is $f(x,t) = 1/2$. Writing out the error shows that it is about 10^{-6}, even on coarse grids.

Diffusion of a Jump

Our next problem concerns a jump in the initial condition, like we get when bringing two metal pieces together, as depicted in Fig. 7.4. However, for simplicity we apply Dirichlet conditions (instead of the more appropriate Neumann conditions) at the boundaries. The problem to be solved reads

$$\frac{\partial u}{\partial t} = \frac{\partial^2 u}{\partial x^2},$$ (7.102)

$$u(0, t) = 0,$$ (7.103)

$$u(1, t) = 1,$$ (7.104)

$$u(x, 0) = \begin{cases} 0, & x \le 0.5, \\ 1, & x > 0.5. \end{cases}$$ (7.105)

Before running any simulations, we should have an idea of what the solution looks like. As time increases, the initial jump in temperature is smoothed out, and for very large times we expect that no further changes in the temperature takes place. This means that as $t \to \infty$, we expect $\partial u / \partial t \to 0$. The PDE is then reduced to the problem $\partial^2 u / \partial x^2 = 0$ with boundary conditions $u(0) = 0$ and $u(1) = 1$. This stationary problem has the solution $u(x) = x$. We therefore expect the jump to diffuse into a straight line.

A sequence of plots is displayed in Fig. 7.14. As we see, this problem is numerically tough to compute. The initial jump triggers a wavy solution, but the waves eventually disappear and a straight line is obtained. Using $\Delta x = 0.005$, i.e., ten times as many spatial points (and 100 times as many time levels, since we keep $\Delta t = \Delta x^2 / 2$), shows that the wavy nature of the solution curves is significantly reduced. As $\Delta x \to 0$, it seems that the wavy nature disappears, so we can conclude that the waves in Fig. 7.14 are numerical artifacts.

In much more complicated PDE problems, such as those solved in industry and scientific research, one can encounter wavy solutions. A fundamental question is then whether the waves are physical or purely numerical. Insight into numerical methods and experience with numerical artifacts as in the present test case constitute important knowledge when judging solutions to complicated PDE problems. Effects that decrease when making the grid finer are very often non-physical effects.

7.4.5 Instability

Let us investigate the numerical solution generated by Algorithm 7.1 in more detail. The quality of the numerical solution will depend on the choice of Δx and Δt. We expect that the smaller these discretization parameters are, the more accurate the solution will be. Looking at the scheme in Algorithm 7.1, we see that Δx and Δt enter the formula only through the fraction $\alpha = \Delta t / \Delta x^2$. The quality of the numerical solution therefore depends on the choice of α. How should we choose α?

Let us start with the problem

$$\frac{\partial u}{\partial t} = \frac{\partial^2 u}{\partial x^2} \quad \text{for } x \in (0, 1), \ t > 0,$$

$$u(0, t) = u(1, t) = 0 \quad \text{for } t > 0,$$

$$u(x, 0) = \sin(3\pi x) \quad \text{for } x \in (0, 1).$$

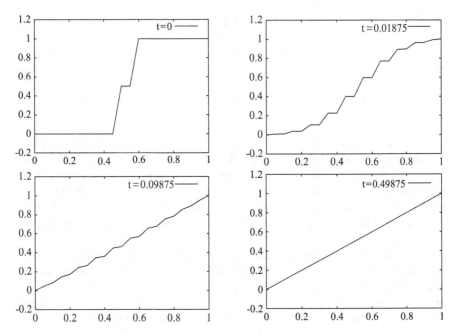

Fig. 7.14 Visualization of the solution $u(x, t_\ell)$ of (7.102)–(7.105) at the time levels $\ell = 1, 16, 80, 400$ (*top left* to *bottom right*); $\Delta x = 0.05$, $\Delta t = \Delta x^2/2$

The analytical solution to this initial-boundary value problem can be shown to be

$$u(x,t) = e^{-9\pi^2 t} \sin(3\pi x). \tag{7.106}$$

In Fig. 7.15 we have graphed this function and the numerical results generated by Algorithm 7.1 at $t = 10$ for various values of the discretization parameters in space and time:

- The solid line represents the analytical solution.
- The dotted line represents results obtained with $\Delta x = 1/9$, $\Delta t = 10/17$, and consequently $\alpha = 0.4765$.
- The dash-dotted line represents numbers generated with $\Delta x = 1/19$, $\Delta t = 10/82$, implying that $\alpha = 0.4402$.
- The dashed line represents approximations obtained by setting $\Delta x = 1/59$, $\Delta t = 10/706$, and thus $\alpha = 0.4931$.

In these cases the scheme produces well-behaved approximations, and the approximations seem to converge toward the solution of the problem as Δt and Δx tend toward zero.

Note that, in each of these plots, $\alpha \leq 0.5$ and Δt is relatively much smaller than Δx. Is this necessary, or can we increase Δt? Unfortunately, if Δt is increased such that $\alpha > 0.5$, then this scheme tends to produce oscillating numbers, and

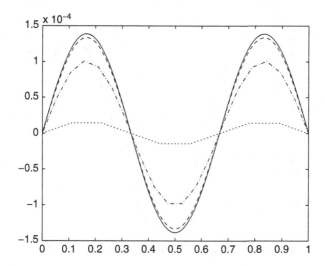

Fig. 7.15 The *solid line* represents the solution (7.106). The *dotted, dash-dotted,* and *dashed lines* are the numerical results implied by Algorithm 7.1 in the cases of $n = 10$ and $m = 17$, with $\alpha = 0.4765$ ($\Delta x = 1/9$), $\alpha = 0.4402$ ($\Delta x = 1/19$), and $\alpha = 0.4931$ ($\Delta x = 1/59$), respectively

eventually the numbers produced do not provide an approximation of the solution of the diffusion equation. This phenomenon is illustrated in Figs. 7.16 and 7.17:

- If, in the case of $\Delta x = 1/59$, Δt is reduced from $10/706$ to $10/681$, and thus α is increased from $\alpha = 0.4931$ to $\alpha = 0.5112$, then oscillations occur in the numerical approximation.
- A further increase of Δt, corresponding to $\alpha = 0.5157$, results in a curve that does not bear any resemblance with the graph of the solution of the problem.

Above we observed that, if $\alpha \leq 0.5$, then our numerical method for the diffusion equation works fine. On the other hand, for discretization parameters Δx and Δt such that $\alpha > 0.5$, this method did not provide acceptable results. Is this a special feature of the problem studied in this example or is this a more general disadvantage of this scheme? To gain further insight, let us perform more experiments.

Now we turn our attention to the diffusion problem

$$\frac{\partial u}{\partial t} = \frac{\partial^2 u}{\partial x^2} \quad \text{for } x \in (0, 1),\ t > 0,$$
$$u(0, t) = u(1, t) = 0 \quad \text{for } t > 0,$$
$$u(x, 0) = x(1 - x) = x - x^2 \quad \text{for } x \in (0, 1).$$

The exact solution to this problem is derived in the next chapter and listed in (8.84) on page 392. How does the numerical solution produced by Algorithm 7.1 compare with the exact solution in this case?

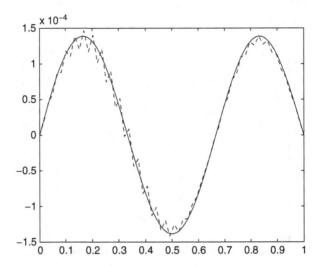

Fig. 7.16 The *dashed line* represents numerical results generated by Algorithm 7.1 with $\alpha = 0.5112$ ($\Delta x = 1/59$). The *solid line* is the graph of the exact solution (7.106)

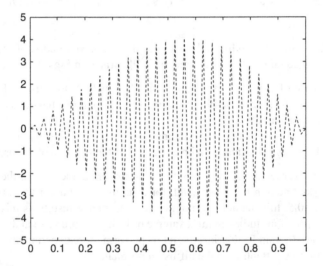

Fig. 7.17 A plot of the numbers generated by Algorithm 7.1 when $\alpha = 0.5157 > 0.5$ ($\Delta x = 1/59$). Observe that for the chosen Δx and Δt values, the method fails to solve the problem under consideration!

In Fig. 7.18 we have plotted the function given by the sum of the first 100 terms of the sine series of the solution (8.84), along with the results from three different numerical experiments. Quick convergence is obtained. Even by using "modest" discretization parameters ($\Delta x = 1/13$ and $\Delta t = 1/34$), the explicit scheme seems to produce accurate approximations. Note that $\alpha \leq 0.5$ for all the parameters used in these experiments.

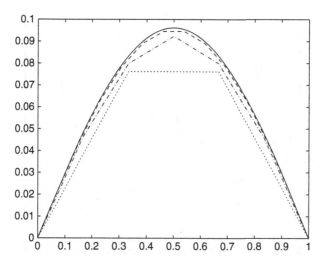

Fig. 7.18 Numerical results in a problem with a sine series solution given in (8.84). The *solid line* represents the first 100 terms in the sine series (8.84). The *dotted, dash-dotted*, and *dashed curves* are the numerical results generated in the cases of $\alpha = 0.3$, $\alpha = 0.45$, and $\alpha = 0.4971$, respectively

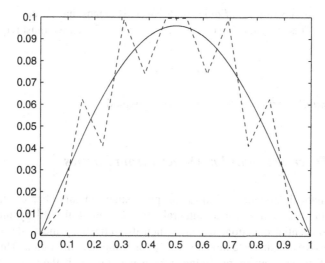

Fig. 7.19 Same problem as in Fig. 7.18, but now the numerical solution corresponds to $\alpha = 0.5828$

Let us fix the number of grid points used in the space dimension to 13, and study the performance of the method as the time step is reduced. Will we observe the same type of behavior as that in Figs. 7.16 and 7.17? Yes, in Figs. 7.19 and 7.20 we used $\alpha = 0.5828$ and $\alpha = 0.6760$, respectively. Clearly, oscillations and instabilities occur. The numerical method we have presented so far to solve the

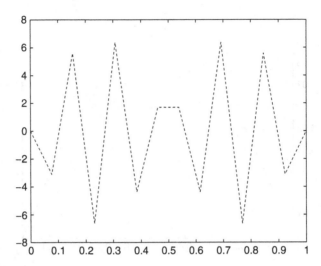

Fig. 7.20 Same problem as in Fig. 7.18, but now the numerical solution corresponds to $\alpha = 0.6760 > 0.5$. The method fails to solve the problem under consideration

diffusion equation is only *conditionally stable*. This means that Δt cannot exceed a critical value. This critical value corresponds to $\alpha = 0.5$, i.e., we must require that

$$\Delta t \le \frac{1}{2}\Delta x^2 . \tag{7.107}$$

This is the *stability criterion* of our numerical method.

7.4.6 A Discrete Algorithm Directly from Physics

The mathematical modeling of diffusion phenomena in this chapter starts with physical principles, expressed as integral formulations, followed by the derivation of a differential equation, which is then discretized by the finite difference method to arrive at a discrete model suitable for computer simulation. This section outlines a different way of reasoning, where we start with the physical principles and go directly to a discrete computational model, without the intermediate differential equation step. Many physicists and engineers prefer such an approach to computerized problem solving.

We focus on the diffusive transport of a substance. Mass conservation in an interval Ω was expressed in Sect. 7.3.1 as

$$\varrho q(a)\Delta t - \varrho q(b)\Delta t + \int_a^b \varrho f \Delta t\, dx = \int_a^b \varrho \Delta c\, dx, \tag{7.108}$$

cf. derivation of (7.12) on page 291. In addition we need Fick's law (7.16):

$$q = -k\frac{\partial c}{\partial x}. \tag{7.109}$$

Let us divide the global domain into intervals of length Δx. That is, if the global domain is the unit interval $(0, 1)$, and we introduce $n+1$ grid points $x_i, i = 0, \ldots, n$, we have $x_i = i\Delta x$. We could apply the integral form (7.108) of mass conservation to one grid cell $[x_i, x_{i+1}]$. However, to get a final discrete algorithm with the same indices as in Algorithm 7.1, we must apply (7.108) to $[x_{i-\frac{1}{2}}, x_{i+\frac{1}{2}}]$. In this case, $a = x_{i-\frac{1}{2}}$ and $b = x_{i+\frac{1}{2}}$:

$$\varrho q_i \Delta t - \varrho q_{i+1}\Delta t + \int_{x_{i-\frac{1}{2}}}^{x_{i+\frac{1}{2}}} \varrho f\Delta t\, dx = \int_{x_{i-\frac{1}{2}}}^{x_{i+\frac{1}{2}}} \varrho\Delta c\, dx.$$

Imagine that this is the mass conservation principle at some time level ℓ such that a possible expression for Δc is $c^{\ell+1} - c^\ell$, the difference in concentration between two time levels. All quantities are then sampled at time level ℓ so we can write

$$\varrho q_i^\ell \Delta t - \varrho q_{i+1}^\ell \Delta t + \int_{x_{i-\frac{1}{2}}}^{x_{i+\frac{1}{2}}} \varrho f(x, t_\ell)\Delta t\, dx = \int_{x_{i-\frac{1}{2}}}^{x_{i+\frac{1}{2}}} \varrho(c^{\ell+1} - c^\ell)dx \qquad . \tag{7.110}$$

When i varies throughout the grid, this equation ensures that mass is conserved in each interval $[x_{i-\frac{1}{2}}, x_{i+\frac{1}{2}}]$.

The second integral over Δc in (7.110) is inconvenient. How should we integrate a function c whose value is known only in the discrete grid points? An obvious idea is to perform a numerical approximation of the integral. The midpoint rule is an attractive candidate, since it involves the c sampled at the point x_i, i.e., we get an expression involving the point values c_i^ℓ and $c_i^{\ell+1}$:

$$\int_{x_{i-\frac{1}{2}}}^{x_{i+\frac{1}{2}}} \varrho(c^{\ell+1} - c^\ell)\, dx \approx \varrho(c_i^{\ell+1} - c_i^\ell)\Delta x.$$

Since we are already approximating integrals, no more errors are introduced if we replace the integral over the known function f by a similar midpoint approximation,

$$\int_{x_{i-\frac{1}{2}}}^{x_{i+\frac{1}{2}}} \varrho f(x, t_\ell)\Delta t\, dx \approx \varrho f_i^\ell \Delta t \Delta x.$$

The mass conservation statement can be written as

$$\varrho q_i^\ell \Delta t - \varrho q_{i+1}^\ell \Delta t + \varrho f_i^\ell \Delta t \Delta x = \varrho(c_i^{\ell+1} - c_i^\ell)\Delta x,$$

or, after dividing by $\varrho \Delta x$ and reordering,

$$c_i^{\ell+1} = c_i^\ell + \Delta t \frac{q_i^\ell - q_{i+1}^\ell}{\Delta x} + \Delta t f_i^\ell. \qquad (7.111)$$

To use (7.111) as an updating formula for $c_i^{\ell+1}$, we need to relate q_i^ℓ and q_{i+1}^ℓ to c^ℓ-values through a discrete version of Fick's law (7.109). Using midpoint differences, we approximate (7.109) as

$$q_i^\ell = \frac{c_i^\ell - c_{i-1}^\ell}{\Delta x}, \quad q_{i+1}^\ell = \frac{c_{i+1}^\ell - c_i^\ell}{\Delta x}.$$

Inserting this discrete Fick's law in (7.111), we arrive at the scheme

$$c_i^{\ell+1} = c_i^\ell + \frac{\Delta t}{\Delta x^2} \left(c_{i-1}^\ell - 2c_i^\ell + c_{i+1}^\ell \right) + \Delta t f_i^\ell. \qquad (7.112)$$

This is exactly the same updating formula as in (7.91). That equation was derived by discretizing the governing PDE by the finite difference method, whereas in the present section we obtained the same scheme by discretizing the basic physical expressions: (a) an integral statement of mass conservation and (b) Fick's law.

Continuous functions, and thereby PDEs, constituted the major tool of analysis of physical problems when the fundamental laws of nature were formulated and studied. If the laws of nature were discovered after the invention of computers, it might have been possible that the derivations in the present section would be the natural way to model nature, without any PDEs. Of course, from a mathematical point of view, the derivations of (7.91) and (7.112) are mathematically equivalent. It is just the need for discretizing integrals instead of derivatives only that makes a difference in the reasoning.

7.4.7 Variable Coefficients

The heat conduction equation (7.82) allows for variable coefficients $\varrho(x)$, $c_v(x)$, and $k(x)$. This is of particular relevance when we study heat conduction in a domain made up of more than one material. We will now learn how to discretize (7.82).

The equation is to be fulfilled at any grid point (x_i, t_ℓ):

$$\varrho(x_i)c_v(x_i)\frac{\partial}{\partial t}u(x_t, t_\ell) = \left[\frac{\partial}{\partial x}\left(k(x)\frac{\partial u}{\partial x} \right) \right]_{x=x_i, t=t_\ell} + f(x_i, t_\ell). \qquad (7.113)$$

The next step is to replace the derivatives by finite differences. For the time derivative, this procedure is the same as for the simplified, constant-coefficient heat conduction equation (7.86), which we treated in Sect. 7.4.1. The only new challenge is the space–derivative term. The idea is to discretize the term in two steps. First we look at the outer first-order derivative

$$\frac{\partial}{\partial x}\phi(x_i, t_\ell),$$

where

$$\phi(x_i, t_\ell) \equiv k(x)\frac{\partial}{\partial x}u(x_i, t_\ell),$$

and thereafter we approximate the inner derivative in ϕ. This is the same procedure as we followed on page 318 when deriving finite difference approximations for second-order derivatives.

The first step involves a centered finite difference around the point x_i:

$$\frac{\partial}{\partial x}\phi(x_i, t_\ell) \approx \frac{\phi^\ell_{i+\frac{1}{2}} - \phi^\ell_{i-\frac{1}{2}}}{\Delta x}.$$

The expression $\phi^\ell_{i+\frac{1}{2}}$ means $\phi(x_{i+\frac{1}{2}}, t_\ell) = \phi(x_i + \frac{1}{2}\Delta x, t_\ell)$. Now, $x_i + \frac{1}{2}\Delta x$ is not a grid point, but this will not turn out to be a problem.

The quantities $\phi^\ell_{i+\frac{1}{2}}$ and $\phi^\ell_{i-\frac{1}{2}}$ involve first-order derivatives and must also be discretized. Using centered finite differences again, we can write

$$\phi^\ell_{i+\frac{1}{2}} = k_{i+\frac{1}{2}}\frac{u^\ell_{i+1} - u^\ell_i}{\Delta x}$$

and

$$\phi^\ell_{i-\frac{1}{2}} = k_{i-\frac{1}{2}}\frac{u^\ell_i - u^\ell_{i-1}}{\Delta x}.$$

The quantity $k_{i+\frac{1}{2}}$ means $k(x_i + \frac{1}{2}\Delta x)$ and is straightforward to evaluate if $k(x)$ is a known function of x. We will assume that for a while.

Using the time-derivative approximation (7.88) and the space-derivative formula derived above, we can write the discrete version of the heat conduction equation (7.82) as

$$\gamma_i \frac{u^{\ell+1}_i - u_i}{\Delta t} = \frac{1}{\Delta x}\left(k_{i+\frac{1}{2}}\frac{u_{i+1} - u_i}{\Delta x} - k_{i-\frac{1}{2}}\frac{u_i - u_{i-1}}{\Delta x}\right) + f^\ell_i, \qquad (7.114)$$

where we have introduced the abbreviation γ_i for $\varrho(x_i)c_v(x_i)$. We assume that we have already computed u at time level ℓ, which means that only $u^{\ell+1}_i$ is an unknown quantity in this equation. Solving with respect to $u^{\ell+1}_i$ gives

$$u^{\ell+1}_i = u^\ell_i + \frac{1}{\gamma_i}\frac{\Delta t}{\Delta x}\left(k_{i+\frac{1}{2}}\frac{u^\ell_{i+1} - u^\ell_i}{\Delta x} - k_{i-\frac{1}{2}}\frac{u^\ell_i - u^\ell_{i-1}}{\Delta x}\right) + \frac{\Delta t}{\gamma_i}f^\ell_i. \qquad (7.115)$$

This is the finite difference scheme corresponding to the PDE (7.82).

The algorithm for solving the discrete version (7.115) of (7.82) can be written as slight modifications of Algorithms 7.1 and 7.3, depending on the choice of boundary conditions.

7.5 Implicit Numerical Methods

The numerical method for diffusion problems from Sect. 7.4 is explicit, meaning that we can compute the value of $u(x, t)$ at a space–time grid point from a formula that is simple to evaluate. In a program, we run through all grid points at all time levels and compute a new u value at each grid point by evaluating the formula.

In many other PDE problems, the values of the unknown function at a new time level are coupled to each other. This yields a linear system of algebraic equations to be solved instead of just evaluating a simple explicit formula. Methods where new values are coupled in a linear system are known as *implicit* methods.

Since explicit methods are much simpler to work with than implicit methods, one can ask why it is not sufficient to always use explicit methods. The answer to this question lies in the instability problem we encountered with the explicit scheme in Sect. 7.4.5: The maximum time step is $\Delta t = \Delta x^2/(2k)$. Say you compute the solution with some value of Δx. Suppose, then, that you need a four times finer resolution in space. Now the time step must be reduced by a factor of 16. To simulate from time zero to some finite time T the overall work increases by a factor of 64. If we could stick to the old value of the time step, the increase in the work would be by a factor of 4 only. (This reasoning may be too simplistic, because increased accuracy in space should normally be accompanied by a smaller time step for increased accuracy in time. Otherwise the total error may be dominated by the temporal error and refinements in space are just a waste of resources.)

One can formulate implicit methods for diffusion equations where there are no stability restrictions on the time step, i.e., any Δt will work. Of course, a large Δt yields inaccurate results, but the solution might exhibit at least correct qualitative features in cases where an explicit scheme produces completely useless results. In particular, if we seek the stationary solution of a diffusion problem as $t \to \infty$ and $u_t \to 0$, one giant time step with an implicit method may be enough to arrive at an accurate solution (!).

7.5.1 The Backward Euler Scheme

Equation (7.87) is our starting point for all finite difference schemes, i.e., we sample the PDE at a space–time point (x_i, t_ℓ). The next step is to replace derivatives by finite differences. This time we use a *backward* difference for the time derivative,

$$\frac{\partial}{\partial t} u(x_i, t_\ell) \approx \frac{u_i^\ell - u_i^{\ell-1}}{\Delta t}. \tag{7.116}$$

For the second-order derivative in space we use (7.89). Inserting these approximations in the PDE yields

$$\frac{u_i^\ell - u_i^{\ell-1}}{\Delta t} = \frac{u_{i-1}^\ell - 2u_i^\ell + u_{i+1}^\ell}{\Delta x^2} + f_i^\ell. \tag{7.117}$$

We assume that $u_i^{\ell-1}$ is known while u_i^ℓ is unknown, $i = 0, \ldots, n$. Note that here we view values at time level ℓ as unknown, while we sought u values at time level $\ell + 1$ in Chap. 7. The reason is because we used a forward difference (involving u_i^ℓ and $u_i^{\ell+1}$) in time in Sect. 7.4.1, while we use a backward difference here (involving u_i^ℓ and $u_i^{\ell-1}$). We could equally well use the backward difference at grid point $(x_i, t_{\ell+1})$. That would result in the equivalent scheme

$$\frac{u_i^{\ell+1} - u_i^\ell}{\Delta t} = \frac{u_{i-1}^{\ell+1} - 2u_i^{\ell+1} + u_{i+1}^{\ell+1}}{\Delta x^2} + f_i^\ell, \tag{7.118}$$

where the unknown values are at time level $\ell + 1$: $u_{i-1}^{\ell+1}$, $u_i^{\ell+1}$, and $u_{i+1}^{\ell+1}$. In the following we stick to (7.117).

The finite difference scheme implied by (7.117) is known as a backward Euler scheme (where "backward" refers to the backward difference in time). Some also call this scheme the fully implicit scheme or the implicit Euler scheme.

7.5.2 The Linear System of Equations

With scheme (7.90), we simply solved for the new unknown. Solving (7.117) with respect to u_i^ℓ gives

$$\left(1 + 2\frac{\Delta t}{\Delta x^2}\right) u_i^\ell = u_i^{\ell-1} + \frac{\Delta t}{\Delta x^2} \left(u_{i-1}^\ell + u_{i+1}^\ell\right) + \Delta t \, f_i^\ell.$$

The problem now is that u_{i-1}^ℓ and u_{i+1}^ℓ are also unknown! It is impossible to derive an explicit formula for u_i^ℓ containing values at the previous time level $\ell - 1$ only. There are three unknown quantities: u_{i-1}^ℓ, u_i^ℓ, and u_{i+1}^ℓ. These three quantities are coupled in scheme (7.117). A fruitful reordering of (7.117) involves collecting all the unknown quantities on the left-hand side and all the known quantities on the right-hand side:

$$-\frac{\Delta t}{\Delta x^2} u_{i-1}^\ell + (1 + 2\alpha) u_i^\ell - \frac{\Delta t}{\Delta x^2} u_{i+1}^\ell = u_i^{\ell-1} + \Delta t \, f_i^\ell.$$

We have in previous parts of the chapter seen that the ratio $\Delta t / \Delta x^2$ appears frequently, so we use our α parameter to represent this ratio. The equation above then reads

$$- \alpha u_{i-1}^{\ell} + (1 + 2\alpha) u_i^{\ell} - \alpha u_{i+1}^{\ell} = u_i^{\ell-1} + \Delta t\, f_i^{\ell} . \qquad (7.119)$$

Let us write down what (7.119) implies in the case we have five grid points ($n = 4$) on $[0, 1]$ and $f = 0$:

$$-\alpha u_0^{\ell} + (1 + 2\alpha) u_1^{\ell} - \alpha u_2^{\ell} = u_1^{\ell-1},$$
$$-\alpha u_1^{\ell} + (1 + 2\alpha) u_2^{\ell} - \alpha u_3^{\ell} = u_2^{\ell-1},$$
$$-\alpha u_2^{\ell} + (1 + 2\alpha) u_3^{\ell} - \alpha u_4^{\ell} = u_3^{\ell-1}.$$

We can assume Dirichlet boundary conditions, say, $u_0^{\ell} = u_4^{\ell} = 0$ for simplicity. Then we have

$$(1 + 2\alpha) u_1^{\ell} - \alpha u_2^{\ell} = u_1^{\ell-1}, \qquad (7.120)$$
$$-\alpha u_1^{\ell} + (1 + 2\alpha) u_2^{\ell} - \alpha u_3^{\ell} = u_2^{\ell-1}, \qquad (7.121)$$
$$-\alpha u_2^{\ell} + (1 + 2\alpha) u_3^{\ell} = u_3^{\ell-1}. \qquad (7.122)$$

This is nothing but a linear system of three equations for u_1^{ℓ}, u_2^{ℓ}, and u_3^{ℓ}. We can write this system in matrix form,

$$Au = b,$$

where

$$A = \begin{pmatrix} 1 + 2\alpha & -\alpha & 0 \\ -\alpha & 1 + 2\alpha & -\alpha \\ 0 & -\alpha & 1 + 2\alpha \end{pmatrix},$$

$$b = \begin{pmatrix} u_1^{\ell-1} \\ u_2^{\ell-1} \\ u_3^{\ell-1} \end{pmatrix}, \quad u = \begin{pmatrix} u_1^{\ell} \\ u_2^{\ell} \\ u_3^{\ell} \end{pmatrix}.$$

To find the new u_i^{ℓ} values we need to solve such a 3×3 system of linear equations, which can be done by Gaussian elimination, hopefully known from a linear algebra course.

As an alternative to forming a linear system where only the u_i^{ℓ} values at the inner grid are unknowns in the system, we can include the boundary points as well. That is, we treat u_i^{ℓ}, $i = 0, 1, 2, 3, 4$, as unknowns and add the boundary conditions as equations:

$$u_0^{\ell} = 0,$$
$$-\alpha u_0^{\ell} + (1 + 2\alpha) u_1^{\ell} - \alpha u_2^{\ell} = u_1^{\ell-1},$$
$$-\alpha u_1^{\ell} + (1 + 2\alpha) u_2^{\ell} - \alpha u_3^{\ell} = u_2^{\ell-1},$$
$$-\alpha u_2^{\ell} + (1 + 2\alpha) u_3^{\ell} - \alpha u_4^{\ell} = u_3^{\ell-1},$$
$$u_4^{\ell} = 0.$$

The matrix system $A u = b$ now has the following matrix A, a right-hand side b, and a vector of unknowns u:

$$A = \begin{pmatrix} 1 & 0 & 0 & 0 & 0 \\ -\alpha & 1+2\alpha & -\alpha & 0 & 0 \\ 0 & -\alpha & 1+2\alpha & -\alpha & 0 \\ 0 & 0 & -\alpha & 1+2\alpha & -\alpha \\ 0 & 0 & 0 & 0 & 1 \end{pmatrix},$$

$$b = \begin{pmatrix} 0 \\ u_1^{\ell-1} \\ u_2^{\ell-1} \\ u_3^{\ell-1} \\ 0 \end{pmatrix}, \quad u = \begin{pmatrix} u_0^{\ell} \\ u_1^{\ell} \\ u_2^{\ell} \\ u_3^{\ell} \\ u_4^{\ell} \end{pmatrix}.$$

In the rest of this chapter, we include the u values at the boundaries in the linear systems.

The more general case, with $n+1$ grid points and Dirichlet boundary conditions, leads to a linear system $A u = b$ with $n+1$ unknowns. The $(n+1) \times (n+1)$ matrix A can be expressed as

$$A = \begin{pmatrix} A_{0,0} & 0 & 0 & \cdots & \cdots & \cdots & \cdots & \cdots & 0 \\ A_{1,0} & A_{1,1} & 0 & \ddots & & & & & \vdots \\ 0 & A_{2,1} & A_{2,2} & A_{2,3} & \ddots & & & & \vdots \\ \vdots & \ddots & & \ddots & \ddots & 0 & & & \vdots \\ \vdots & & \ddots & \ddots & \ddots & \ddots & \ddots & & \vdots \\ \vdots & & & 0 & A_{i,i-1} & A_{i,i} & A_{i,i+1} & \ddots & \vdots \\ \vdots & & & & \ddots & \ddots & \ddots & \ddots & 0 \\ \vdots & & & & & \ddots & \ddots & \ddots & A_{n-1,n} \\ 0 & \cdots & \cdots & \cdots & \cdots & 0 & A_{n,n-1} & A_{n,n} \end{pmatrix}. \quad (7.123)$$

The components $A_{i,j}$ follow from (7.119). Let us explain how we derive the $A_{i,j}$ expression in detail. We start with a general row, say, the row with index i. This row arises from the (7.119), and the matrix entries correspond to the coefficients in front of the unknowns in this equation:

$$A_{i,i-1} = -\alpha, \quad (7.124)$$
$$A_{i,i+1} = -\alpha, \quad (7.125)$$
$$A_{i,i} = 1 + 2\alpha. \quad (7.126)$$

The first and last rows are different since these correspond to the boundary conditions. In case of Dirichlet conditions,

$$u(0, t) = D_0(t), \quad u(1, t) = D_1(t),$$

we get

$$A_{0,0} = 1, \tag{7.127}$$
$$A_{0,1} = 0, \tag{7.128}$$
$$A_{n-1,n} = 0, \tag{7.129}$$
$$A_{n,n} = 1. \tag{7.130}$$

Similarly, we must derive the elements on the right-hand side b:

$$b_0 = D_0(t), \tag{7.131}$$
$$b_n = D_1(t), \tag{7.132}$$
$$b_i = u_i^{\ell-1} + \Delta t \, f_i^{\ell}, \tag{7.133}$$

with $i = 1, \ldots, n - 1$.

To solve the linear system at time level ℓ, we need to fill A and b in a program with the expressions above and then call some Gaussian elimination routine to solve the system. Algorithm 7.4 summarizes the details. By the notation $u_i^{\ell} \leftarrow u$, we indicate that the values of the solution vector of the linear system, u, are to be inserted in the data structure holding the numerical values u_i^{ℓ} at the grid points. Very often this will be the same data structure in a program (the vector holding the solution at the new time level).

Algorithms 7.1 and 7.4 solve the same PDE problem, but apply different numerical methods. There are more steps and more complicated operations (such as filling matrices and solving linear systems) in Algorithm 7.4 compared with Algorithm 7.1. The advantage of Algorithm 7.4 is that there are no restrictions on the choice of Δt, i.e., Algorithm 7.4 is stable for any choice of Δt or α.

7.5.3 Solution of Tridiagonal Linear Systems

An important step in Algorithm 7.4 is the solution of the linear system of equations $Au = b$. When A is an $(n+1) \times (n+1)$ matrix, a standard Gaussian elimination procedure requires on the order of n^3 operations to compute the solution. The explicit Algorithm 7.1 involves just on the order of n operations to compute a new solution at a new time level. That is, the implicit method will be about n^2 computationally heavier than the explicit scheme!

Fortunately, we can reduce the work in the Gaussian elimination procedure to the order of n. The key idea is to observe that most of the elements in A are zeros. In

Algorithm 7.4

Implicit Method for the Diffusion Equation with $u(0, t) = u(1, t) = 0$.

SET INITIAL CONDITION:
$u_i^0 = I(x_i)$, for $i = 0, \ldots, n$
$\alpha = \Delta t / \Delta x^2$
$t_0 = 0$
for $\ell = 1, 1, \ldots, m$
    $t_\ell = t_{\ell-1} + \Delta t$
    FILL MATRIX AND RIGHT-HAND SIDE:
    $A_{i,j} = 0$ *for* $i, j = 0, 1, \ldots, n$
    $A_{0,0} = 1$
    $A_{n,n} = 1$
    $b_0 = D_0(t_\ell)$
    $b_n = D_1(t_\ell)$
    for $i = 1, \ldots, n - 1$
        $A_{i,i-1} = -\alpha$, $A_{i,i+1} = -\alpha$, $A_{i,i} = 1 + 2\alpha$
        $b_i = u_i^{\ell-1} + \Delta t \, f_i^\ell$
    solve the linear system $A u = b$ wrt. u
    $u_i^\ell \leftarrow u$

fact, there are at most three entries in each row that can be different from zero. It is possible to utilize this fact in the Gaussian elimination procedure and construct a much more efficient solution algorithm. This algorithm works with the nonzero elements in A only.

Since all entries in row number i are zero, except for the diagonal entry $A_{i,i}$ and its two "neighbors" $A_{i,i-1}$ and $A_{i,i+1}$, we realize that the nonzero entries in A appear on the diagonal, the subdiagonal and the superdiagonal, see (7.123). We say that the coefficient matrix A has a *tridiagonal* structure and that A is tridiagonal. Obviously, it suffices to store the three diagonals only in the computer's memory. This can save much memory compared with storing the complete matrix with $(n + 1)^2$ entries. But even more important is the fact that we can reduce work in solving $A u = b$ from the order of n^3 to the order of n.

To derive a tailored Gaussian elimination algorithm for tridiagonal matrices, we take a look at (7.123). Imagine that we start the standard Gaussian elimination procedure.[18] We first eliminate $A_{i,0}$ from $i = 1, \ldots, n$. However, we know that $A_{i,0} = 0$ for $i = 2, \ldots, n$, so it suffices to eliminate $A_{1,0}$ only. For a general column number k, we just eliminate $A_{i,i+1}$, and note that the rest of the column is zero, i.e., $A_{i,k} = 0$ for $k = i + 2, \ldots, n$.

After having turned A into an upper triangular form by the Gaussian elimination procedure, we find u from the so-called back substitution. In addition, in this process, we can work with the nonzeroes only of the upper triangular matrix.

---

[18] We assume that you know this procedure, but if not, you can just accept the end result of this section: Algorithm 7.5.

Algorithm 7.5

Solution of a Tridiagonal Linear System.

$p_i = A_{i-1,i}$ *for* $i = 1, \ldots, n$
$q_i = A_{i,i}$ *for* $i = 0, \ldots, n$
$r_i = A_{i,i+1}$ *for* $i = 0, \ldots, n - 1$
ELIMINATION TO UPPER TRIANGULAR MATRIX:
$d_0 = q_0$
$c_0 = b_0$
for $k = 1, \ldots, n$:
    $m_k = p_k/d_{k-1}$
    $d_k = q_k - m_k r_{k-1}$
    $c_k = b_k - m_k c_{k-1}$
BACK SUBSTITUTION:
$u_n = c_n/d_n$
for $k = n - 1, n - 2, \ldots, 0$:
    $u_k = (c_k - r_k u_{k+1})/d_k$

We can now formulate the algorithm for solving a linear system $Au = b$, where A is tridiagonal. The three diagonals of the A matrix are supposed to be stored as three vectors: The diagonal is stored in q_i, $i = 0, \ldots, n$; the subdiagonal is stored in p_i, $i = 1, \ldots, n$; and the superdiagonal is stored in r_i, $i = 0, \ldots, n - 1$. Given the right-hand side b_i, $i = 0, \ldots, n$, Algorithm 7.5 can then be used to compute the solution u_i, $i = 0, \ldots, n$.

Calling some function implementation of Algorithm 7.5 to solve for u_i^ℓ, $i = 0, \ldots, n$ actually implies unnecessary work: Since the coefficient matrix A is constant in time, we can perform the elimination (also known as the factorization) step once and for all and only call the back substitution procedure at each time level. That is, we split Algorithm 7.5 into two parts: the elimination (factorization) part and the back-substitution part. Initially, at $t = 0$, we perform the elimination and store the vectors c_i and d_i. At each time level we pass c_i and d_i together with b_i and r_i to some function implementing the back substitution algorithm. The two steps are incorporated in Algorithm 7.6, which is a refined version of Algorithm 7.4. In the line $c, d = \text{factorize}(p, q, r)$, we mean that factorize is a function that takes the p, q, and r vectors as arguments and produces the c and d vectors as a result. A similar notation is used in the backsubstitution step.

7.5.4 Comparing the Explicit and Implicit Methods

Let us compare the implicit Algorithm 7.6 and the explicit Algorithm 7.1 with respect to computational efforts. The computation of the initial condition and storing new values in u_i^- are identical steps in the two algorithms, where the work is proportional to n. The implicit algorithm needs an additional initialization step in

Algorithm 7.6

Implicit Method for the Diffusion Equation with $u(0, t) = D_0(t)$ and $u(1, t) = D_1(t)$.

SET INITIAL CONDITION:
$u_i^0 = I(x_i)$, for $i = 0, \ldots, n$
$\alpha = \Delta t / \Delta x^2$
$t_0 = 0$
COMPUTE CONSTANT COEFFICIENT MATRIX:
$A_{i,j} = 0$ for $i, j = 0, 1, \ldots, n$
$A_{0,0} = 1$
$A_{n,n} = 1$
for $i = 1, \ldots, n - 1$
$\quad A_{i,i-1} = -\alpha, A_{i,i+1} = -\alpha, A_{i,i} = 1 + 2\alpha$
FACTORIZE MATRIX:
$c, d = \text{factorize}(p, q, r)$
PERFORM TIME SIMULATION:
for $\ell = 1, 1, \ldots, m$
$\quad t_\ell = t_{\ell-1} + \Delta t$
$\quad$ FILL RIGHT-HAND SIDE:
$\quad b_0 = D_0(t_\ell)$
$\quad b_n = D_1(t_\ell)$
$\quad$ for $i = 1, \ldots, n - 1$
$\quad\quad b_i = u_i^{\ell-1} + \Delta t \, f_i^\ell$
$\quad$ SOLVE LINEAR SYSTEM:
$\quad u = \text{backsubstitution}(b, r, c, d)$
$\quad u_i^\ell \leftarrow u$

which we factor the matrix. The amount of work is proportional to n, and since we perform the operations only once, the work is "drowned" in a long time simulation. We therefore do not consider this part. What matters, is what we do in the time loop. The explicit scheme contains $n + 1$ calls to $f(x, t)$, in addition to n multiplications, and $3(n + 1)$ additions/subtractions. We do not bother with the difference between $n + 1$ and n, and say that the work consists of n calls and $3n$ additions/subtractions. The explicit scheme also has n calls to $f(x, t)$, n multiplications and n additions to form the right-hand side, and n subtractions, n multiplications, and n divisions in the back-substitution procedure.

We can conclude that the work in the explicit and implicit methods is about the same. Which algorithm is best depends on whether we can choose a larger time step Δt in the implicit algorithm and achieve the same accuracy as that produced by the maximum time step in the explicit method. It can be shown that the errors in both the explicit and implicit methods are of the form

$$C_1 \Delta x^2 + C_2 \Delta t,$$

where C_1 and C_2 are constants depending on the exact solution $u(x,t)$. It makes sense to keep both error terms at the same level, which means that we should choose Δt to be proportional to Δx^2. Therefore, if we refine the spatial grid (reduce Δx), we should also reduce Δt. In other words, we keep $\alpha = \Delta t / \Delta x^2$ constant in a series of experiments where Δx is reduced. With the implicit method we can choose $\alpha > 1/2$ if we want, but to obtain accuracy, we still have the unfavorable situation where doubling the number of grid points requires reducing the time step by a factor of four. In other words, both methods should apply $\Delta t = \alpha \Delta x^2$, the only difference being that the implicit method allows for any α, while the explicit method will typically apply $\alpha = 0.5$. This implies that the implicit method is 2α times more efficient than the explicit method. We strongly emphasize that these considerations are very rough.

The discussions above are a bit disappointing: We derived a more complicated implicit scheme, but it is not much more efficient than the explicit scheme. It would be nice to have an unconditionally stable implicit scheme that allows larger time steps in a way that makes it more efficient than the explicit scheme. This is the topic of the next section.

7.5.5 The Crank–Nicolson Scheme

It is easy to increase the accuracy of the explicit and implicit schemes by employing more accurate finite differences. In both the explicit and implicit methods we used one-sided differences in time. A central difference can decrease the error in time to be proportional to Δt^2 instead of just Δt. The idea of the following so-called *Crank–Nicolson scheme* is to apply a central derivative in time.

We approximate the PDE at the spatial point x_i and the time point $t_{\ell-1/2}$. A centered difference at this time point is

$$\left[\frac{\partial u}{\partial t}\right]_i^{\ell-1/2} \approx \frac{u_i^\ell - u_i^{\ell-1}}{\Delta t}.$$

We need to approximate the second derivative,

$$\left[\frac{\partial^2 u}{\partial x^2}\right]_i^{\ell-1/2} \approx \frac{u_{i-1}^{\ell-1/2} - 2u_i^{\ell-1/2} + u_{i+1}^{\ell-1/2}}{\Delta x^2},$$

A problem arises here: $u_{i-1}^{\ell-1/2}$, $u_i^{\ell-1/2}$, and $u_{i+1}^{\ell-1/2}$ are not quantities at our integer steps in time. One way out of this problem is to approximate a quantity at time level $\ell - 1/2$ by an arithmetic average of the quantity at time levels ℓ and $\ell - 1$. For example,

$$u_i^{\ell-1/2} \approx \frac{1}{2}\left(u_i^{\ell-1} + u_i^\ell\right).$$

Using such an average for $u_{i-1}^{\ell-1/2}$, $u_i^{\ell-1/2}$, and $u_{i+1}^{\ell-1/2}$ leads to the finite difference approximation

$$\left[\frac{\partial^2 u}{\partial x^2}\right]_i^{\ell-1/2} \approx \frac{1}{2}\frac{u_{i-1}^{\ell-1} - 2u_i^{\ell-1} + u_{i+1}^{\ell-1}}{\Delta x^2} + \frac{1}{2}\frac{u_{i-1}^{\ell} - 2u_i^{\ell} + u_{i+1}^{\ell}}{\Delta x^2}.$$

Setting the first-derivative in time equal to the second-derivative in space plus the source term $f(x_i, t_{\ell-1/2})$ yields the resulting finite difference scheme:

$$\frac{u_i^{\ell} - u_i^{\ell-1}}{\Delta t} = \frac{1}{2}\frac{u_{i-1}^{\ell} - 2u_i^{\ell} + u_{i+1}^{\ell}}{\Delta x^2} + \frac{1}{2}\frac{u_{i-1}^{\ell-1} - 2u_i^{\ell-1} + u_{i+1}^{\ell-1}}{\Delta x^2} + f_i^{\ell-1/2}. \quad (7.134)$$

The quantities at time level $\ell - 1$ are considered known. There are then three coupled unknown quantities at time level ℓ, just as in the implicit backward Euler scheme. We reorder the equation so that the unknowns appear on the left-hand side and the known quantities on the right-hand side:

$$-\frac{\alpha}{2}u_{i-1}^{\ell} + (1 + \alpha) u_i^{\ell} - \frac{\alpha}{2}u_{i+1}^{\ell} = \frac{\alpha}{2}u_{i-1}^{\ell-1} + (1 - \alpha)u_i^{\ell-1} + \frac{\alpha}{2}u_{i+1}^{\ell-1} + \Delta t \, f_i^{\ell-1/2}. \quad (7.135)$$

The new unknown quantities are coupled to each other, as in the previous implicit scheme. We must therefore compute the new unknowns by solving a linear system. The coefficient matrix of this system has the same form as in (7.123), but the matrix entries and right-hand side entries have different formulas:

$$A_{0,0} = 1, \quad (7.136)$$
$$A_{0,1} = 0, \quad (7.137)$$
$$A_{n-1,n} = 0, \quad (7.138)$$
$$A_{n,n} = 1, \quad (7.139)$$
$$A_{i,i-1} = -\frac{\alpha}{2}, \quad (7.140)$$
$$A_{i,i+1} = -\frac{\alpha}{2}, \quad (7.141)$$
$$A_{i,i} = 1 + \alpha, \quad (7.142)$$
$$b_0 = D_0(t_\ell), \quad (7.143)$$
$$b_n = D_1(t_\ell), \quad (7.144)$$
$$b_i = \alpha u_{i-1}^{\ell-1} + (1 - 2\alpha)u_i^{\ell-1} + \alpha u_{i+1}^{\ell-1} + \Delta t \, f_i^{\ell-1/2}. \quad (7.145)$$

In these formulas, the index i runs from 1 to $n - 1$.

Algorithms 7.4 and 7.6 can again be used to solve the problem. Only the formulas for the matrix and right-hand side entries need to be changed.

Like the backward Euler scheme, the Crank–Nicolson scheme just derived is unconditionally stable, i.e., the solution does not exhibit large non-physical oscillations if Δt exceeds a critical value. But is the Crank–Nicolson scheme more efficient than the explicit and implicit Euler schemes? To answer this question we must look at the error in the Crank–Nicolson scheme. It can be shown that the error now is

$$E = C_1 \Delta x^2 + C_2 \Delta t^2,$$

where C_1 and C_2 are constants that depend on the behavior of the exact solution. The temporal error is now of order Δt^2, in contrast to just Δt for the explicit and implicit Euler schemes. This means that the error in the Crank–Nicolson scheme approaches zero fast as we reduce Δt. To keep the two error terms approximately the same size, we should choose Δt proportional to Δx: $\Delta t = \beta \Delta x$ for some constant β.

Consider a computation on a grid with a given Δx, performed with all three schemes. Suppose we then halve Δx. In the implicit and Euler schemes we should divide Δt by four, and in the explicit method this is required because of stability reasons. In the Crank–Nicoloson scheme we only need to divide Δt by two. That is, we can halve the work compared to the other two methods. This makes the Crank–Nicolson scheme more efficient. Again, we must emphasize that these considerations are very rough.

7.5.6 The θ Scheme

In the previous sections we derived three different methods to solve the diffusion PDE numerically: (1) the explicit forward Euler scheme, (2) the implicit backward Euler scheme, and (3) the implicit Crank–Nicolson scheme. It is possible to derive a generalized scheme that simplifies to these three simpler schemes. This generalized scheme is called the θ scheme, because of a parameter θ used to weight contributions from time level $\ell - 1$ versus time level ℓ. The θ scheme can be written as

$$\frac{u_i^\ell - u_i^{\ell-1}}{\Delta t} = \alpha\theta\frac{u_{i-1}^\ell - 2u_i^\ell + u_{i+1}^\ell}{\Delta x^2} + \alpha(1-\theta)\frac{u_{i-1}^{\ell-1} - 2u_i^{\ell-1} + u_{i+1}^{\ell-1}}{\Delta x^2} + f_i^*,$$
(7.146)

where

$$f_i^* = \theta f_i^\ell + (1-\theta) f_i^{\ell-1}.$$

Again, we collect the unknowns on the left-hand side and the known quantities on the right-hand side:

$$-\alpha\theta u_{i-1}^\ell + (1 + 2\alpha\theta)\, u_i^\ell - \alpha\theta u_{i+1}^\ell =$$
$$\alpha(1-\theta)u_{i-1}^{\ell-1} + (1 - 2\alpha(1-\theta))u_i^{\ell-1} + \alpha(1-\theta)u_{i+1}^{\ell-1} +$$
$$\Delta t\left(\theta f_i^\ell + (1-\theta) f_i^{\ell-1}\right).$$
(7.147)

Algorithm 7.7

*The θ Scheme for the Diffusion Equation with $u(0, t) = D_0(t)$
and $u(1, t) = D_1(t)$.*

SET INITIAL CONDITION:
$u_i^0 = I(x_i), \quad for\ i = 0, \ldots, n$
$\alpha = \Delta t / \Delta x^2$
$t_0 = 0$
COMPUTE CONSTANT COEFFICIENT MATRIX:
$A_{i,j} = 0$ for $i, j = 0, 1, \ldots, n$
$A_{0,0} = 1$
$A_{n,n} = 1$
$for\ i = 1, \ldots, n - 1$
$\quad A_{i,i-1} = -\alpha\theta, \ A_{i,i+1} = -\alpha\theta$
$\quad A_{i,i} = 1 + 2\alpha\theta$
FACTORIZE MATRIX:
$c, d = $ factorize(p, q, r)
$for\ \ell = 1, 1, \ldots, m$
$\quad t_\ell = t_{\ell-1} + \Delta t$
$\quad$ FILL RIGHT-HAND SIDE:
$\quad b_0 = D_0(t_\ell)$
$\quad b_n = D_1(t_\ell)$
$\quad for\ i = 1, \ldots, n - 1$
$\quad\quad b_i = \alpha(1 - \theta)u_{i-1}^{\ell-1} + (1 - 2\alpha(1 - \theta))u_i^{\ell-1} + \alpha(1 - \theta)u_{i+1}^{\ell-1} +$
$\quad\quad\quad \Delta t\,(\theta f_i^\ell + (1 - \theta)f_i^{\ell-1})$
$\quad$ SOLVE LINEAR SYSTEM:
$\quad u = $ backsubstitution(b, r, c, d)
$\quad u_i^\ell \leftarrow u$

Looking at this scheme, we realize that if we set $\theta = 0$, we recover the explicit forward Euler scheme.[19] Setting $\theta = 1$, we get the implicit backward Euler scheme, and setting $\theta = 1/2$, we arrive at the Crank–Nicolson scheme. Therefore, in a program it can be wise to implement the θ scheme and then just choose a θ value to recover one of the three specialized schemes.

The value $\theta = 0$ is a special case, since then we have a diagonal matrix system and there is hence no need to invoke a factorization and back-substitution procedure. In programs we might use the θ scheme for all implicit schemes ($\theta \neq 0$) and offer a separate and simpler implementation for the explicit scheme ($\theta = 0$).

Finally, we summarize our numerical methods for the diffusion PDE by listing the most general computational procedure: Algorithm 7.7, expressing the θ scheme.

---

[19] Scheme (7.147) with $\theta = 0$ equals (7.90) if in the latter we exchange ℓ with $\ell - 1$ and $\ell + 1$ with ℓ.

7.6 Exercises

Exercise 7.1. Suppose we have a long tube, modeled as the interval $[0, L]$, filled with water. We inject droplets of ink in the middle of the tube at certain discrete time events $t_i = iT$, $i = 1, 2, \ldots$. Argue why the following source function is a reasonable model for such a type of injection:

$$f(x,t) = \begin{cases} K, & L/2 - \Delta L \le x \le L/2 + \Delta L, \ iT - \Delta t \le t \le iT + \Delta t, \ i = 1, 2, \ldots \\ 0 & \text{otherwise} \end{cases}$$

Here, K is a constant reflecting the mount of mass injected. Hint: Make sketches of (o, better yet, animate) the function $f(x,t)$.

Set up a one-dimensional PDE model with appropriate boundary and initial conditions for this problem. How do you expect the solution to develop in time?   ◇

Exercise 7.2. What kind of temperature problem can be described by the PDE

$$\frac{\partial T}{\partial t} = \frac{\partial^2 T}{\partial x^2},$$

with boundary conditions $k\frac{\partial T}{\partial x} = h_T(T - T_0 - A \sin(\omega t))$ at $x = 0$ and $T(1000, t) = T_0$, and initial condition $T(x, 0) = T_0$?   ◇

Exercise 7.3. What kind of flow can be described by the PDE

$$\frac{\partial v}{\partial t} = \frac{\partial^2 v}{\partial x^2},$$

with boundary conditions $v(0, t) = 0$ and $v(H, t) = A \sin(\omega t)$, and initial condition $v(x, 0) = 0$?   ◇

Exercise 7.4. Suppose we have a diffusion problem on $[0, L]$:

$$\frac{\partial u}{\partial t} = k\frac{\partial^2 u}{\partial x^2} + A \sin^2(\omega t) \exp\left(-(x - x_0)^2\right),$$
$$u(x, t) = c_0, \quad x = 0, L,$$
$$u(x, 0) = c_0.$$

Use c_0 as scale for u and a time scale as in the previous example (just omit ϱ and c_v from the t_c expression). Show that the scaled version of the PDE may be written as (dropping bars)

$$\frac{\partial u}{\partial t} = \frac{\partial^2 u}{\partial x^2} + \delta \sin^2(\omega \frac{L^2}{k} t) \exp\left(-L(x - x_0)^2\right),$$

where

$$\delta = \frac{AL^2}{kc_0}$$

is a dimensionless number.   ◇

Exercise 7.5. Derive an algorithm for solving (7.86) with Robin boundary conditions:

$$-\frac{\partial u}{\partial x} = h_T(u - u_s), \quad x = 0,$$

$$\frac{\partial u}{\partial x} = h_T(u - u_s), \quad x = 1.$$

(Hint: Follow the reasoning used for incorporating Neumann boundary conditions.)   ◇

Exercise 7.6. Show that $u(x,t) = ax + bt + c$, where a, b, and c are constants, is an exact solution of both the diffusion equation *and* the numerical scheme, provided that the boundary conditions and the source term are adjusted to fit this solution. That is, a program should compute $u_i^\ell = ai\,\Delta x + b\ell\,\Delta t + c$ exactly (to machine precision).   ◇

Exercise 7.7. Suppose that we want to test a computer program that implements non-homogeneous Dirichlet conditions of the type

$$u(0,t) = D_0(t), \quad u(1,t) = D_1(t),$$

and non-homogeneous Neumann conditions of the type

$$\frac{\partial}{\partial x}u(0,t) = h_0(t), \quad \frac{\partial}{\partial x}u(1,t) = h_1(t).$$

To this end, we want to compare the numerical solution with an exact solution of the problem. Show that the function

$$u(x,t) = e^{-at}\sin(b(x - c))$$

is a solution of the diffusion equation, where a, b, and c are constants. The nice feature of this solution is that it obeys certain time-dependent non-homogeneous boundary values by choosing certain values of the constants a, b, and c. Use the freedom in choosing a, b, and c to find an exact solution of a diffusion problem with a non-homogeneous Dirichlet condition at $x = 0$ and a non-homogeneous Neumann condition at $x = 1$.   ◇

Exercise 7.8. We consider a scaled diffusion PDE without source terms,

$$\frac{\partial u}{\partial t} = \frac{\partial^2 u}{\partial x^2}.$$

Suppose we manage to find two exact solutions $v(x,t)$ and $w(x,t)$ to this equation. Show that the linear combination

$$\hat{u}(x,t) = av(x,t) + bw(x,t)$$

is also a solution of the equation for arbitrary values of the constants a and b. Set up boundary and initial conditions that $\hat{u}$ fulfills (hint: simply evaluate v and w at the boundary and at $t = 0$).

Show that

$$u(x,t;k) = e^{-\pi^2 k^2 t} \sin(\pi k x) \tag{7.148}$$

is a solution of the diffusion PDE without source terms for any value of k. Choose $v(x,t) = u(x,t;1)$ and $w(x,t) = u(x,t;100)$ and form the linear combination

$$\hat{u}(x,t) = v(x,t) + 0.1w(x,t).$$

Sketch the function $\hat{u}(x,0)$ from an understanding of how a small-amplitude, rapidly oscillating sine function $w(x,t)$ is added to the half-wave sine function $v(x,t)$. Set up the complete initial-boundary value problem for $\hat{u}$, and adjust a program for the diffusion equation such that the program can compute this $\hat{u}$. Choose $\Delta t = \Delta x^2/2$. Find the time $t = T$ when there are no more visible tracks of the rapidly oscillating part of the initial condition. Explain that the solution you observe as graphical output of the program is in accordance with the analytical expression for $\hat{u}$.   ◇

Exercise 7.9. Algorithm 7.1 is easy to implement in a computer program if the computer language has arrays whose first index is 0. All languages with some inheritance from C (e.g., C++, Java, C#, Perl, Python, Ruby) have this feature. Some languages (Matlab, for instance) use index 1 for the first element in an array. You should then rewrite Algorithm 7.1 so that the index in space (usually called i in this book) runs from 1 to $n + 1$ and not from 0 to n. (This rewrite might sound trivial, but a simple change from i to $i + 1$ very often causes errors in programs.)   ◇

Exercise 7.10. Write a program that implements Algorithm 7.1 in your favorite computer language. Note that if the arrays in your language cannot start at 0, you should rewrite the algorithm as suggested in Exercise 7.9. Use test problems from Sect. 7.4.4 to test your program.   ◇

Exercise 7.11. The exercise is similar to Exercise 7.10, but now you should implement Algorithm 7.2.   ◇

Exercise 7.12. Use the program from Exercise 7.11 to solve the diffusion problem

$$\frac{\partial u}{\partial t} = 10\frac{\partial^2 u}{\partial x^2} + x, \quad x \in (0, L),\ t > 0,$$
$$u(0,t) = 2, \quad t > 0,$$
$$u(L,t) = 1, \quad t > 0,$$
$$u(x,0) = 1.$$

Find, in particular, the value $u(L/2, 4)$. (Hint: Scale the problem first.)   ◇

Exercise 7.13. The exercise is similar to Exercise 7.10, but now you should implement Algorithm 7.3.   ◇

Exercise 7.14. Modify Algorithm 7.3 so that there can be a non-homogeneous Neumann condition at $x = 0$: $\partial u/\partial x = N_0(t)$. Implement the algorithm and verify the implementation.   ◇

Exercise 7.15. Suppose we want to solve (7.82) with scheme (7.115). Extend Algorithm 7.1 to this case. Find a test problem for your program using the method of manufactured solutions.   ◇

Exercise 7.16. Extend the algorithm in Exercise 7.15 to handle Neumann conditions $\partial u/\partial x = 0$ at both ends.   ◇

Exercise 7.17. Assume that the problem to be solved by Algorithm 7.6 has Neumann conditions $\partial u/\partial x = 0$ at the end $x = 1$. Modify the algorithm accordingly.   ◇

Exercise 7.18. Use your favorite programming language and implement Algorithm 7.6. (If arrays in your language cannot start at index 0, you need to rewrite the algorithm first, so that i runs from 1 to $n + 1$ instead of from 0 to n.) Verify your implementation.   ◇

Exercise 7.19. Assume that the problem to be solved by Algorithm 7.6 has a variable diffusion coefficient as explained in Sect. 7.4.7. Modify the algorithm to handle this extension of the problem.   ◇

Exercise 7.20. Write a program, in a programming language of your choice, that implements Algorithm 7.7. Set up at least two test problems to test that the program works correctly.   ◇

Exercise 7.21. Use the program from Exercise 7.20 to solve the problem (7.102)–(7.105). Apply two grid spacings, $\Delta x = 0.1$ and $\Delta x = 0.01$. First, run the explicit method ($\theta = 0$) with the maximum Δt for these two cases. Then run the implicit backward Euler scheme ($\theta = 1$) with $\Delta t = 0.5\Delta x^2$ and $\Delta t = 10\Delta x^2$. Repeat the latter tests with the Crank–Nicolson scheme ($\theta = 0.5$). Create animations of the solutions in each case and write a report about how the various methods solve this quite challenging numerical problem.   ◇

Exercise 7.22. Modify Algorithm 7.7 such that one can choose between non-homogeneous Dirichlet or Neumann conditions at the boundary points.   ◇

Exercise 7.23. Write extensions in the algorithm from Exercise 7.22 such that a variable diffusion coefficient is allowed.   ◇

Exercise 7.24. Make a program implementing the algorithm from Exercise 7.23. Set up a series of test cases for checking that one can handle Dirichlet and Neumann boundary conditions as well as a variable diffusion coefficient (use the method of manufactured solutions to test the latter feature).   ◇

Exercise 7.25. Suppose we want to solve the diffusion equation in the limit where $\partial u/\partial t \to 0$. This gives rise to the Poisson equation:

$$-\frac{d^2u}{dx^2} = f \text{ in } [0, 1].$$

There is no initial condition associated with this equation, because there is no evolution in time – the unknown function u is just a function of space: $u = u(x)$. However, the equation is associated with boundary conditions, say, $u(0) = u(1) = 0$.

(a) Replace the second-derivative by a finite difference approximation. Explain that this gives rise to a (tridiagonal) linear system, exactly as for the implicit backward Euler scheme.

(b) Compare the equations from (a) with the equations generated by the backward Euler scheme. Show that the former arises in the limit as $\Delta t \to \infty$ in the latter.

(c) Construct an analytical solution of the Poisson equation when f is constant.

(d) The result from (b) tells us that we can take one very long time step in a program implementing the backward Euler scheme and then arrive at the solution of the Poisson equation. Demonstrate, by using a program, that this is the case for the test problem from (c).

⬦

7.7 Projects

7.7.1 Diffusion of a Jump

We look at the physical problem of bringing two pieces of metal together, as depicted in Fig. 7.4 on page 282, where the initial temperatures of the pieces are different. Heat conduction will smooth out the temperature difference, as shown in Fig. 7.5 on page 282.

(a) Formulate a one-dimensional initial-boundary value problem for this heat conduction application. The initial condition must reflect the temperature jump, and the boundaries are insulated.

(b) Scale the problem. Hint: Use the same scales as we did for (7.52)–(7.55).
    Many find modeling and scaling difficult, so the resulting scaled problem is listed here for convenience:

$$\frac{\partial u}{\partial t} = \frac{\partial^2 u}{\partial x^2},$$

$$\frac{\partial u}{\partial x} = 0, \quad x = 0, 1,$$

$$u(x, 0) = \begin{cases} 0, \ x \leq 0.5, \\ 1, \ x > 0.5. \end{cases}$$

(c) Write a computer program that solves the scaled problem by the explicit forward Euler scheme.

(d) Show that

$$u(x,t) = e^{-\pi^2 t} \cos(\pi x)$$

is a solution of the scaled problem if the initial condition is set to $u(x,0) = \cos(\pi x)$. Implement this initial condition, run the code, and plot the maximum error as a function of time. Argue why you think the code is free of bugs.

(e) Turn to the step function as the initial condition in the code. State clearly the expected qualitative behavior of the evolution of the temperature, and comment plots to provide confidence in the computed results.

(f) Find from computer experiments the time T it takes to reach a stationary (constant) solution.

(g) You are now asked to plot the physical time T (with dimension) it takes to achieve a constant temperature in the two metal pieces as a function of the heat conduction coefficient and the total length of the pieces. You do not need to produce the plot, but you need to explain how you will solve the problem. (Hint: There is no need for simulations beyond the single run from (f) above! And for those who are too accurate with the mathematics and give up because T is infinite, remember that scientific computing is about approximations.)

7.7.2 Periodical Injection of Pollution

This project concerns the periodic injection of mass in a diffusive transport problem. One important application can be a factory that launches pollution into the air in a periodic manner. The goal of the study is to see how the pollution spreads in time and how areas at some distance from the source are affected.

We assume that the pollution spreads equally in all three space dimensions. Since the source of pollution, i.e. thepipe at the factory, is localized in a very small area, it is reasonable to presume that the concentration depends on time and the distance from the source. This means that the problem exhibits spherical symmetry, with the origin at the source. In spherical coordinates the problem has only two independent variables: time t and the radial distance r from the source.

The diffusion PDE expressed in spherical coordinates reads

$$\frac{\partial}{\partial t}c(r,t) = k\frac{1}{r^2}\left(r^2\frac{\partial}{\partial r}c(r,t)\right) + f(r,t). \tag{7.149}$$

(a) The source term $f(r,t)$ is used to model how pollution is injected into the air. We assume that the factory is operative for 8 h/day. During operation, $f = K$, where K is a constant reflecting the amount of pollution that enters the air per unit time. The source f is also localized; during operation $f = K$ in a small area $r \leq r_0$, where r_0 is typically the radius of the factory pipe. When the factory is not in operation, no pollution enters the air ($f = 0$).

Find a mathematical specification of the function $f(r, t)$. Implement f as a function in a program. (Hint: Exercise 7.1 provides some ideas.)

(b) At distances far from the source ($r = 0$) we assume that there is no pollution. Mathematically, we can express this as

$$\lim_{r \to \infty} c(r, t) = 0.$$

Infinite domains are inconvenient in computer models, so we implement the condition at a distance $r = L$, where L is sufficiently large. Initially, we assume that there is no pollution in the air. Set up a complete initial-boundary value problem for this physical application.

(c) Construct an explicit finite difference scheme for (7.149). Hint: Use the ideas from Sect. 7.4.7.

(d) Set up a complete algorithm for computing $c(r, t)$.

(e) Implement the algorithm from (d) above in a computer program.

(f) Suggest a series of test problems with the purpose of checking that the program implementation is correct.

(g) Choosing values for the constants K, k, L, r_0, Δt, and the total simulation time can be challenging. Scaling will help to make the choices of input data significantly simpler. Scale the problem, using $t_c = r_0^2/k$ as the time scale, r_0 as the space scale, and K as the scale for f. Show that the scaled PDE takes the form (dropping the bars in scaled quantities)

$$\frac{\partial c}{\partial t} = \frac{1}{r^2} \frac{\partial}{\partial r} \left(r^2 \frac{\partial c}{\partial r} \right) + \alpha f(r, t),$$

where α is a dimensionless constant that equals unity if we choose the scale for the concentration as $t_c K$. Set up the complete scaled problem, and explain how you can use the program developed for the unscaled problem by setting the input parameters K, k, L, and so on, to suitable values.

(h) Perform numerical experiments to determine an appropriate location of the "inifinite" boundary (L/r_0). The principle is that the solution should be approximately unaffected by moving the boundary further toward infinity.

(i) Create an animation of the scaled concentration. The animation should contain at least ten periods (days) of simulation.

(j) Improve the animation by plotting both the f and c functions such that one can see the effect of the pollution injection on c.

(k) Discuss the validity of this statement: "The movie created in (j) is the complete solution of the problem".

7.7.3 Analyzing Discrete Solutions

We consider the problem solved by Algorithm 7.1. One can show[20] that the following discrete function is an exact analytical solution of the computations in Algorithm 7.1:

$$u_i^\ell = A \left(1 - \frac{4\Delta t}{\Delta x^2} \sin^2 \frac{\pi k \Delta x}{2} \right)^\ell \sin(\pi k(i - 1)\Delta x) . \tag{7.150}$$

The parameters A and k are constants that can be freely chosen.

(a) Insert the expression (7.150) in the finite difference scheme in Algorithm 7.1 and show that it fulfills this discrete equation. Check also that the boundary conditions $u_1 = u_n = 0$ are fulfilled. What is the corresponding initial condition?

The present diffusion problem, where we have a mathematical expression for the numerical solution, is ideal for testing the correctness of a program implementing Algorithm 7.1: We write out u_i^ℓ and check that it coincides with the solution (7.150) to machine precision, regardless of what Δx is (but recall that $\alpha \le 0.5$).

(b) We have exact mathematical solutions of the PDE and the finite difference scheme given by the expressions (7.148) and (7.150), respectively. Comparing these two solutions, we see that both are of the form

$$u(x_i, t_\ell) = A\xi^\ell \sin(\pi k(i - 1)\Delta x),$$

with

$$\xi = \xi_{PDE} \equiv e^{-\pi^2 k^2 \Delta t}$$

in the PDE case and

$$\xi = \xi_\Delta \equiv 1 - \frac{4\Delta t}{\Delta x^2} \sin^2 \frac{\pi k \Delta x}{2}$$

in the discrete case. The error in the numerical solution is therefore contained in the difference $e = \xi_{PDE} - \xi_\Delta$:

$$e(k, \Delta x, \Delta t) = e^{-\pi^2 k^2 \Delta t} - 1 + \frac{4\Delta t}{\Delta x^2} \sin^2 \frac{\pi k \Delta x}{2} .$$

Find Taylor series expansions of the functions e^y and $\sin^2 y$, and use these power expansions in the expressions for e. Explain that the largest (so-called

---

[20] The derivation is, in principle, quite simple. Both the PDE and the finite difference scheme allow solutions of the form $u(x, t) = A\xi^\ell \sin(kx)$, where ℓ counts time levels, and A, ξ, and k are parameters to be adjusted by inserting the expression in the PDE or the finite difference scheme.

leading) terms in the power expansion of e are proportional to Δt and Δx^2. We often write this as

$$e = \mathcal{O}(\Delta t) + \mathcal{O}(\Delta x^2)$$

and say that the finite difference scheme is of first order in time (Δt^1) and second order in space (Δx^2).

(c) From a specific solution (7.148) of a diffusion PDE, we realize that the values of the solution decrease in time. Show that the exact solution (7.150) of the corresponding numerical problem does not necessarily decrease in time; the solution can increase if

$$\Delta t > \frac{1}{2}\Delta x^2 .$$

Such an increase will be amplified with time, and we cannot allow this if the numerical solution will have the same qualitative features as the solution of the underlying PDE. We therefore say that the numerical solution is unstable if $\Delta t > \frac{1}{2}\Delta x^2$. A different way of achieving the same stability or instability condition is given in the next chapter.

7.7.4 Compare Different Scalings

You are advised to work through Project 7.7.1 before attacking the present project. We consider the same physical problem, but with different boundary conditions. Let the boundary conditions be changed to a fixed temperature of U_a at the left boundary and known heat flow of Q_b at the right boundary. The scaling of this initial-boundary value problem was discussed in Sect. 7.3.5. Two scales for the unknown function were suggested, (7.73), referred to here as scale 1, and (7.79), referred to as scale 2.

(a) Implement the scaled problem in the computer code from Project 7.7.1. Introduce an input parameter for switching between scale 1 and 2. Run problems with $\beta = 1$, $\beta \gg 1$ with scale 1, and $\beta \ll 1$ with scale 2. As commented on in Sect. 7.3.5, the latter two simulations demonstrate that the scaling fails, in the sense that the magnitude of the solution is not about unity. The purpose of this project is to analyze this observation a bit more by looking into analytical expressions.

(b) As time increases, we expect the solution to stabilize and change little with time, since there are no time-dependent boundary conditions or source terms in our present model. Mathematically this means that we expect

$$\lim_{t \to \infty} \frac{\partial u}{\partial t} = 0 .$$

In this limit, we have a *stationary solution* $u_s(x)$ of the problem. The stationary solution does not depend on the initial condition, i.e., the initial condition is forgotten as $t \rightarrow \infty$. Justify this assertion by plotting the stationary solution corresponding to the step initial condition and the initial condition $u = U_a$.

(c) Formulate a PDE and boundary conditions to be fulfilled by u_s. Write the scaled version of the problem in b), using the scales 1 and 2.

(d) Solve analytically for the scaled stationary solution (integrate twice and apply the boundary conditions to determine the integration constants). Show that scale 1 yields temperatures that are much greater than unity if $\beta \gg 1$.

Chapter 8
Analysis of the Diffusion Equation

In Chap. 7 we studied several aspects of the theory of diffusion processes. We saw how these equations arise in models of several physical phenomena and how they can be approximately solved by suitable numerical methods. The analysis of diffusion equations is a classic subject of applied mathematics and of scientific computing. Its impact on the field of partial differential equations (PDEs) has been very important, both from a theoretical and practical point of view. The purpose of this chapter is to dive somewhat deeper into this field and thereby increase our understanding of this important topic.

Our aim is to study several mathematical properties of the diffusion problem

$$u_t = u_{xx} \quad \text{for } x \in (0, 1), \, t > 0, \tag{8.1}$$

$$u(0, t) = u(1, t) = 0 \quad \text{for } t > 0, \tag{8.2}$$

$$u(x, 0) = f(x) \quad \text{for } x \in (0, 1), \tag{8.3}$$

where f is a given initial condition defined on the unit interval $(0, 1)$. Here we have introduced a different notation for the derivative than what was used in Chap. 7: u_x is short-hand for $\partial u/\partial x$, u_{xx} is short-hand for $\partial^2 u/\partial x^2$, and u_t is short-hand for $\partial u/\partial t$. The reason for introducing this notation is twofold: we save some writing space when writing the derivatives, and it is a widely used notation in the mathematical analysis of PDEs. Note that, if not stated otherwise, we will consider a problem with homogeneous Dirichlet boundary conditions, cf. (8.2).

The three main subjects of this chapter are

- To investigate whether or not several physical properties of diffusion are satisfied by the mathematical model (8.1)–(8.3)
- To derive a procedure, commonly referred to as "separation of variables", for computing the analytical solution of the diffusion equation
- To analyze the stability properties of an explicit numerical method, scheme (7.91) in Chap. 7, for solving (8.1)–(8.3)

Prior to reading this chapter the reader should study Chapter 7 in detail.

A. Tveito et al., *Elements of Scientific Computing*, Texts in Computational Science and Engineering 7, DOI 10.1007/978-3-642-11299-7_8,
© Springer-Verlag Berlin Heidelberg 2010

8.1 Properties of the Solution

The problem (8.1)–(8.3) is not trivial. We are seeking a function u of the space variable x and time t, i.e., $u = u(x, t)$, such that the first order derivative of u with respect to t equals the second order derivative of u with respect to x. In addition, this function must satisfy the homogeneous boundary condition (8.2) and the initial condition (8.3). It is by no means obvious[1] that such a function exists. Furthermore, if a solution exists, is it unique?

This situation is more the rule than the exception in science. We are faced with a challenging problem and we do not know if we can solve it. Questions such as existence and uniqueness of a solution and the solvability[2] of the problem are very often open. Our ultimate goal is of course to compute the solution, provided that it exists, of the problem. However, instead of aiming at this goal straight away we will follow the methods of science: Start by deriving simple properties of the problem, slowly and stepwise increase our knowledge and finally, if in the position to do so, determine the problem's solution.

We will now derive a series of properties that a solution of (8.1)–(8.3) must satisfy. These properties will do the following:

- Increase our knowledge in a stepwise manner.
- Help us determine whether or not this is a good model for the physical process that it is supposed to describe.
- Give us a set of properties that the numerical approximations of u should satisfy. This is useful information for designing appropriate numerical schemes for the diffusion equation and in the debugging process of a computer code developed to solve it.
- Provide us with a set of tools suitable for analyzing the diffusion equation. This methodology will hopefully be applicable for understanding a broader class of PDEs, that is, a class of problems related to (8.1)–(8.3) and which may include equations for which it is impossible to derive an explicit solution formula.

Throughout this section we will assume that u is a smooth[3] solution of (8.1)–(8.3).

8.1.1 Energy Arguments

Consider the diffusion equation (8.1), the boundary condition (8.2) and the initial condition (8.3). Recall that this is a prototype of a model for, e.g., the heat evolution

---

[1] Except for well-trained mathematicians.

[2] The question of whether or not we are capable of computing the solution (or at least an approximation of it).

[3] The partial derivatives of all orders of u with respect to x and t are assumed to be continuous.

in a medium occupying the closed unit interval $[0, 1]$ along the x-axis. Our goal is to study the behavior of the heat distribution u as t increases.

It turns out that by applying basic techniques from mathematical analysis we can derive an interesting property for a function $E_1(t)$ defined by

$$E_1(t) = \int_0^1 u^2(x, t)\, dx \quad \text{for } t \geq 0. \tag{8.4}$$

Note that $E_1(t)$ is a function of time only! At the present stage it might be difficult to understand the reason for studying this function. However, we will see below that it turns out to be very useful.

If we multiply the left- and right-hand sides of the diffusion equation (8.1) by u it follows that

$$u_t u = u_{xx} u \quad \text{for } x \in (0, 1),\ t > 0.$$

By the chain rule for differentiation, we observe that

$$\frac{\partial}{\partial t} u^2 = 2u u_t,$$

and hence

$$\frac{1}{2} \frac{\partial}{\partial t} u^2 = u_{xx} u \quad \text{for } x \in (0, 1),\ t > 0.$$

Next, if we integrate this equation with respect to x and apply the rule of integration by parts,[4] it follows that u must satisfy

$$\frac{1}{2} \int_0^1 \frac{\partial}{\partial t} u^2(x, t)\, dx = \int_0^1 u_{xx}(x, t) u(x, t)\, dx \tag{8.5}$$

$$= u_x(1, t) u(1, t) - u_x(0, t) u(0, t) - \int_0^1 u_x(x, t) u_x(x, t)\, dx$$

$$= - \int_0^1 u_x^2(x, t)\, dx \quad \text{for } t > 0,$$

where the last equality is a consequence of the boundary condition (8.2).

Recall that we assumed that u is a smooth solution of the diffusion equation. This implies that we can interchange the order of integration and derivation in (8.5), that is,

$$\frac{\partial}{\partial t} \int_0^1 u^2(x, t)\, dx = -2 \int_0^1 u_x^2(x, t)\, dx \quad \text{for } t > 0, \tag{8.6}$$

---

[4] Recall that

$$\int_a^b u'(x) v(x)\, dx = [u(x)v(x)]_a^b - \int_a^b u(x) v'(x)\, dx.$$

see, e.g., [28] for further details. Equation (8.6) shows that

$$E_1'(t) = -2 \int_0^1 u_x^2(x,t)\,dx \quad \text{for } t > 0,$$

which in turns implies that

$$E_1'(t) \le 0.$$

Thus E is a non-increasing function of time t, i.e.,

$$E_1(t_2) \le E_1(t_1) \quad \text{for all } t_2 \ge t_1 \ge 0,$$

and in particular

$$\int_0^1 u^2(x,t)\,dx \le \int_0^1 u^2(x,0)\,dx = \int_0^1 f^2(x)\,dx \quad \text{for } t > 0.$$
$$(8.7)$$

This inequality shows that the integral of the square of the solution of the diffusion equation is bounded by the integral of the square of the initial temperature distribution f. From a physical point of view, this property seems to be reasonable. There are no source terms[5] present in our model problem (8.1) and the temperature at $x = 0$ and $x = 1$ is kept at zero for all time $t > 0$, see (8.2). Therefore, it seems reasonable that the temperature $u(x,t)$ at any point x will approach zero as time increases. Thus, (8.7) is in agreement with our intuition.[6]

8.1.2 A Bound on the Derivative

We will now consider one more "energy argument". Recall that $u(x,t)$ can represent the temperature at position x at time t. This means that u_x represents the "speed" at which heat flows along the x-axis. From a physical point of view, it seems to be reasonable that this speed can somehow be bounded by the properties of the initial condition f, see (8.3). We will now show by mathematical methods that this is indeed the case. To this end, we define a function $E_2(t)$ by

---

[5] That is, the function f in (7.1) from Chap. 7 is identical to zero. This function is frequently referred to as a source term.

[6] The alert reader may wonder why this section was titled "Energy Arguments"? There are no references to energy, since it is used in the theory for heat transfer problems in physics, above. In fact, there is no strong relation between $E_1(t)$ and the energy present in the medium for which (8.1)–(8.3) is modeling the heat evolution. We will not dwell upon this issue. However, it should be mentioned that in some models given by PDEs, functions similar to E_1 do represent the energy present in the underlying physical process. This has led to broader use of the term in mathematics; it is commonly applied to investigations of functions similar to and of the form of E_1.

$$E_2(t) = \int_0^1 u_x^2(x,t)\, dx.$$

Our analysis of E_2 is similar to that of E_1. However, instead of multiplying the diffusion equation (8.1) by u, as we did in the analysis leading to inequality (8.7), we will now multiply it by u_t and integrate with respect to x. Consequently, u must satisfy the equation

$$\int_0^1 u_t^2(x,t)\, dx = \int_0^1 u_{xx}(x,t) u_t(x,t)\, dx.$$

Integration by parts leads to

$$\int_0^1 u_t^2(x,t)\, dx = [u_x(x,t) u_t(x,t)]_0^1 - \int_0^1 u_x(x,t) u_{tx}(x,t)\, dx$$

$$= u_x(0,t) u_t(0,t) - u_x(1,t) u_t(1,t) - \int_0^1 u_x(x,t) u_{xt}(x,t)\, dx,$$

where we have used the basic property that we can change the order of differentiation[7] of u with respect to x and t. The chain rule for differentiation implies that

$$\frac{\partial}{\partial t} u_x^2 = 2 u_x u_{xt},$$

and hence

$$\int_0^1 u_t^2(x,t)\, dx = u_x(0,t) u_t(0,t) - u_x(1,t) u_t(1,t) - \int_0^1 u_x(x,t) u_{xt}(x,t)\, dx$$

$$= u_x(0,t) u_t(0,t) - u_x(1,t) u_t(1,t) - \frac{1}{2} \int_0^1 \frac{\partial}{\partial t} u_x^2(x,t)\, dx.$$

As above, we can interchange the order of integration and differentiation and thereby conclude that

$$\frac{1}{2} \frac{\partial}{\partial t} \int_0^1 u_x^2(x,t)\, dx = - \int_0^1 u_t^2(x,t)\, dx - u_x(0,t) u_t(0,t) + u_x(1,t) u_t(1,t).$$

(8.8)

Note that (8.8) contains the derivatives $u_t(0,t)$ and $u_t(1,t)$ of u with respect to time t at the boundary of the solution domain $(0,1)$. What do we know about these quantities? They are not present in our model problem (8.1)–(8.3). However, according to the boundary condition (8.2), $u(0,t)$ and $u(1,t)$ are equal to zero for all $t > 0$. Thus, $u(0,t)$ and $u(1,t)$ are constant with respect to time and we therefore conclude that

---

[7] We assumed that u is smooth.

$$u_t(0, t) = u_t(1, t) = 0 \quad \text{for } t > 0. \tag{8.9}$$

Equations (8.8) and (8.9) imply that

$$E_2'(t) = \frac{\partial}{\partial t} \int_0^1 u_x^2(x, t) \, dx = -2 \int_0^1 u_t^2(x, t) \, dx \leq 0,$$

and thus E_2 is a non-increasing function with respect to time t,

$$E_2(t_2) \leq E_2(t_1) \quad \text{for all } t_2 \geq t_1 \geq 0. \tag{8.10}$$

$$\boxed{\int_0^1 u_x^2(x, t) \, dx \leq \int_0^1 u_x^2(x, 0) \, dx = \int_0^1 f_x^2(x) \, dx \quad \text{for } t > 0.}$$
$$\tag{8.11}$$

Consider an object with a non-uniform temperature distribution at time $t = 0$. Assume that the temperature at the boundary of this object is kept constant with respect to both time and space.[8] We have all experienced how the temperature distribution in such cases will tend toward a uniform distribution; for sufficiently large t, a constant temperature throughout the medium is reached. Furthermore, the speed at which the temperature differences are evened out in the object, will decay with time. Inequality (8.10) shows, in a somewhat modified form, that this latter property is fulfilled by our model problem.

8.1.3 Stability

We will now investigate some interesting consequences of the bound (8.7) for the solution u of our model problem (8.1)–(8.3). Recall that f represents the temperature distribution in the medium at time $t = 0$. In real-world simulations, such initial states of the system under consideration will in many cases be based on physical measurements. These measurements will always contain errors. It is impossible to measure the temperature at every point in a medium with 100% accuracy. Consequently, it becomes important to investigate whether or not small changes in f will introduce major changes in the solution u of the problem. In mathematics this is referred to as a question of stability: Is the solution u of (8.1)–(8.3) stable with respect to perturbations in the initial condition f?

---

[8] The temperature is constant with respect to the spatial position along the boundary.

Let us consider a problem with a modified initial condition,

$$v_t = v_{xx} \quad \text{for } x \in (0, 1), \ t > 0, \tag{8.12}$$
$$v(0, t) = v(1, t) = 0 \quad \text{for } t > 0, \tag{8.13}$$
$$v(x, 0) = g(x) \quad \text{for } x \in (0, 1). \tag{8.14}$$

If g is close to f, will v be approximately equal to u? To answer this question we will study the difference e between u and v, that is, we will analyze the function

$$e(x, t) = u(x, t) - v(x, t) \quad \text{for } x \in [0, 1], \ t \geq 0.$$

From (8.1)–(8.3) and (8.12)–(8.14), we find that

$$e_t = (u - v)_t = u_t - v_t = u_{xx} - v_{xx} = (u - v)_{xx} = e_{xx},$$

and furthermore

$$e(0, t) = u(0, t) - v(0, t) = 0 - 0 = 0 \quad \text{for } t > 0,$$
$$e(1, t) = u(1, t) - v(1, t) = 0 - 0 = 0 \quad \text{for } t > 0,$$
$$e(x, 0) = u(x, 0) - v(x, 0) = f(x) - g(x) \quad \text{for } x \in (0, 1).$$

So, interestingly, e solves the diffusion equation with homogeneous Dirichlet boundary conditions at $x = 0, 1$ and with initial condition

$$h = f - g,$$

that is, e satisfies

$$e_t = e_{xx} \quad \text{for } x \in (0, 1), \ t > 0,$$
$$e(0, t) = e(1, t) = 0 \quad \text{for } t > 0,$$
$$e(x, 0) = f(x) - g(x) \quad \text{for } x \in (0, 1).$$

In particular, this means that e must satisfy the properties derived above for the model problem (8.1)–(8.3). Hence, by inequality (8.7), we conclude that

$$\int_0^1 e^2(x, t) \, dx \leq \int_0^1 h^2(x) \, dx \quad \text{for } t > 0,$$

or

$$\int_0^1 (u(x, t) - v(x, t))^2 \, dx \leq \int_0^1 (f(x) - g(x))^2 \, dx \quad \text{for } t > 0. \tag{8.15}$$

This means that if g is close to f then the integral of $(u - v)^2$, at any time $t > 0$, must be small. This, in turn, implies that the region where v differs significantly from u must be small (see Exercise 8.13). Hence, we conclude that minor changes in the initial condition of (8.1)–(8.3) will not alter its solution significantly. The problem is stable!

8.1.4 Uniqueness

We will now prove that (8.1)–(8.3) can have at most one smooth solution. Assume that both u and v are smooth solutions of this problem; that is, u solves (8.1)–(8.3) and v solves (8.12)–(8.14), with $g(x) = f(x)$ for all x in $(0, 1)$. Then (8.15) implies that

$$\int_0^1 (u(x, t) - v(x, t))^2 \, dx = 0 \quad \text{for } t > 0.$$

Since u and v are both smooth, it follows that the function $(u - v)^2$ is continuous. Furthermore,

$$(u(x, t) - v(x, t))^2 \geq 0 \quad \text{for all } x \in [0, 1] \text{ and } t \geq 0,$$

and we can therefore conclude that

$$u(x, t) = v(x, t) \quad \text{for all } x \in [0, 1] \text{ and } t \geq 0,$$

see Exercise 8.12.

> The problem (8.1)–(8.3) can have at most one smooth solution.

8.1.5 Maximum Principles

Consider a steel rod positioned along the unit interval $[0, 1]$. Suppose that the temperature of this rod at any position $x \in (0, 1)$ is given by the function $f(x)$ at time $t = 0$. That is, f represents the initial temperature distribution in the rod. Furthermore, assume that the temperature at the endpoints $x = 0$ and $x = 1$ of the rod at time $t > 0$ is determined by the functions $g_1 = g_1(t)$ and $g_2 = g_2(t)$, respectively, and that the rod is isolated elsewhere.[9] Under these circumstances the temperature $u(x, t)$ is, as we have seen in Chap. 7, governed by the following set of equations:

---

[9] Heat can only leave and enter the rod through its endpoints.

$$u_t = u_{xx} \quad \text{for } x \in (0, 1),\, t > 0, \tag{8.16}$$
$$u(0, t) = g_1(t) \text{ and } u(1, t) = g_2(t) \quad \text{for } t > 0, \tag{8.17}$$
$$u(x, 0) = f(x) \quad \text{for } x \in (0, 1), \tag{8.18}$$

for an appropriate choice of scales.

8.1.6 Physical Considerations

Let us first, for the sake of simplicity, consider the case of $g_1(t) = g_2(t) = 0$ for all $t > 0$, i.e., the temperature at the end points of the rod is kept at zero. In this case, we know from physical experience that the temperature at every point of the rod will approach zero as time increases. Furthermore, heat will not accumulate at any point. On the contrary, this is a diffusion process and the heat will thus be smoothed out in time. Consequently, it seems reasonable that the temperature at any point x and time t cannot exceed the maximum value of the initial temperature distribution f and the zero temperature kept at the endpoints of the rod. In mathematical symbols this reads

$$u(x, t) \le \max(\max_x f(x), 0) \quad \text{for all } x \in [0, 1],\, t > 0.$$

In a similar manner we can argue that the temperature throughout the rod at any time cannot be less than the minimum value of the initial temperature distribution f and the temperature at the end points of the rod, i.e.,

$$u(x, t) \ge \min(\min_x f(x), 0) \quad \text{for all } x \in [0, 1],\, t > 0.$$

We can carry this discussion one step further by allowing non-zero temperatures on the boundary. In such more general cases we would also, due to the physical interpretation of heat conduction, expect the highest and lowest temperatures to appear either initially or at the boundary of the rod. For example, assume that $f(x) = 0$ for $x \in (0, 1)$ and that $g_1(t) = g_2(t) = c$, where c is a positive constant, for all $t > 0$. Then the temperature throughout the rod will increase with time from zero to c, and the temperature cannot, at any time or position, be higher than c or lower than zero!

8.1.7 Analytical Considerations

Above we argued from a "physical" point of view that the maximum and minimum temperature of the rod is obtained either at the boundaries or initially. What about the mathematical model (8.16)–(8.18)? Will a solution of this problem satisfy this property? The purpose of the present section is to investigate whether or not this is the case.[10]

In this section we will need some of the basic properties of the maxima of functions defined on closed sets. These results are, for the sake of easy reference and completeness, recaptured without any derivations or proofs. The reader unfamiliar with these concepts and properties should consult a suitable introductory mathematics text.

Let us first consider functions of a single variable x. Suppose $q(x)$ is a smooth function defined on the closed unit interval $[0, 1]$. Assume that q achieves its maximum value at an interior point[11] x^*, i.e.,

$$x^* \in (0, 1) \quad \text{and} \quad q(x) \leq q(x^*) \text{ for all } x \in [0, 1],$$

then q must satisfy

$$q'(x^*) = 0, \tag{8.19}$$
$$q''(x^*) \leq 0, \tag{8.20}$$

see any introductory text on calculus. Consequently, if a smooth function $h(x)$, defined on $[0, 1]$, is such that

$$h'(x) \neq 0 \quad \text{for all } x \in (0, 1)$$

or

$$h''(x) > 0 \quad \text{for all } x \in (0, 1),$$

then h must achieve[12] its maximum value at one of the endpoints of the unit interval. Hence, we conclude that

$$h(0) \geq h(x) \quad \text{for all } x \in [0, 1] \tag{8.21}$$

or

$$h(1) \geq h(x) \quad \text{for all } x \in [0, 1], \tag{8.22}$$

must hold. This means that

$$h(x) \leq \max(h(0), h(1)) \quad \text{for all } x \in [0, 1].$$

---

[10] From a modeling point of view this is an important question. High-quality models must, of course, fulfill the basic properties of the underlying physical process!

[11] Any point p in the open interval $(0, 1)$ is referred to as an interior point of the closed interval $[0, 1]$.

[12] Recall that a smooth function defined on a closed interval is bounded. Furthermore, such a function will always achieve both its maximum and minimum value within this interval, see e.g., Apostol [4].

Furthermore, if (8.21) holds, then h must satisfy

$$h'(0) \leq 0, \tag{8.23}$$

and if (8.22) is the case, then it follows that

$$h'(1) \geq 0. \tag{8.24}$$

Further information on this important topic can be found in your favorite calculus textbook. It is important that you get this right. In order to illuminate the properties (8.19), (8.20), (8.23), and (8.24), we recommend the reader make some plots of smooth functions defined on the closed unit interval!

Recall that the unknown function u in (8.16)–(8.18) is a function of two variables. Are similar properties as those stated above valid in this case? Yes, indeed: The same results hold with respect to each of the variables!

Consider a smooth function v of the spatial position $x \in [0, 1]$ and time $t \in [0, T]$, i.e., $v = v(x, t)$. More precisely, we assume that the partial derivatives of all orders of v with respect x and t are continuous, and that

$$v : \overline{\Omega}_T \to \mathbb{R},$$

where

$$\Omega_T = \{(x, t) \,|\, 0 < x < 1 \text{ and } 0 < t < T\}, \tag{8.25}$$
$$\partial\Omega_T = \{(x, 0) \,|\, 0 \leq x \leq 1\} \cup \{(1, t) \,|\, 0 \leq t \leq T\} \cup \{(x, T) \,|\, 0 \leq x \leq 1\}$$
$$\cup \{(0, t) \,|\, 0 \leq t \leq T\}, \tag{8.26}$$

and

$$\overline{\Omega}_T = \Omega_T \cup \partial\Omega_T. \tag{8.27}$$

That is, $\overline{\Omega}_T$ denotes the closure of Ω_T and thus forms a closed set. The reader should draw a figure illustrating these sets!

Remember that a continuous function defined on a closed set always achieves both its maximum and minimum value within this set, see, e.g., Apostol [4]. Assume that $(x^*, t^*) \in \Omega_T$, an interior point, is a maximum point for v in $\overline{\Omega}_T$, i.e.,

$$v(x, t) \leq v(x^*, t^*) \quad \text{for all } (x, t) \in \overline{\Omega}_T.$$

Then, as in the single-variable case, v must satisfy

$$v_x(x^*, t^*) = 0 \quad \text{and} \quad v_t(x^*, t^*) = 0, \tag{8.28}$$
$$v_{xx}(x^*, t^*) \leq 0 \quad \text{and} \quad v_{tt}(x^*, t^*) \leq 0. \tag{8.29}$$

This means that if $w = w(x, t)$ is a smooth function defined on $\overline{\Omega}_T$ such that

$$w_x(x,t) \neq 0 \quad \text{or} \quad w_t(x,t) \neq 0 \quad \text{for all } (x,t) \in \Omega_T, \tag{8.30}$$

or

$$w_{xx}(x,t) > 0 \quad \text{or} \quad w_{tt}(x,t) > 0 \quad \text{for all } (x,t) \in \Omega_T, \tag{8.31}$$

then the maximum value of w must be attained at the boundary $\partial \Omega_T$ of Ω_T. This observation is, as we will see below, the main ingredient of the analysis presented in this section. Note also that, analogously to (8.24), if the maximum is achieved for $t = T$, say, at (x^*, T), then w must satisfy the inequality[13]

$$w_t(x^*, T) \geq 0. \tag{8.32}$$

As mentioned above, we will not go through any derivations of these properties for the maxima of smooth functions defined on closed sets. Further details can be found in, e.g., Apostol [4] or Edwards and Penney [13]. Instead we will now turn our attention toward the heat equation and, more specifically, to how these results can be utilized to increase our insight into this problem.

Assume that u is a smooth solution of (8.16)–(8.18). For an arbitrary time $T > 0$ we want to study the behavior of u on the closed set $\overline{\Omega}_T$ defined in (8.27). Assume that u achieves its maximum value at an interior point $(x^*, t^*) \in \Omega_T$. From (8.16), it follows that

$$u_t(x^*, t^*) = u_{xx}(x^*, t^*),$$

and by (8.28) we conclude that

$$u_{xx}(x^*, t^*) = 0.$$

Thus we can not use (8.29) directly to exclude the possibility that u attains its maximum value at an interior point. We have to take a detour!

To this end we define a family of auxiliary functions $\{v_\epsilon\}_{\epsilon>0}$ by

$$v^\epsilon(x,t) = u(x,t) + \epsilon x^2 \quad \text{for } \epsilon > 0, \tag{8.33}$$

where u solves (8.16)–(8.18). Note that

$$v_t^\epsilon(x,t) = u_t(x,t), \tag{8.34}$$
$$v_{xx}^\epsilon(x,t) = u_{xx}(x,t) + 2\epsilon > u_{xx}(x,t).$$

Now, if v^ϵ achieves its maximum at an interior point, say $(x^*, t^*) \in \Omega_T$, then property (8.28) implies that

$$v_t^\epsilon(x^*, t^*) = 0,$$

---

[13] Similar properties must hold at any maximum point located at the boundary $\partial \Omega_T$ of Ω_T, i.e., at maximum points with coordinates of the form $(x^*, 0)$, $(0, t^*)$ or $(1, t^*)$. However, in the present analysis we will only need the inequality (8.32).

and consequently

$$u_t(x^*, t^*) = 0.$$

Next we can apply the diffusion equation (8.16) to conclude that

$$u_{xx}(x^*, t^*) = u_t(x^*, t^*) = 0,$$

and therefore

$$v_{xx}^\epsilon(x^*, t^*) = u_{xx}(x^*, t^*) + 2\epsilon = 2\epsilon > 0.$$

This violates property (8.29), which must hold at a maximum point of v^ϵ. Hence, we conclude that v^ϵ must attain its maximum value at the boundary of Ω_T, i.e.,

$$v^\epsilon(x, t) \leq \max_{(y,s) \in \partial \Omega_T} v^\epsilon(y, s) \quad \text{for all } (x, t) \in \overline{\Omega}_T.$$

Can v^ϵ reach its maximum value at time $t = T$ and for $x^* \in (0, 1)$? No, for the following reasons this can not be the case. Assume that (x^*, T), with $0 < x^* < 1$, is such that

$$v^\epsilon(x, t) \leq v^\epsilon(x^*, T) \quad \text{for all } (x, t) \in \overline{\Omega}_T.$$

Then, according to property (8.32),

$$v_t^\epsilon(x^*, T) \geq 0,$$

which in turn implies that

$$u_t(x^*, T) \geq 0,$$

see (8.34). Consequently, since u satisfies the diffusion equation,

$$u_{xx}(x^*, T) = u_t(x^*, T) \geq 0,$$

and it follows that

$$v_{xx}^\epsilon(x^*, t^*) = u_{xx}(x^*, t^*) + 2\epsilon \geq 0 + 2\epsilon > 0.$$

This contradicts property (8.29) that must be fulfilled at such a maximum point. This means that

$$v^\epsilon(x, t) \leq \max_{(y,s) \in \Gamma_T} v^\epsilon(y, s) \quad \text{for all } (x, t) \in \overline{\Omega}_T, \tag{8.35}$$

where

$$\Gamma_T = \{(0, t) \mid 0 \leq t \leq T\} \cup \{(x, 0) \mid 0 \leq x \leq 1\} \cup \{(1, t) \mid 0 \leq t \leq T\}.$$

Let

$$M = \max\left(\max_{t \geq 0} g_1(t), \max_{t \geq 0} g_2(t), \max_{x \in (0,1)} f(x)\right)$$

and

$$\Gamma = \{(0,t) \,|\, t \geq 0\} \cup \{(x,0) \,|\, 0 \leq x \leq 1\} \cup \{(1,t) \,|\, t \geq 0\}.$$

Then, since $\Gamma_T \subset \Gamma$, (8.35) and the definition (8.33) of v^ϵ imply that

$$v^\epsilon(x,t) \leq \max_{(y,s)\in\Gamma} v^\epsilon(y,s) \leq M + \epsilon \quad \text{for all } (x,t) \in \overline{\Omega}_T.$$

Next, again by (8.33), it follows that

$$u(x,t) \leq v^\epsilon \quad \text{for all } (x,t) \in \overline{\Omega}_T,$$

and consequently

$$u(x,t) \leq M + \epsilon \quad \text{for all } (x,t) \in \overline{\Omega}_T \text{ and all } \epsilon > 0.$$

This inequality is valid for all $\epsilon > 0$, and, hence, it follows that

$$u(x,t) \leq M \quad \text{for all } (x,t) \in \overline{\Omega}_T. \tag{8.36}$$

In the argument leading to (8.36), the end "point" $T > 0$ of the time interval $[0,T]$, in which we studied the behavior of the solution of the diffusion equation, was arbitrary, cf. also definitions (8.25)–(8.27). Thus, we can choose T as large as we need! This means that this inequality must hold for any $t > 0$, i.e.,

$$u(x,t) \leq M \quad \text{for all } x \in [0,1], \, t > 0. \tag{8.37}$$

8.1.8 The Minimum Principle

We have now shown that the solution u of our model problem (8.16)–(8.18) is bounded by the maximum value M attained by the boundary conditions g_1 and g_2 and the initial temperature distribution f. Can we derive a similar property for the minimum value achieved by u? That is, can we prove that the temperature $u(x,t)$, at any point (x,t), is larger than the minimum value achieved by the boundary and initial conditions? From a physical point of view, this property seems to be reasonable. Let us now have a look at the mathematics. Consider the function w defined by

$$w(x,t) = -u(x,t) \quad \text{for all } x \in [0,1], \, t > 0.$$

Note that

$$w_t = (-u)_t = -u_t = -u_{xx} = (-u)_{xx} = w_{xx},$$

and consequently w satisfies the diffusion equation. Moreover,

$$w(0,t) = -g_1(t) \text{ and } w(1,t) = -g_2(t) \quad \text{for } t > 0,$$
$$w(x,0) = -f(x) \quad \text{for } x \in (0,1),$$

and thus, from the analysis presented above, we find that

$$w(x,t) \leq \max\left(\max_{t\geq 0}(-g_1(t)),\ \max_{t\geq 0}(-g_2(t)),\ \max_{x\in(0,1)}(-f(x))\right)$$
$$= -\min\left(\min_{t\geq 0} g_1(t),\ \min_{t\geq 0} g_2(t),\ \min_{x\in(0,1)} f(x)\right) \quad \text{for all } x \in [0,1],\ t > 0,$$

or

$$u(x,t) \geq \min\left(\min_{t\geq 0} g_1(t),\ \min_{t\geq 0} g_2(t),\ \min_{x\in(0,1)} f(x)\right) \quad \text{for all } x \in [0,1],\ t > 0.$$

8.1.9 Summary

Above we have derived maximum and minimum principles for the solution u of the problem

$$u_t = u_{xx} \quad \text{for } x \in (0,1),\ t > 0, \tag{8.38}$$
$$u(0,t) = g_1(t) \text{ and } u(1,t) = g_2(t) \quad \text{for } t > 0, \tag{8.39}$$
$$u(x,0) = f(x) \quad \text{for } x \in (0,1). \tag{8.40}$$

More precisely, we have shown that the maximum and minimum value of u is either obtained initially or at the boundary of the solution domain.

A smooth solution of the problem (8.38)–(8.40) must satisfy the bound

$$m \leq u(x,t) \leq M \quad \text{for all } x \in [0,1],\ t > 0, \tag{8.41}$$

where

$$m = \min\left(\min_{t\geq 0} g_1(t),\ \min_{t\geq 0} g_2(t),\ \min_{x\in(0,1)} f(x)\right), \tag{8.42}$$

$$M = \max\left(\max_{t\geq 0} g_1(t),\ \max_{t\geq 0} g_2(t),\ \max_{x\in(0,1)} f(x)\right). \tag{8.43}$$

8.1.10 Uniqueness Revisited

In Sect. 8.1.3 we used the energy bound (8.7) to prove that the model problem (8.1)–
(8.3) can have at most one smooth solution. We will now see that it is also possible
to utilize (8.41)–(8.43) to prove this.

Suppose that both u and v are smooth functions satisfying (8.16)–(8.18). Let

$$e(x,t) = u(x,t) - v(x,t) \quad \text{for } x \in [0,1],\ t \geq 0,$$

and note that

$$e_t = (u-v)_t = u_t - v_t = u_{xx} - v_{xx} = (u-v)_{xx} = e_{xx}.$$

Thus, e satisfies the diffusion equation. Furthermore,

$$
\begin{aligned}
e(0,t) &= u(0,t) - v(0,t) = g_1(t) - g_1(t) = 0 \quad \text{for } t > 0,\\
e(1,t) &= u(1,t) - v(1,t) = g_2(t) - g_2(t) = 0 \quad \text{for } t > 0,\\
e(x,0) &= u(x,0) - v(x,0) = f(x) - f(x) = 0 \quad \text{for } x \in (0,1),
\end{aligned}
$$

and consequently (8.41)–(8.43) imply that

$$0 \leq e(x,t) \leq 0 \quad \text{for all } x \in [0,1],\ t \geq 0,$$

or

$$u(x,t) - v(x,t) = 0 \quad \text{for all } x \in [0,1],\ t \geq 0.$$

> The problem (8.16)–(8.18) can have at most one smooth solution.

8.2 Separation of Variables and Fourier Analysis

The great French mathematician Jean Baptiste Joseph Fourier (1768–1830) did
important mathematical work on the theory of heat. He invented[14] a new framework
for studying the diffusion equation. Fourier's work has had an enormous impact, not
only on the theory of PDEs, but on several branches of modern mathematics. If you
decide to continue to study mathematics you will most certainly run into theories
that somehow can be tracked back to Fourier's original work on heat conduction.

---

[14] Are new mathematical theories inventions or discoveries? It seems like neither mathematicians
nor philosophers are able to agree upon this. The interested reader can consult the books [12] and
[11] for a general discussion of the "nature of mathematics".

In the days of Fourier there were of course no computers available. So he developed a technique for computing the solution of the diffusion equation using only paper and pencil. We will now go through the three basic steps in this classic methodology.

8.2.1 Separation of Variables

To begin with, we will focus on the PDE (8.1) and the boundary condition (8.2) separately and try to determine functions that satisfy these two equations. That is, we are trying to find functions satisfying

$$u_t = u_{xx} \quad \text{for } x \in (0, 1), \, t > 0, \tag{8.44}$$

$$u(0, t) = u(1, t) = 0 \quad \text{for } t > 0. \tag{8.45}$$

We will show below how to take care of the initial condition (8.3). Note that the zero function, $u(x, t) = 0$ for all $x \in [0, 1]$ and $t > 0$, satisfies these two equations. However, we want to find non-trivial solutions of this problem.

The basic idea is to seek a solution of (8.44) and (8.45) that can be expressed as a product of two univariate functions. Thus, we make the ansatz[15]

$$u(x, t) = X(x)T(t), \tag{8.46}$$

where $X = X(x)$ and $T = T(t)$ only depend on x and t, respectively.[16]

If a function of the form (8.46) is supposed to solve (8.44) and (8.45), then

$$u(0, t) = 0 = X(0)T(t) \quad \text{and} \quad u(1, t) = 0 = X(1)T(t).$$

Consequently, $X(x)$ must satisfy

$$X(0) = 0 \quad \text{and} \quad X(1) = 0, \tag{8.47}$$

provided that $T(t) \neq 0$ for $t > 0$.

Next, by plugging the ansatz (8.46) for $u(x, t)$ into the diffusion equation (8.44), we find that X and T must satisfy the relation

$$X(x)T'(t) = X''(x)T(t) \quad \text{for all } x \in (0, 1), t > 0.$$

---

[15] *Ansatz* is the German word for *beginning*. It is frequently used in the mathematical literature. By an ansatz, we mean a "guess". More precisely, we guess that the solution of a particular problem is in a specific form. Thereafter, we investigate whether or not, and under what circumstances, this guess actually forms a solution of the problem.

[16] We are trying to split u into a product of two univariate functions and the technique is therefore referred to as "separation of variables".

This means, if we assume that $X(x) \neq 0$ for $x \in (0, 1)$ and $T(t) \neq 0$ for $t > 0$, that

$$\frac{T'(t)}{T(t)} = \frac{X''(x)}{X(x)} \quad \text{for all } x \in (0, 1), t > 0. \tag{8.48}$$

The left-hand side of this equation only depends on time t and the right-hand side only depends on the space coordinate x. This is interesting, and very important! What does it mean? Well, it means that if a function u of the form (8.46) satisfies (8.44) and (8.45), then there must exist a constant λ such that

$$\frac{T'(t)}{T(t)} = \lambda, \tag{8.49}$$

$$\frac{X''(x)}{X(x)} = \lambda. \tag{8.50}$$

It is important that you understand this. Take a few minutes to think about it! You may also want to do Exercise 8.11.

Note that (8.49) and (8.50) are two ordinary differential equations.[17] In these two equations there are three unknowns; two functions T, X and a constant λ. Indicating that there might be more than one solution to this system. This is indeed the case. There are infinitely many solutions!

We have already encountered (8.49) in Chap. 2 in connection with models for exponential growth of rabbit populations. Its solution, see Chap. 2, is given in terms of the exponential function

$$T(t) = ce^{\lambda t}, \tag{8.51}$$

where c is a constant. In this case there is no initial condition present. Thus, this function satisfies (8.49) for all constants c.

Next, we turn our attention toward (8.50). This equation can also be written on the form

$$X''(x) = \lambda X(x), \tag{8.52}$$

with the boundary conditions

$$X(0) = 0 \text{ and } X(1) = 0, \tag{8.53}$$

see (8.47). Thus, we are seeking a function such that the second-order derivative of this function equals the function itself multiplied by a constant. From introductory courses in calculus we know that

$$\sin'(x) = \cos(x),$$
$$\cos'(x) = -\sin(x),$$

---

[17] There are no partial derivatives present!

and therefore

$$\sin''(x) = -\sin(x),$$
$$\cos''(x) = -\cos(x).$$

This means that both $\sin(x)$ and $\cos(x)$ solve the differential equation (8.52) with $\lambda = -1$. However,

$$\sin(0) = 0 \text{ and } \sin(1) \neq 0,$$
$$\cos(0) = 1 \text{ and } \cos(1) \neq 0,$$

so neither of these functions satisfies the boundary conditions (8.53).

Can we somehow modify these functions in order to obtain solutions of (8.52)–(8.53)? Yes. Recall that for any integer k

$$\sin(k\pi) = 0,$$

and furthermore

$$\sin''(k\pi x) = -k^2 \pi^2 \sin(k\pi x).$$

This means that, for $k = \ldots, -2, -1, 0, 1, 2, \ldots$,

$$X(x) = X_k(x) = \sin(k\pi x), \tag{8.54}$$
$$\lambda = \lambda_k = -k^2 \pi^2 \tag{8.55}$$

satisfy the differential equation (8.52) and the boundary condition (8.53).

What about the cosine function? From trigonometry we know that

$$\cos(k\pi + \pi/2) = 0,$$

provided that k is an integer. Thus,

$$X(x) = \cos(k\pi x + \pi/2), \tag{8.56}$$
$$\lambda = -k^2 \pi^2, \tag{8.57}$$

also generates a solution of (8.52)–(8.53) for every integer k. However, by trigonometric identities, it follows that

$$\cos(k\pi x + \pi/2) = \sin(k\pi x + \pi) = -\sin(k\pi x) = \sin(-k\pi x).$$

This means that (8.56) and (8.57) do not add any new solutions to the problem (8.52) and (8.53); they are contained in the set of solutions given by (8.54) and (8.55).

Let us summarize our findings. We conclude from the analysis above, and (8.46), (8.51), (8.54), and (8.55), that the functions

$$u_k(x,t) = c_k e^{-k^2 \pi^2 t} \sin(k\pi x) \quad \text{for } k = \ldots, -2, -1, 0, 1, 2, \ldots \qquad (8.58)$$

satisfy both the diffusion equation (8.44) and the boundary condition (8.45). Here, c_k represents an arbitrary constant, see (8.51). At this point we recommend that you do Exercise 8.1.

Example 8.1. Consider the problem

$$u_t = u_{xx} \quad \text{for } x \in (0, 1), \, t > 0, \qquad (8.59)$$
$$u(0, t) = u(1, t) = 0 \quad \text{for } t > 0, \qquad (8.60)$$
$$u(x, 0) = \sin(\pi x) \quad \text{for } x \in (0, 1). \qquad (8.61)$$

In view of the theory developed in Sect. 8.2.1, it follows that

$$u(x, t) = e^{-\pi^2 t} \sin(\pi x)$$

satisfies (8.59) and (8.60). Furthermore,

$$u(x, 0) = e^{-\pi^2 0} \sin(\pi x) = \sin(\pi x),$$

and thus this is the unique smooth solution of this problem. ∎

Example 8.2. Our second example is

$$u_t = u_{xx} \quad \text{for } x \in (0, 1), \, t > 0,$$
$$u(0, t) = u(1, t) = 0 \quad \text{for } t > 0,$$
$$u(x, 0) = 7 \sin(5\pi x) \quad \text{for } x \in (0, 1).$$

Setting $k = 5$ and $c_5 = 7$ in (8.58), we find that

$$u(x, t) = 7e^{-25\pi^2 t} \sin(5\pi x)$$

satisfies the diffusion equation and the boundary condition of this problem. Furthermore,

$$u(x, 0) = 7e^{-25\pi^2 0} \sin(5\pi x) = 7 \sin(5\pi x)$$

and hence the initial condition is also satisfied. ∎

8.2.2 Super-Positioning

Above we saw how we could use the method of separation of variables to construct infinitely many solutions of the two equations

$$u_t = u_{xx} \quad \text{for } x \in (0, 1),\ t > 0, \tag{8.62}$$
$$u(0, t) = u(1, t) = 0 \quad \text{for } t > 0. \tag{8.63}$$

Our goal now is to show that we can use the solutions given in (8.58) to generate even further solutions of this problem. More precisely, it turns out that any linear combination of the functions in (8.58) will satisfy (8.62) and (8.63). This is a consequence of the linearity of the diffusion equation. Let us have a closer look at this matter.

Assume that both v_1 and v_2 satisfy (8.62) and (8.63). Let

$$w = v_1 + v_2,$$

and observe that

$$w_t = (v_1 + v_2)_t = (v_1)_t + (v_2)_t = (v_1)_{xx} + (v_2)_{xx} = (v_1 + v_2)_{xx} = w_{xx}.$$

Furthermore,

$$w(0, t) = v_1(0, t) + v_2(0, t) = 0 + 0 = 0 \quad \text{for } t > 0,$$

and

$$w(1, t) = v_1(1, t) + v_2(1, t) = 0 + 0 = 0 \quad \text{for } t > 0,$$

and hence w is also a solution of these two equations.

Next, for an arbitrary constant a_1 consider the function

$$p(x, t) = a_1 v_1(x, t).$$

Since a_1 is a constant and v_1 satisfy the diffusion equation it follows that

$$p_t = (a_1 v_1)_t = a_1(v_1)_t = a_1(v_1)_{xx} = (a_1 v_1)_{xx} = p_{xx},$$

and furthermore

$$p(0, t) = a_1 v_1(0, t) = 0 \quad \text{for } t > 0,$$
$$p(1, t) = a_1 v_1(1, t) = 0 \quad \text{for } t > 0.$$

Hence, we conclude that p satisfies both (8.62) and (8.63).

These properties of w and p show that any linear combination of solutions of (8.62) and (8.63) will also satisfy these equations. That is, if v_1 and v_2 satisfy (8.62) and (8.63), then any function of the form

$$a_1 v_1 + a_2 v_2,$$

where a_1 and a_2 are arbitrary constants, also solves[18] this problem for all constants a_1 and a_2.

What does this mean for the set of solutions given in (8.58)? Well, it means that any linear combination of these functions will form a solution of the problem. In mathematical terms this is expressed as follows. For any, possibly infinite, sequence of numbers

$$\ldots, c_{-2}, c_{-1}, c_0, c_1, c_2, \ldots$$

such that the series of functions

$$\sum_{k=-\infty}^{-1} c_k e^{-k^2 \pi^2 t} \sin(k \pi x) + \sum_{k=0}^{\infty} c_k e^{-k^2 \pi^2 t} \sin(k \pi x)$$

converges,[19] the function

$$u(x, t) = \sum_{k=-\infty}^{-1} c_k e^{-k^2 \pi^2 t} \sin(k \pi x) + \sum_{k=1}^{\infty} c_k e^{-k^2 \pi^2 t} \sin(k \pi x) \qquad (8.64)$$

defines a formal solution of (8.62) and (8.63).

We close this discussion with a simplifying observation. It turns out that, without any loss of generality, we can remove the first sum from the formula given in (8.64). This is so because

$$\sin(-y) = -\sin(y) \quad \text{for all } y \in \mathbb{R}$$

and, consequently, for any positive integer k,

$$c_{-k} e^{-k^2 \pi^2 t} \sin(-k \pi x) + c_k e^{-k^2 \pi^2 t} \sin(k \pi x) = (c_k - c_{-k}) e^{-k^2 \pi^2 t} \sin(k \pi x).$$

Thus, the set of solutions of the diffusion equation (8.62) and the boundary condition (8.63) that we have constructed in this section can be written in the following form: For any sequence of numbers

$$c_1, c_2, \ldots,$$

the function

$$u(x, t) = \sum_{k=1}^{\infty} c_k e^{-k^2 \pi^2 t} \sin(k \pi x) \qquad (8.65)$$

---

[18] This is a typical feature of linear problems.

[19] From a rigorous mathematical point of view there are several open questions that we have not addressed here. What does it mean that a series of functions converges? Can we differentiate term-wise? And so forth? Prior to answering these questions we don't know whether or not (8.64) forms a solution of (8.62) and (8.63). These questions are beyond the scope of this introductory text and we will not dwell upon them. Instead we will treat (8.64) as a "formal solution" of the problem. Further details can be found in, e.g., [28].

defines a formal solution of (8.62) and (8.63), provided that the series in (8.65) converges.

In this section we have added simple functions, given in terms of the sine function, to obtain more general solutions of our problem. This technique is often referred to as super-positioning, and hence the title of this section.

8.2.3 Fourier Series and the Initial Condition

On the one hand, in Sect. 8.1.3 we showed that our model problem (8.1)–(8.3) can have at most one smooth solution. On the other hand, in Sect. 8.2.2 we constructed infinitely many functions satisfying the diffusion equation (8.1) and the boundary condition (8.2). This means that, for a given initial condition f,

- There is at most one function on the form (8.65) that fulfills the initial condition (8.3), or
- None of these functions satisfies this condition.

The purpose of this section is to clarify the role of the initial condition f in Fourier's approach to heat transfer problems. This is a difficult issue and in this introductory text we can only scratch the surface of this field.[20] To gain further insight, we will start our investigations with a few examples.

8.2.4 Some Simple Examples

In examples 1 and 2 we saw how the method of separation of variables can be used to compute the analytical solution of our model problem in some relatively simple cases. Let us now consider some examples in which it is necessary to apply the super-positioning technique developed above.

Example 8.3. In this example we consider a problem with an initial condition given by two sine modes

$$u_t = u_{xx} \quad \text{for } x \in (0, 1), \, t > 0, \tag{8.66}$$
$$u(0, t) = u(1, t) = 0 \quad \text{for } t > 0, \tag{8.67}$$
$$u(x, 0) = 2.3 \sin(3\pi x) + 10 \sin(6\pi x) \quad \text{for } x \in (0, 1). \tag{8.68}$$

Consider the general formula (8.65) for a formal solution of the diffusion equation (8.66) and the boundary condition (8.67). Let

$$c_k = 0 \quad \text{for } k \neq 3 \text{ and } k \neq 6,$$
$$c_3 = 2.3 \text{ and } c_6 = 10.$$

---

[20] The theory for general Fourier series and Hilbert spaces.

Then we find that the unique smooth solution of this problem is given by the formula

$$u(x,t) = 2.3e^{-9\pi^2 t}\sin(3\pi x) + 10e^{-36\pi^2 t}\sin(6\pi x).$$

Indeed, the reader should verify by hand that this function satisfies the three equations (8.66), (8.67), and (8.68).

■

Example 8.4. Let us determine a formula for the solution of the following problem:

$u_t = u_{xx}$ for $x \in (0, 1)$, $t > 0$,

$u(0, t) = u(1, t) = 0$ for $t > 0$,

$u(x, 0) = 20\sin(\pi x) + 8\sin(3\pi x) + \sin(67\pi x) + 1002\sin(10^4\pi x)$ for $x \in (0, 1)$.

This is easily accomplished by setting

$$c_k = 0 \quad \text{for } k \neq 1, 3, 67, 10^4,$$
$$c_1 = 20,\ c_3 = 8,\ c_{67} = 1,\ c_{10^4} = 1002,$$

in formula (8.65). The reader should verify that the solution of this problem is given by

$$u(x,t) = 20e^{-\pi^2 t}\sin(\pi x) + 8e^{-9\pi^2 t}\sin(3\pi x)$$
$$+ e^{-(67\pi)^2 t}\sin(67\pi x) + 1002e^{-(10^4\pi)^2 t}\sin(10^4\pi x)$$

■

8.2.5 *Initial Conditions Given by a Sum of Sine Functions*

The technique used in the last two examples is generalized in a straightforward manner to handle cases where the initial condition consists of any finite number of sine modes. This can be expressed in mathematical terms as follows.

Let S be any finite set of positive integers and consider an initial condition of the form

$$f(x) = \sum_{k \in S} c_k \sin(k\pi x),$$

where

$$\{c_k\}_{k \in S}$$

are arbitrary given constants. In this case our model problem takes the form

$$u_t = u_{xx} \quad \text{for } x \in (0, 1),\ t > 0,$$
$$u(0, t) = u(1, t) = 0 \quad \text{for } t > 0,$$
$$u(x, 0) = \sum_{k \in S} c_k \sin(k\pi x) \quad \text{for } x \in (0, 1),$$

and our goal is to show that

$$u(x,t) = \sum_{k \in S} c_k e^{-k^2 \pi^2 t} \sin(k \pi x)$$

solves this problem.

Since the sum is finite we can differentiate term-wise, that is,

$$u_t(x,t) = \sum_{k \in S} \frac{\partial}{\partial t} \left(c_k e^{-k^2 \pi^2 t} \sin(k \pi x) \right) = \sum_{k \in S} -k^2 \pi^2 c_k e^{-k^2 \pi^2 t} \sin(k \pi x),$$

and

$$u_{xx}(x,t) = \sum_{k \in S} \frac{\partial^2}{\partial x^2} \left(c_k e^{-k^2 \pi^2 t} \sin(k \pi x) \right) = \sum_{k \in S} -k^2 \pi^2 c_k e^{-k^2 \pi^2 t} \sin(k \pi x),$$

and we see that this function satisfies the diffusion equation. Moreover, by elementary properties of the sine function, it follows that

$$u(0,t) = \sum_{k \in S} c_k e^{-k^2 \pi^2 t} \sin(0) = 0,$$

$$u(1,t) = \sum_{k \in S} c_k e^{-k^2 \pi^2 t} \sin(k \pi) = 0,$$

and thus the boundary conditions are fulfilled. Finally, since

$$u(x,0) = \sum_{k \in S} c_k e^0 \sin(k \pi x) = \sum_{k \in S} c_k \sin(k \pi x) = f(x),$$

we conclude that we have found the unique smooth solution of this problem.

Can we develop this technique one step further? What about initial conditions defined in terms of infinite series of sine functions? This means that the initial condition is of the form

$$f(x) = \sum_{k=1}^{\infty} c_k \sin(k \pi x) \quad \text{for } x \in (0,1),$$

where the sum on the right-hand side is an infinite series. Assume that this series converges for all $x \in (0,1)$ and $t \geq 0$. Consider the formal solution

$$u(x,t) = \sum_{k=1}^{\infty} c_k e^{-k^2 \pi^2 t} \sin(k \pi x) \tag{8.69}$$

of our model problem (8.1)–(8.3). Under what circumstances will this function define a "proper" solution of this problem? This issue is beyond the scope of this text,[21] and thus we will not dwell upon this subject. Instead, we will assume that these formal solutions are well behaved.

Note that if this series allows for term-wise differentiation, i.e.,

$$u_t(x,t) = \sum_{k=1}^{\infty} \frac{d}{dt}\left(c_k e^{-k^2\pi^2 t}\sin(k\pi x)\right),$$

$$u_{xx}(x,t) = \sum_{k=1}^{\infty} \frac{d^2}{dx^2}\left(c_k e^{-k^2\pi^2 t}\sin(k\pi x)\right),$$

then (8.69) defines the unique smooth solution of our model problem. The argument for this property is completely analogous to that used in the case of an initial condition given in terms of a finite sine series discussed above, and it is left as an exercise to the reader.

8.2.6 Computing Fourier Sine Series

So far we have seen that if the initial condition f is given by a sum, finite or infinite, of sine functions that vanish at $x = 0$ and at $x = 1$, then we can construct, at least formally, solutions of our model problem (8.1)–(8.3). But very few functions are given in terms of sine series. What about fairly simple initial temperature distributions? For example,

$$f(x) = x(1-x) \quad \text{for } x \in (0,1),$$

or

$$f(x) = 10 \quad \text{for } x \in (0,1).$$

So, the state of affairs is really bad. We cannot even handle what, from a physical point of view, most people would characterize as the simplest case, an initially uniform temperature distribution! How can we extend the present theory to handle such cases? Fourier himself asked this question and pursued the subject with great interest.[22] Our goal now is to develop a more general approach to this problem.

Faced with the fact that very few functions are given in terms of sine series, what can we do? Well, we can sort of take the "opposite view" to handle this problem. More precisely, for a given initial condition f we might ask if we can find constants $c_1, c_2, c_3, \ldots$ such that

---

[21] It is an important topic in more advanced texts on Fourier analysis, see, e.g., Weinberger [29]

[22] As is so often the case with breakthroughs in science, Fourier's work on this subject caused a lot of controversy. Fourier's results were basically correct but his arguments contained errors. He even had problems with getting his work accepted for publication!

$$f(x) = \sum_{k=1}^{\infty} c_k \sin(k\pi x) \quad \text{for } x \in (0, 1).\tag{8.70}$$

If this is the case, then we can use the methodology developed above to solve our problem. Fourier realized this and it eventually made him find what today is referred to as Fourier analysis.

Let us have a closer look at (8.70). It is important to note that, in this equation, f is a given function and the constants $c_1, c_2, c_3, \ldots$ are the unknowns that we want to determine. Thus, this is one equation with infinitely many unknowns! How can we solve this problem? Is it possible to solve? We will now see that a basic property of the sine function leads to an amazingly simple solution of this problem.

Let k and l be two positive integers and consider the integral

$$\int_0^1 \sin(k\pi x) \sin(l\pi x)\, dx.$$

Recall the trigonometric identity that, for any real numbers a and b,

$$\sin(a)\sin(b) = \frac{1}{2}\left(\cos(a - b) - \cos(a + b)\right).\tag{8.71}$$

By applying this identity, the reader can verify in a straightforward manner that

$$\int_0^1 \sin(k\pi x) \sin(l\pi x)\, dx = \begin{cases} 0 & k \neq l, \\ 1/2 & k = l, \end{cases}\tag{8.72}$$

see Exercise 8.5. How can this property be exploited in the present situation? Ideally, we would like to have an explicit formula for the constants $c_1, c_2, c_3, \ldots$ in (8.70). It turns out that this can easily be accomplished by applying (8.72). If we multiply the left- and right-hand sides of (8.70) by $\sin(l\pi x)$ and integrate, we find that

$$\int_0^1 f(x) \sin(l\pi x)\, dx = \int_0^1 \left(\sum_{k=1}^{\infty} c_k \sin(k\pi x)\right) \sin(l\pi x)\, dx.$$

Let us assume that we can interchange the order of integration and summation,[23]

$$\int_0^1 f(x) \sin(l\pi x)\, dx = \sum_{k=1}^{\infty} c_k \int_0^1 \sin(k\pi x) \sin(l\pi x)\, dx,$$

$$= \frac{1}{2} c_l,$$

---

[23] This depends on the convergence properties of the involved series. The topic is beyond the scope of this book.

where the last equality is a consequence of the property (8.72) of the sine function. Thus we have derived an amazingly simple formula for the coefficients in (8.70),

$$c_k = 2 \int_0^1 f(x) \sin(k\pi x)\, dx \quad \text{for } k = 1, 2, \ldots . \tag{8.73}$$

It is important that you get this right. Our argument shows that if the function f can be written in the form (8.70) then the coefficients must satisfy (8.73). It does *not* provide any information about which functions can be expressed as a series of sine functions. This topic is treated in more advanced texts on Fourier analysis, see, e.g., Weinberger [29].

Roughly speaking, every "well-behaved"[24] function can be written in the form (8.70), with Fourier coefficients given by (8.73). We will not pursue this question any further. Throughout this text every initial condition f will be such that the Fourier sine series converges and that (8.70) holds.

8.2.7 Summary

Let us summarize our findings so far.

The Fourier coefficients for a well-behaved function $f(x)$, $x \in (0, 1)$, is defined by

$$c_k = 2 \int_0^1 f(x) \sin(k\pi x)\, dx \quad \text{for } k = 1, 2, \ldots , \tag{8.74}$$

and the associated Fourier sine series by

$$f(x) = \sum_{k=1}^{\infty} c_k \sin(k\pi x) \quad \text{for } x \in (0, 1). \tag{8.75}$$

Furthermore, if f satisfies (8.75), then

$$u(x, t) = \sum_{k=1}^{\infty} c_k e^{-k^2 \pi^2 t} \sin(k\pi x) \tag{8.76}$$

---

[24] Typically, continuous functions and functions with a finite number of jump discontinuities.

defines a formal[a] solution of the problem

$$u_t = u_{xx} \quad \text{for } x \in (0, 1),\ t > 0, \tag{8.77}$$
$$u(0, t) = u(1, t) = 0 \quad \text{for } t > 0, \tag{8.78}$$
$$u(x, 0) = f(x) \quad \text{for } x \in (0, 1). \tag{8.79}$$

---

[a] Recall that we have not analyzed the convergence properties of this series properly. Under what circumstances will the Fourier series of a function converge toward the correct limit? Can we differentiate term-wise? Will the limit, if it exists, satisfy the diffusion equation? Thus we refer to (8.76) as a "formal solution".

Note that, for a given positive integer N, we can approximate the formal solution (8.76) by the Nth partial sum u_N of the Fourier series, i.e.,

$$u(x, t) \approx u_N(x, t) = \sum_{k=1}^{N} c_k e^{-k^2\pi^2 t} \sin(k\pi x). \tag{8.80}$$

This is important from a practical point of view. By hand, or on a computer, we can, in general, not compute an infinite sum of sine functions!

8.2.8 More Examples

We will now consider some slightly more advanced examples illuminating the theory developed above.

Example 8.5. We want to determine the Fourier sine series of the constant function

$$f(x) = 10 \quad \text{for } x \in (0, 1).$$

From formula (8.74) we find that

$$c_k = 2 \int_0^1 f(x) \sin(k\pi x)\, dx = 2 \int_0^1 10 \sin(k\pi x)\, dx$$
$$= 20 \left[-\frac{1}{k\pi} \cos(k\pi x) \right]_0^1 = -\frac{20}{k\pi}(\cos(k\pi) - 1)$$
$$= \begin{cases} 0 & \text{if k is even,} \\ \frac{40}{k\pi} & \text{if k is odd.} \end{cases}$$

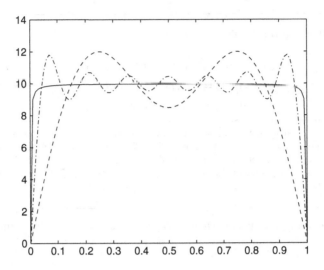

Fig. 8.1 The first two (*dashed line*), seven (*dash-dotted line*), and 100 (*solid line*) terms of the Fourier sine series of the function $f(x) = 10$

Thus we find that the Fourier sine series of this function is

$$f(x) = 10 = \sum_{k=1}^{\infty} \frac{40}{(2k-1)\pi} \sin((2k-1)\pi x) \quad \text{for } x \in (0,1), \qquad (8.81)$$

see (8.75).

In Fig. 8.1 we have plotted the Nth partial sum of this series for $N = 2, 7, 100$.

∎

The alert reader might have noticed a subtle difficulty in the Fourier analysis presented above. Every sine mode, $\sin(k\pi x)$, is zero at $x = 0$ and at $x = 1$, provided that k is an integer. Consequently, any function g given by a Fourier sine series, finite or infinite,

$$g(x) = \sum_{k=1}^{\infty} c_k \sin(k\pi x),$$

will also have the property that

$$g(0) = g(1) = 0.$$

Thus, (8.75) cannot, in general, be extended to hold at the end points of the closed unit interval $[0, 1]$; convergence toward $f(x)$ is only obtained for $x \in (0, 1)$. In addition, the convergence tends to be slow close to the endpoints of the interval, see Fig. 8.1. On the other hand, if f is a well-behaved function satisfying

$$f(0) = f(1) = 0, \qquad (8.82)$$

then its Fourier sine series will converge quickly toward $f(x)$ for all x in the closed interval $[0, 1]$. Finally note that, even if (8.82) does not hold, (8.76) provides the correct formal solution of the problem (8.77)–(8.79). We will not dwell upon this issue. Further details on this advanced topic can, e.g., be found in Weinberger [29].

Example 8.6. Above we mentioned that we were not in a position to solve our model problem for even the simplest possible initial condition, i.e., we could not solve (8.77)–(8.79) if f were a constant ($\neq 0$). This is no longer the case. According to (8.76) and (8.81),

$$u(x,t) = \sum_{k=1}^{\infty} \frac{40}{(2k-1)\pi} e^{-(2k-1)^2 \pi^2 t} \sin((2k-1)\pi x) \qquad (8.83)$$

is the formal solution of

$$u_t = u_{xx} \quad \text{for } x \in (0, 1), \ t > 0,$$
$$u(0,t) = u(1,t) = 0 \quad \text{for } t > 0,$$
$$u(x,0) = 10 \quad \text{for } x \in (0, 1).$$

In Figs. 8.2 and 8.3 we have graphed the function given by the 100th partial sum of the series (8.83) at time $t = 0.5$ and $t = 1$, respectively. Note that the solution decays rapidly. This is in accordance with the physics of the underlying heat transfer problem: Recall that the temperature at the boundary of the solution domain is kept

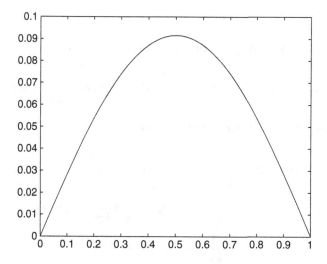

Fig. 8.2 A plot of the function given by the sum of the first 100 terms of the series defining the formal solution of the problem studied in Example 8.6. The figure shows a snapshot of this function at time $t = 0.5$

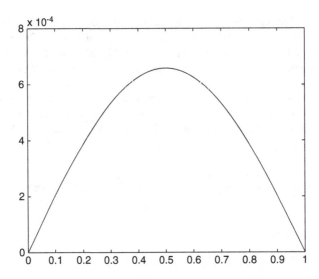

Fig. 8.3 A plot of the function given by the sum of the first 100 terms of the series defining the formal solution of the problem studied in Example 8.6. The figure shows a snapshot of this function at time $t = 1$

at zero for all time. Thus, we expect that the temperature throughout the interval $[0, 1]$ will approach zero as t increases!

∎

Example 8.7. Let us compute the Fourier coefficients $\{c_k\}$ of the function

$$f(x) = x(1 - x) = x - x^2$$

mentioned above. According to formula (8.74),

$$c_k = 2 \int_0^1 (x - x^2) \sin(k\pi x)\,dx = 2 \int_0^1 x \sin(k\pi x)\,dx - 2 \int_0^1 x^2 \sin(k\pi x)\,dx,$$

where, by applying integration by parts, we find that

$$\int_0^1 x \sin(k\pi x)\,dx = \left[-x\frac{1}{k\pi}\cos(k\pi x)\right]_0^1 - \int_0^1 -\frac{1}{k\pi}\cos(k\pi x)\,dx$$

$$= -\frac{1}{k\pi}\cos(k\pi) + \frac{1}{k\pi}\left[\frac{1}{k\pi}\sin(k\pi x)\right]_0^1$$

$$= -\frac{1}{k\pi}\cos(k\pi) = \frac{(-1)^{k+1}}{k\pi},$$

and

$$\int_0^1 x^2 \sin(k\pi x)\,dx = \left[-x^2\frac{1}{k\pi}\cos(k\pi x)\right]_0^1 - \int_0^1 -2x\frac{1}{k\pi}\cos(k\pi x)\,dx$$

$$= -\frac{1}{k\pi}\cos(k\pi) + \frac{2}{k\pi}\left[x\frac{1}{k\pi}\sin(k\pi x)\right]_0^1$$

$$-\frac{2}{k\pi}\int_0^1 \frac{1}{k\pi}\sin(k\pi x)\,dx$$

$$= -\frac{1}{k\pi}\cos(k\pi) + \frac{2}{(k\pi)^2}\left[\frac{1}{k\pi}\cos(k\pi x)\right]_0^1$$

$$= -\frac{1}{k\pi}\cos(k\pi) + \frac{2}{(k\pi)^3}\cos(k\pi) - \frac{2}{(k\pi)^3}$$

$$= \frac{(-1)^{k+1}}{k\pi} + \frac{2(-1)^k}{(k\pi)^3} - \frac{2}{(k\pi)^3}.$$

Thus

$$c_k = 2\left(\frac{2}{(k\pi)^3} - \frac{2(-1)^k}{(k\pi)^3}\right) = \begin{cases} 0 & \text{if } k \text{ is even,} \\ \frac{8}{(k\pi)^3} & \text{if } k \text{ is odd,} \end{cases}$$

and, at least formally,[25] we find that

$$f(x) = x - x^2 = \sum_{k=1}^{\infty}\left(\frac{8}{((2k-1)\pi)^3}\right)\sin((2k-1)\pi x).$$

In Fig. 8.4 we have plotted the functions defined by the first, the first seven, and the first 100 terms of this series. Note that

$$f(0) = f(1) = 0,$$

and consequently it seems that f is accurately approximated by as few as seven terms of its Fourier series, see the discussion following Example 8.5. ∎

Example 8.8. We want to solve the problem

$$u_t = u_{xx} \quad \text{for } x \in (0,1), \ t > 0,$$
$$u(0,t) = u(1,t) = 0 \quad \text{for } t > 0,$$
$$u(x,0) = x(1-x) = x - x^2 \quad \text{for } x \in (0,1).$$

By the formula derived in the previous example we find that this problem can be written in the form

---

[25] We have not proved that the series converges toward the correct limit!

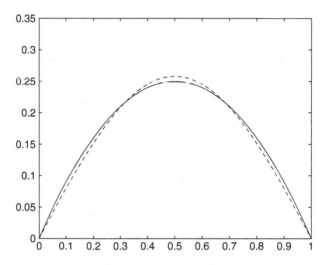

Fig. 8.4 The first (*dashed line*), the first seven (*dash-dotted line*), and the first 100 (*solid line*) terms of the Fourier sine series of the function $f(x) = x - x^2$. It is impossible to distinguish between the figures representing the first seven and the first 100 terms. They are both accurate approximations of $x - x^2$

$$u_t = u_{xx} \quad \text{for } x \in (0,1),\ t > 0,$$
$$u(0,t) = u(1,t) = 0 \quad \text{for } t > 0,$$
$$u(x,0) = \sum_{k=1}^{\infty} \left(\frac{8}{((2k-1)\pi)^3} \right) \sin((2k-1)\pi x) \quad \text{for } x \in (0,1),$$

and then it follows by (8.76) that

$$u(x,t) = \sum_{k=1}^{\infty} \left(\frac{8}{((2k-1)\pi)^3} \right) e^{-(2k-1)^2\pi^2 t} \sin((2k-1)\pi x) \qquad (8.84)$$

defines a formal solution of it. As in Example 8.6 we observe that this solution decays as time t increases, see Figs. 8.5 and 8.6. This is in agreement with our physical intuition of the problem under consideration.

■

8.2.9 Analysis of an Explicit Finite Difference Scheme

In Sect. 7.4.5 we found that the α parameter in the explicit finite difference scheme from Algorithms 7.1–7.3 must be chosen to be less than one-half. In the present section we will give an alternative theoretical explanation why the explicit scheme

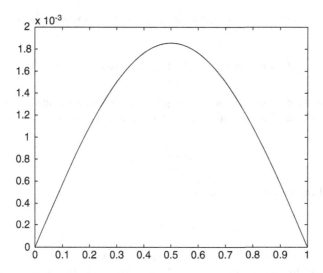

Fig. 8.5 A snapshot at time $t = 0.5$ of the function given by the sum of the first 100 terms of the series defining the formal solution of the problem studied in Example 8.8

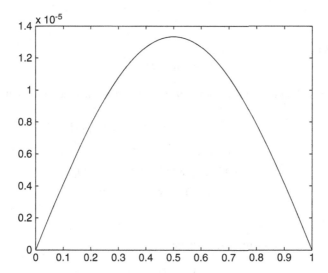

Fig. 8.6 A snapshot at time $t = 1$ of the function given by the sum of the first 100 terms of the series defining the formal solution of the problem studied in Example 8.8

becomes unstable for $\alpha > 1/2$. The reasoning will be based on a discrete analog to the maximum principle, worked out in Sect. 8.1.5.

Due to the boundary condition (8.2), the bounds of (8.41)–(8.43) take the form

$$\min\left(\min_x f(x), 0\right) \leq u(x,t) \leq \max\left(\max_x f(x), 0\right) \quad \text{for all } x \in (0,1) \text{ and } t \geq 0,$$

where u denotes the solution of our model problem (8.1)–(8.3). In particular, this means that

$$|u(x,t)| \leq \max_{Y} |f(x)| \quad \text{for all } x \in (0,1) \text{ and } t \geq 0.$$

Let us see if we can prove a similar property for the approximations generated by the explicit scheme. To this end, assume that Δt and Δx satisfy

$$\alpha = \frac{\Delta t}{\Delta x^2} \leq \frac{1}{2}.$$

Note that, if this condition holds, then

$$1 - 2\alpha \geq 0, \tag{8.85}$$

and this turns out to be the key point in the present analysis.

To simplify the notation, we introduce the symbol $\bar{u}^\ell$ for the maximum of the absolute value of the discrete approximation at time step t_ℓ, that is,

$$\bar{u}^\ell = \max_i |u_i^\ell| \quad \text{for } \ell = 0, \ldots, m,$$

and note that

$$\bar{u}^0 = \max_i |f(x_i)|.$$

If (8.85) holds, then it follows from (7.91) and the triangle inequality that

$$\begin{aligned}
|u_i^{\ell+1}| &= |\alpha u_{i-1}^\ell + (1 - 2\alpha)u_i^\ell + \alpha u_{i+1}^\ell| \\
&\leq |\alpha u_{i-1}^\ell| + |(1 - 2\alpha)u_i^\ell| + |\alpha u_{i+1}^\ell| \\
&= \alpha |u_{i-1}^\ell| + (1 - 2\alpha)|u_i^\ell| + \alpha |u_{i+1}^\ell| \\
&\leq \alpha \bar{u}^\ell + (1 - 2\alpha)\bar{u}^\ell + \alpha \bar{u}^\ell \\
&= \bar{u}^\ell
\end{aligned} \tag{8.86}$$

for $i = 2, \ldots, n - 1$. Moreover, we have

$$u_1^{\ell+1} = u_n^{\ell+1} = 0,$$

and consequently, since (8.86) is valid for $i = 2, \ldots, n - 1$,

$$\max_i |u_i^{\ell+1}| \leq \bar{u}^\ell,$$

or

$$\bar{u}^{\ell+1} \leq \bar{u}^\ell.$$

Finally, by a straightforward induction argument we conclude that

$$\bar{u}^{\ell+1} \le \bar{u}^0 = \max_i |f(x_i)|,$$

which is the result we were seeking.

Assume that the discretization parameters Δt and Δx satisfy

$$\alpha = \frac{\Delta t}{\Delta x^2} \le \frac{1}{2}. \qquad (8.87)$$

Then the approximations generated by the explicit scheme (7.91) satisfy the bound

$$\max_i |u_i^\ell| \le \max_i |f(x_i)| \quad \text{for } \ell = 0, \dots, m, \qquad (8.88)$$

where f is the initial condition in the model problem (8.1)–(8.3).

Thus, if (8.87) is satisfied then the numbers generated by (7.91) will not "blow up"; they will always be bounded by the magnitude of the initial condition. Furthermore, it seems that the scheme produces accurate approximations of the solution of the diffusion equation, provided that Δt and Δx are small. On the other hand, if this condition is not satisfied, then the experiments presented above indicate that this method does not work. Thus, the scheme is only conditionally stable!

8.2.10 Consequences of the Stability Criterion

Let us have a closer look at the implications of this stability condition. Assume that we use 11 grid points in the space dimension, i.e., $n = 11$ (corresponding to $\Delta x = 0.1$). Then Δt must satisfy $\Delta t \le 0.005$, provided that $T = 1$. This means that the number of time steps m must be at least 200. More generally, m and n must satisfy

$$m \ge 2T(n-1)^2.$$

Hence, the number of time steps needed increases rapidly with the number of grid points used in the space dimension.

Assume that we want to compute an approximation of the solution of the diffusion equation in the time interval [0,1], i.e., $T = 1$. Then, for $n = 101$, m must satisfy $m \ge 20,000$, and in the case of $n = 1,001$ at least $2 \cdot 10^6$ time steps must be taken! For the present model problem this causes no problems on modern computers. However, in many realistic simulation processes, involving two or three space dimensions and large solution domains, explicit schemes of this type tend to "stall" due to the large number of time steps needed. To solve such problems implicit methods must be applied, since they do not demand similar severe restrictions on the discretization parameters. We will define and analyze implicit schemes in later chapters.

8.2.11 Exercises

Exercise 8.1. Let c be an arbitrary constant and k an arbitrary integer. Define the function $u = u(x,t)$ by

$$u(x,t) = ce^{-k^2\pi^2 t} \sin(k\pi x) \quad \text{for } x \in [0, 1], \ t > 0.$$

(a) Compute u_t and u_{xx} and verify that

$$u_t = u_{xx}.$$

(b) Verify that
$$u(0, t) = u(1, t) = 0 \quad \text{for } t > 0.$$

(c) Show that
$$u(x,0) = c \sin(k\pi x) \quad \text{for } x \in (0, 1).$$

◇

Exercise 8.2.   (a) Use formulas (8.74) and (8.75) to compute the Fourier sine series of the function
$$f(x) = 1 \quad \text{for } x \in (0, 1).$$

(b) Write a computer program that can plot the function given by the Nth partial sum of the Fourier series of f. Make plots for $N = 3, 7, 80, 100$.
(c) Find a formal solution of the problem

$$u_t = u_{xx} \quad \text{for } x \in (0, 1), \ t > 0,$$
$$u(0, t) = u(1, t) = 0 \quad \text{for } t > 0,$$
$$u(x,0) = f(x) \quad \text{for } x \in (0, 1),$$

see (8.76)–(8.79).
(d) Write a computer program that can graph the function defined by the first N terms of the series of the formal solution in (c). The program should take N and the time t at which you want to graph the approximate solution, as input parameters. Make plots for $N = 3, 7, 80, 100$, at $t = 0.25$ and $t = 2$.
(e) Redo (a)–(d) but this time apply the initial condition

$$f(x) = x \quad \text{for } x \in (0, 1).$$

◇

Exercise 8.3.　(a) Compute the Fourier coefficients $\{c_k\}$ of the function

$$f(x) = x^2 - x^3 \quad \text{for } x \in (0, 1).$$

(b) Use Fourier's method to compute a formal solution of the following problem

$$u_t = u_{xx} \quad \text{for } x \in (0, 1),\ t > 0, \tag{8.89}$$
$$u(0, t) = u(1, t) = 0 \quad \text{for } t > 0, \tag{8.90}$$
$$u(x, 0) = x^2 - x^3 \quad \text{for } x \in (0, 1). \tag{8.91}$$

(c) Derive an explicit scheme for (8.89)–(8.91).
(d) Write a computer program that implements the scheme in (c).
(e) Use the program from (d) to graph an approximation of

$$u(x, 2) \quad \text{for } x \in [0, 1]. \tag{8.92}$$

(f) Consider the formal solution that you derived in (b). Write a computer program that uses the first 50 terms of this sine series to compute an approximation of $u(x, 2)$ for $x \in [0, 1]$.
(g) Use the computer code that you developed in (f) to graph a second approximation of (8.92).

◇

Exercise 8.4.　(a) Compute the Fourier sine series of the function

$$f(x) = e^x \quad \text{for } x \in (0, 1).$$

(b) Write a function[26] that, for a given positive integer N and $x \in (0, 1)$, computes the sum of the first N terms of the Fourier sine series in (a).
(c) Write a computer program that calls the function in (b) and plots the partial sum of the Fourier series. The code should take N as an input parameter. Make plots for $N = 5, 10, 50$.

◇

Exercise 8.5. Let k and l be positive integers. Apply the identity (8.71) to prove that

$$\int_0^1 \sin(k\pi x) \sin(l\pi x)\, dx = \begin{cases} 0 & k \neq l, \\ 1/2 & k = l. \end{cases}$$

◇

Exercise 8.6. So far we have only considered the method of separation of variables for the diffusion equation on the spatial interval $[0, 1]$. What about other intervals, say, $[0, L]$, where L is a positive constant? Can we generalize the theory developed above to handle such cases? We will now investigate this question.

---

[26] Computer code!

(a) Let l and k be two positive integers. Show that

$$\int_0^L \sin\left(\frac{k\pi x}{L}\right) \sin\left(\frac{l\pi x}{L}\right) dx = \begin{cases} 0 & k \neq l, \\ \frac{L}{2} & k = l. \end{cases}$$

(b) Assume that f is a function defined on $(0, L)$. We want to compute the Fourier sine coefficients of this function on this interval. That is, we want to compute constants $c_1, c_2, \ldots$ such that

$$f(x) = \sum_{k=1}^{\infty} c_k \sin\left(\frac{k\pi x}{L}\right) \quad \text{for } x \in (0, L).$$

Derive the formula

$$c_k = \frac{2}{L} \int_0^L f(x) \sin\left(\frac{k\pi x}{L}\right) dx.$$

(c) Show by differentiation that the function

$$u(x, t) = ce^{-(k\pi/L)^2 t} \sin\left(\frac{k\pi x}{L}\right),$$

where c is an arbitrary constant, satisfies the diffusion equation

$$u_t = u_{xx} \quad \text{for } x \in (0, L), \ t > 0.$$

(d) Compute the Fourier sine series of the function

$$f(x) = e^x \quad \text{for } x \in (0, L),$$

and find a formal solution of the problem

$$u_t = u_{xx} \quad \text{for } x \in (0, L), \ t > 0, \tag{8.93}$$
$$u(0, t) = u(L, t) = 0 \quad \text{for } t > 0, \tag{8.94}$$
$$u(x, 0) = e^x \quad \text{for } x \in (0, L). \tag{8.95}$$

(e) Write a computer program that graphs an approximation of the formal solution that you derived in (d). The program should take L, T, and N as input parameters and plot the N partial sum $u_N(x, T)$, $x \in (0, L)$, of the Fourier series of the formal solution, see (8.80).

(f) Derive and implement an explicit finite difference scheme for (8.93)–(8.95). The program should take the discretization parameters Δx and Δt, and the time T (the time at which we want to evaluate the solution) as input parameters. The code should plot an approximation of $u(x, T)$ for $x \in (0, L)$. Compare the results with those obtained in (e).

◇

Exercise 8.7. In order to determine the Fourier coefficients of a function $f(x)$, $x \in (0, 1)$, we have to compute integrals of the form

$$c_k = \int_0^1 f(x) \sin(k\pi x)\, dx \quad k = 1, 2 \ldots,$$

cf. (8.74) and (8.75). For many functions f, this can be tedious, difficult, and in some cases impossible to do by hand. This is the case for functions such as

$$f(x) = e^{-x^2},$$
$$f(x) = \sin(e^{\cos(x)}).$$

However, we can use the numerical techniques developed in Chap. 1 to compute approximations of the Fourier coefficients. In this exercise we will write a computer program that utilizes the trapezoidal rule to do so.

(a) Write a function that takes x as an input parameter and returns the number e^{-x^2}.
(b) Write a function *TrapRuleFourier* that takes the number of intervals n to use in the trapezoidal rule, and a positive integer k as input parameters. The function should call the function developed in (a) and return an approximation of the Fourier coefficient c_k for the function e^{-x^2}.
(c) Write a program that computes the Nth partial sum of the Fourier series of e^{-x^2}. It should take the integer N as input and plot the approximation.
(d) Run the program in (c) with $N = 10,100, 1{,}000$ and compare the graph with that of e^{-x^2}.
(e) Apply Simpson's rule instead of the trapezoidal rule and redo (b)–(d).
(f) Use formula (8.80) and the code developed in (a)–(e) to write a program that computes an approximation of the formal solution of the problem

$$u_t = u_{xx} \quad \text{for } x \in (0, 1), \, t > 0,$$
$$u_x(0, t) = u_x(1, t) = 0 \quad \text{for } t > 0,$$
$$u(x, 0) = e^{-x^2} \quad \text{for } x \in (0, 1).$$

The program should graph an approximation of $u(x, 3)$ for $x \in (0, 1)$.
(g) Modify your program so that it computes an approximation of the formal solution of

$$u_t = u_{xx} \quad \text{for } x \in (0, 1), \, t > 0,$$
$$u_x(0, t) = u_x(1, t) = 0 \quad \text{for } t > 0,$$
$$u(x, 0) = \sin(e^{\cos(x)}) \quad \text{for } x \in (0, 1).$$

◇

Exercise 8.8. Consider the problem

$$v_t + Kv = v_{xx} \quad \text{for } x \in (0,1), \, t > 0,$$
$$v(0,t) = v(1,t) = 0 \quad \text{for } t > 0,$$
$$v(x,0) = f(x) \quad \text{for } x \in (0,1),$$

where K is a constant and f is a given initial condition. Let u denote the solution of (8.1)–(8.3). Show that

$$v(x,t) = e^{-Kt}u(x,t).$$

⋄

Exercise 8.9. In the text above we showed that the numerical approximations generated by the explicit scheme (7.91) satisfy the bound

$$\max_i |u_i^\ell| \leq \max_i |f(x_i)| \quad \text{for } \ell = 0,\ldots,m. \tag{8.96}$$

The purpose of the present exercise is to prove a somewhat stronger result.

We will assume throughout this exercise that the discretization parameters have been chosen such that

$$\alpha = \frac{\Delta t}{\Delta x^2} \leq \frac{1}{2}. \tag{8.97}$$

Let the function $G : \mathbb{R}^3 \to \mathbb{R}$ be defined by

$$G(U_-, U, U_+) = \alpha U_- + (1 - 2\alpha)U + \alpha U_+.$$

(a) Show that the approximations generated by the scheme (7.91) satisfy

$$u_i^{\ell+1} = G(u_{i-1}^\ell, u_i^\ell, u_{i+1}^\ell).$$

(b) Prove that

$$\frac{\partial G}{\partial U_-} \geq 0, \quad \frac{\partial G}{\partial U} \geq 0, \quad \frac{\partial G}{\partial U_+} \geq 0,$$

provided that the discretization parameters satisfy (8.97).

For notational purposes we introduce the symbols $\bar{u}_{\max}^\ell$ and $\bar{u}_{\min}^\ell$ for the maximum and minimum values, respectively, of the discrete approximations at time step t_ℓ, i.e.,

$$\bar{u}_{\max}^\ell = \max_i u_i^\ell \quad \text{for } \ell = 0,\ldots,m,$$

and

$$\bar{u}_{\min}^\ell = \min_i u_i^\ell \quad \text{for } \ell = 0,\ldots,m.$$

(c) Apply the monotonicity property derived in (b) to prove that

$$u_i^{\ell+1} \leq \bar{u}_{\max}^\ell \quad \text{for } i = 2,\ldots,n-1 \text{ and } \ell = 0,\ldots,m,$$

and use this inequality to conclude that

$$\bar{u}_{max}^{\ell+1} \leq \bar{u}_{max}^{\ell} \quad \text{for } \ell = 1, \ldots, m.$$

(d) Show that

$$\bar{u}_{max}^{\ell} \leq \max\left(\max_x f(x), 0\right).$$

(e) Prove in a similar manner that

$$\bar{u}_{min}^{\ell} \geq \min\left(\min_x f(x), 0\right).$$

From the inequalities derived in (d) and (e) we conclude that

$$\min\left(\min_x f(x), 0\right) \leq u_i^{\ell} \leq \max\left(\max_x f(x), 0\right) \quad \text{for } i = 1, \ldots, n \text{ and } \ell = 0, \ldots, m,$$

$$(8.98)$$

which is a discrete analog to the maximum principle (8.41)–(8.43) valid for the continuous problem (8.1)–(8.3).

(f) Above we mentioned that we would derive a stronger bound for the numerical approximations of the solution of the diffusion equation than inequality (8.96). Explain why (8.98) is a stronger result than (8.96), i.e., use (8.98) to derive (8.96).

◇

Exercise 8.10. The purpose of this exercise is to give a slightly different argument for the property (8.7) than in the text above. Recall that the starting point of our analysis leading to this inequality was to multiply the left- and right-hand sides of the diffusion equation by u. In this exercise we ask you to start off by differentiating the function $E_1(t)$, defined in (8.4), with respect to t and apply the differential equation (8.1), integration by parts, the boundary condition (8.2), and so on, in order to derive (8.7).

◇

Exercise 8.11. Assume that $g(t)$ is a function of $t > 0$ and that $f(x)$ is a function of $x \in (0, 1)$. Prove that if

$$g(t) = f(x) \quad \text{for all } x \in (0, 1) \text{ and } t > 0,$$

then there must exist a constant λ such that

$$g(t) = \lambda \text{ for all } t > 0,$$

and

$$f(x) = \lambda \text{ for all } x \in (0, 1).$$

◇

Exercise 8.12. Let $f : (0, 1) \to \mathbb{R}$ be a continuous function. Show that if

$$\int_0^1 f(x)\,dx = 0$$

and

$$f(x) \geq 0 \text{ for } x \in (0, 1),$$

then $f(x) = 0$ for all $x \in (0, 1)$.   ◇

Exercise 8.13.   (a) Assume that q is a non-negative function defined on the unit interval $[0, 1]$ and that the integral of q is small. That is, $q(x) \geq 0$ for $x \in [0, 1]$ and

$$\int_0^1 q(x)\,dx \leq \epsilon,$$

where ϵ is a small positive number. We want to study the size of the set where q is "fairly" large. To this end, consider the set

$$S = \{x \in [0, 1];\ q(x) \geq \sqrt{\epsilon}\}.$$

We will assume that q is such that S consists of a countable disjoint union of intervals, i.e.,

$$S = \bigcup_{i=1}^{\infty} [a_i, b_i], \quad (a_i, b_i) \bigcap (a_j, b_j) = \emptyset \text{ for } i \neq j.$$

The length $|S|$ of S is defined in a straightforward manner as

$$|S| = \sum_{i=0}^{\infty} (b_i - a_i).$$

Show that

$$|S| \leq \sqrt{\epsilon}.$$

(b) Solve the problem (8.1)–(8.3) in the case of

$$u(x, 0) = f(x) = 8.1 \sin(3\pi x).$$

(c) Solve the problem (8.12)–(8.14) in the case of

$$v(x, 0) = g(x) = 8 \sin(3\pi x).$$

(d) Compute

$$\int_0^1 (f(x) - g(x))^2\,dx.$$

(e) Compute

$$\int_0^1 (u(x,t) - v(x,t))^2 \, dx$$

for $t = 10^{-4}$ and $t = 10^{-2}$.

(f) Define the set S by

$$S = \{x \in [0, 1]; \; |u(x, 10^{-2}) - v(x, 10^{-2})| \geq 0.213\}.$$

Show that

$$|S| \leq 0.046.$$

$\diamond$

8.3 Projects

8.3.1 Neumann Boundary Conditions

In this chapter we have so far only considered problems with Dirichlet boundary conditions. That is, the value of the unknown function in the PDE is specified at the boundary, see (8.2).

In Chap. 7 we encountered a second type of boundary conditions, namely, Neumann conditions. In this case the derivative, or flux, of the unknown function u in the differential equation is prescribed at the boundary; that is, assuming that the solution domain is the unit interval $(0, 1)$,

$$u_x(0, t) \text{ and } u_x(1, t) \text{ for } t > 0$$

are given. The purpose of this project is to reconsider the theory developed above for this kind of boundary conditions. Let us therefore consider the following model problem:

$$u_t = u_{xx} \quad \text{for } x \in (0, 1), \; t > 0, \tag{8.99}$$
$$u_x(0, t) = u_x(1, t) = 0 \quad \text{for } t > 0, \tag{8.100}$$
$$u(x, 0) = f(x) \quad \text{for } x \in (0, 1), \tag{8.101}$$

where f is a given initial condition.

For given positive integers m and n we define

$$\Delta t = \frac{T}{m},$$
$$\Delta x = \frac{1}{n-1},$$
$$x_i = (i-1)\Delta x \quad \text{for } i = 1, \ldots, n,$$
$$t_\ell = \ell \Delta t \quad \text{for } \ell = 0, \ldots, m,$$

where $[0, T]$ is the time interval in which we want to discretize (8.99)–(8.101).

(a) Use the discretization technique presented in Chap. 7 for the boundary condition (8.100) to derive an explicit scheme for this problem.
(b) Write a computer program that implements the scheme in (a).
(c) Recall the bound (8.7) valid in the case of homogeneous Dirichlet boundary conditions. Explain why we can compare the size of the sum

$$\Delta x \left(\frac{1}{2}(u_1^m)^2 + \sum_{i=2}^{n-1}(u_i^m)^2 + \frac{1}{2}(u_n^m)^2 \right) \tag{8.102}$$

with that of

$$\Delta x \left(\frac{1}{2}(f(x_1))^2 + \sum_{i=2}^{n-1}(f(x_i))^2 + \frac{1}{2}(f(x_n))^2 \right) \tag{8.103}$$

in order to test experimentally whether or not a similar inequality holds in the case of homogeneous Neumann boundary conditions. Let

$$f(x) = e^x x(1 - x) \quad \text{for } x \in (0, 1).$$

and compute the sums (8.102) and (8.103) for a series of different discretization parameters Δx and Δt. What are your experiments indicating?
(d) Show that

$$\int_0^1 u^2(x, t) \, dx \le \int_0^1 u^2(x, 0) \, dx = \int_0^1 f^2(x) \, dx \quad \text{for } t \ge 0.$$

(e) A physical interpretation of the homogeneous Neumann boundary condition (8.100) is that heat can neither enter nor leave the body for which (8.99)–(8.101) is modeling the heat evolution. Thus it seems reasonable that the "total" heat present in the body will be constant with respect to time t.
Apply the definition of the initial condition f given in (c). For various values of the discretization parameters Δx and Δt, compute the sums

$$\Delta x \left(\frac{1}{2}u_1^m + \sum_{i=2}^{n-1}u_i^m + \frac{1}{2}u_n^m \right)$$

and

$$\Delta x \left(\frac{1}{2}f(x_1) + \sum_{i=2}^{n-1}f(x_i) + \frac{1}{2}f(x_n) \right).$$

What do you observe?

(f) Show that

$$\int_0^1 u(x,t)\,dx = \int_0^1 f(x)\,dx \quad \text{for } t \geq 0.$$

(g) If heat can neither leave nor enter a body Ω ($= (0,1)$) then the temperature throughout Ω will approach a constant temperature C as time increases. Thus, our conjecture is that

$$u(x,t) \to C \quad \text{as } t \to \infty \tag{8.104}$$

for all $x \in (0,1)$. Furthermore, it seems reasonable that C equals the average of the initial temperature distribution f.

Let

$$f(x) = 3.14 + \cos(2\pi x) \quad \text{for } x \in (0,1)$$

and define

$$C = \int_0^1 f(x)\,dx.$$

Design a numerical experiment suitable for testing the hypothesis (8.104). Perform several experiments and comment on your results.

We will now modify the method of separation of variables, presented in Sect. 8.2, to handle Neumann boundary conditions. It turns out that this can be accomplished in a rather straight forward manner.

(h) Apply the ansatz

$$u(x,t) = X(x)T(t)$$

to (8.99)–(8.100) and show that this leads to the two eigenvalue problems

$$T'(t) = \lambda T(t), \tag{8.105}$$
$$X''(x) = \lambda X(x), \quad X'(0) = 0 \text{ and } X'(1) = 0, \tag{8.106}$$

where λ is a constant.

(i) Show that

$$T(t) = ce^{\lambda t}$$

solves (8.105) for any constant c.

(j) Show that

$$X(x) = \cos(k\pi x)$$

satisfies (8.106) for $k = 0, 1, 2, \ldots$

(k) Use the results obtained in (i) and (j) and with the super-positioning principle to conclude that any convergent series of the form

$$u(x,t) = c_0 + \sum_{k=1}^{\infty} c_k e^{-k^2\pi^2 t} \cos(k\pi x)$$

defines a formal solution of (8.99) and (8.100).

(l) Prove that

$$\int_0^1 \cos(k\pi x)\cos(l\pi x)\,dx = \begin{cases} 0 & k \neq l, \\ 1/2 & k = l > 0, \\ 1 & k = l = 0. \end{cases}$$

(m) Let f be a smooth function defined on $(0, 1)$. We want to compute the Fourier cosine series of this function; that is, we want to determine constants $c_0, c_1, \ldots$ such that

$$f(x) = c_0 + \sum_{k=1}^{\infty} c_k \cos(k\pi x). \tag{8.107}$$

Show that if (8.107) holds, then

$$c_0 = \int_0^1 f(x)\,dx$$

and

$$c_k = 2\int_0^1 f(x)\cos(k\pi x)\,dx \quad \text{for } k = 1, 2, \ldots.$$

(n) Let f be defined as in question (g) and compute the solution of (8.99)–(8.101) by hand.

(o) Use the solution formula derived in (n) to show that

$$\lim_{t \to \infty} u(x, t) = \int_0^1 f(y)\,dy \quad \text{for all } x \in [0, 1],$$

cf. your numerical experiments in (g).

8.3.2 Variable Coefficients

In this project we will consider a problem with a variable thermal conductivity k. This means that $k = k(x)$ is a function of the spatial position x. For the sake of simplicity, we will assume that k is smooth and that there exist positive numbers m and M such that

$$0 < m \leq k(x) \leq M \quad \text{for } x \in [0, 1].$$

In this case, the model for the heat conduction takes the form (see Chap. 7)

$$u_t = (ku_x)_x \quad \text{for } x \in (0, 1),\ t > 0, \tag{8.108}$$
$$u(0, t) = u(1, t) = 0 \quad \text{for } t > 0, \tag{8.109}$$
$$u(x, 0) = f(x) \quad \text{for } x \in (0, 1), \tag{8.110}$$

where f represents a given initial temperature distribution.

For given positive integers n and m we define the discretization parameters Δx and Δt by

$$\Delta x = \frac{1}{n-1}$$

and

$$\Delta t = \frac{T}{m},$$

where T represents the endpoint of the time interval $[0, T]$ in which we want to discretize the model problem (8.108)–(8.110). As in the text above, the grid points are given by

$$x_i = (i-1)\Delta x \quad \text{for } i = 1,\ldots,n,$$
$$t_\ell = \ell\Delta t \quad \text{for } \ell = 0,\ldots,m,$$

and in addition we introduce the notation

$$x_{i+1/2} = x_i + \Delta x/2 \quad \text{for } i = 1,\ldots,n-1.$$

Let

$$k_{i+1/2} = k(x_{i+1/2}),$$
$$u_i^\ell \approx u(x_i, t_\ell), \quad \text{for } i = 1,\ldots,n-1 \text{ and } \ell = 0,1,\ldots,m,$$

and consider the approximations

$$u_t(x_i, t_\ell) \approx \frac{u_i^{\ell+1} - u_i^\ell}{\Delta t}, \tag{8.111}$$

$$(ku_x)_x(x_i, t_\ell) \approx \frac{(ku_x)(x_{i+1/2}, t_\ell) - (ku_x)(x_{i-1/2}, t_\ell)}{\Delta x}, \tag{8.112}$$

$$(ku_x)(x_{i+1/2}, t_\ell) \approx k_{i+1/2}\frac{u_{i+1}^\ell - u_i^\ell}{\Delta x}, \tag{8.113}$$

$$(ku_x)(x_{i-1/2}, t_\ell) \approx k_{i-1/2}\frac{u_i^\ell - u_{i-1}^\ell}{\Delta x}. \tag{8.114}$$

(a) Use the approximations (8.111)–(8.114) to derive the scheme

$$u_i^{\ell+1} = \alpha k_{i-1/2}u_{i-1}^\ell + \left(1 - \alpha(k_{i-1/2} + k_{i+1/2})\right)u_i^\ell + \alpha k_{i+1/2}u_{i+1}^\ell \tag{8.115}$$

for $i = 2,\ldots,n-1$ and $\ell = 0,\ldots,m-1$. The boundary conditions (8.109) and the initial condition (8.110) are handled by setting

$$u_1^\ell = u_n^\ell = 0 \quad \text{for } \ell = 1, \ldots, m,$$
$$u_i^0 = f(x_i) \quad \text{for } i = 1, \ldots, n.$$

Write a computer program that implements this scheme.

(b) Recapture the argument leading to the bound (8.88). Prove that the numerical approximations generated by (8.115) satisfy

$$\max_i |u_i^\ell| \le \max_i |f(x_i)| \quad \text{for } \ell = 0, \ldots, m,$$

provided that

$$\Delta t \le \frac{1}{2M} \Delta x^2. \tag{8.116}$$

(c) Our goal now is to investigate whether or not the energy bound (8.7), valid for the model problem (8.1)–(8.3), holds in the present case. Let

$$k(x) = 1 + x,$$

$$f(x) = x(1 - x),$$

and $T = 1$. For several choices of the discretization parameters Δx and Δt, use the scheme (8.115) to compute approximate solutions of the model problem (8.108)–(8.110). Make sure that the condition (8.116) is satisfied. In each experiment you should compute

$$\Delta x \left(\frac{1}{2}(u_1^m)^2 + \sum_{i=2}^{n-1}(u_i^m)^2 + \frac{1}{2}(u_n^m)^2 \right) \tag{8.117}$$

and

$$\Delta x \left(\frac{1}{2}(f(x_1))^2 + \sum_{i=2}^{n-1}(f(x_i))^2 + \frac{1}{2}(f(x_n))^2 \right). \tag{8.118}$$

Explain why we can use (8.117) and (8.118) to test experimentally whether or not an inequality of the form (8.7) seems to be valid in the present case. What are your experiments indicating?

(d) Modify the argument leading to (8.7) to the present problem and show that

$$\int_0^1 u^2(x, t)\, dx \le \int_0^1 u^2(x, 0)\, dx = \int_0^1 f^2(x)\, dx \quad \text{for } t \ge 0.$$

(e) In Sect. 8.1.1 we established a bound for the first order derivative of the solution of the heat equation (8.1)–(8.3), cf. inequality (8.11). Is a similar property valid for the problem (8.108)–(8.110)? More precisely, we will analyze whether the bound

$$\int_0^1 k(x)u_x^2(x,t)\,dx \le \int_0^1 k(x)u_x^2(x,0)\,dx = \int_0^1 k(x)f_x^2(x)\,dx \quad \text{for } t \ge 0$$

$$(8.119)$$

holds.

Design a series of numerical experiments suitable for testing this conjecture. Apply the definitions of f, k and T given in (c). According to your experiments, is (8.119) true?

(f) Recapture the mathematical derivation of inequality (8.11), adapt that argument to the present case and prove that (8.119) must hold.

(g) Use the result obtained in (e) to show that

$$\int_0^1 u_x^2(x,t)\,dx \le \frac{M}{m}\int_0^1 u_x^2(x,0)\,dx = \frac{M}{m}\int_0^1 f_x^2(x)\,dx \quad \text{for } t \ge 0.$$

Chapter 9
Parameter Estimation and Inverse Problems

We have seen how mathematical models can be expressed in terms of differential equations. For example,

- Exponential growth,

$$r'(t) = ar(t) \quad \text{for } t > 0, \tag{9.1}$$
$$r(0) = r_0. \tag{9.2}$$

- Logistic growth,

$$r'(t) = ar(t)\left(1 - \frac{r(t)}{R}\right) \quad \text{for } t > 0, \tag{9.3}$$
$$r(0) = r_0. \tag{9.4}$$

- Diffusion,

$$u_t = (ku_x)_x \quad \text{for } x \in (0, 1), \, t > 0, \tag{9.5}$$
$$u(0, t) = u(1, t) = 0 \quad \text{for } t > 0, \tag{9.6}$$
$$u(x, 0) = f(x) \quad \text{for } x \in (0, 1). \tag{9.7}$$

Such equations can be used to analyze the underlying phenomena and to predict the behavior of various systems, which in turn can assist engineers solve practical problems.

In order to use such models we must somehow assign suitable values to the involved parameters:

- r_0 and a in the model for exponential growth
- r_0, a, and R in the model for logistic growth
- $f(x)$ and $k(x)$ in the model for diffusion

For some problems this can be accomplished by measuring the parameters directly. For example, one could perform physical experiments with a steel rod to record the diffusion coefficient k present in (9.5). This approach is unfortunately not always feasible. For population models it may, e.g., be almost impossible to measure the

A. Tveito et al., *Elements of Scientific Computing*, Texts in Computational Science and Engineering 7, DOI 10.1007/978-3-642-11299-7_9,
© Springer-Verlag Berlin Heidelberg 2010

growth rate a or the carrying capacity R. Instead, as was discussed in Sect. 5.2, these quantities must be estimated from historical data (commonly referred to as observation data), i.e. from recordings of the size of the population in the past. One may therefore refer to the latter method as indirect.

The purpose of this chapter is to shed some light onto parameter estimation problems. We will limit our discussion to the indirect approach. The task of estimating the size of parameters often leads to difficult equations, which may not have a unique solution, depending continuously on the observation data. Due to this fact, and the practical importance of the matter, this is currently an active research field.

9.1   Parameter Estimation in Exponential Growth

We will consider the estimation of the growth rate a and the initial condition r_0 in (9.1) and (9.2) from a slightly different perspective than that presented in Sect. 5.2. The purpose is to define a rather general approach that can be applied to a wide class of problems.

The solution of (9.1) and (9.2) depends on a and r_0, and we employ the notation

$$r(t; a, r_0) = r_0 e^{at} \tag{9.8}$$

to emphasize this fact. Let us reconsider the example analyzed in Sect. 5.2.1. That is, we want to use the total world population from 1950 to 1955, reported in Table 5.2, to determine a and r_0 and thereafter use the resulting model to estimate the population in the year 2000.

Let us set $t = 0$ at 1950, such that $t = 1$ corresponds to 1951, $t = 2$ corresponds to 1952, and so forth. If the population growth from 1950 to 1955 was exponential, then it would be possible to find real numbers a and r_0 such that, see Table 5.2,

$$
\begin{aligned}
1950: \quad & r(0; a, r_0) = 2.555 \cdot 10^9, \\
1951: \quad & r(1; a, r_0) = 2.593 \cdot 10^9, \\
1952: \quad & r(2; a, r_0) = 2.635 \cdot 10^9, \\
1953: \quad & r(3; a, r_0) = 2.680 \cdot 10^9, \\
1954: \quad & r(4; a, r_0) = 2.728 \cdot 10^9, \\
1955: \quad & r(5; a, r_0) = 2.780 \cdot 10^9,
\end{aligned}
$$

or

$$
\begin{aligned}
r_0 &= 2.555 \cdot 10^9, & (9.9) \\
r_0 e^{a} &= 2.593 \cdot 10^9, & (9.10) \\
r_0 e^{2a} &= 2.635 \cdot 10^9, & (9.11) \\
r_0 e^{3a} &= 2.680 \cdot 10^9, & (9.12)
\end{aligned}
$$

$$r_0 e^{4a} = 2.728 \cdot 10^9, \tag{9.13}$$
$$r_0 e^{5a} = 2.780 \cdot 10^9. \tag{9.14}$$

We thus have six equations, but only two unknowns a and r_0. The number of people on our planet did not, of course, grow precisely exponentially during this period, and one cannot therefore expect there to exist numbers a and r_0 satisfying (9.9)–(9.14). Instead we have to be content with trying to estimate a and r_0 such that these equations are approximately satisfied.

To this end, consider the function

$$J(a, r_0) = \frac{1}{2} \sum_{t=0}^{t=5} (r(t; a, r_0) - d_t)^2$$
$$= \frac{1}{2} \sum_{t=0}^{t=5} (r_0 e^{at} - d_t)^2,$$

where

$$d_0 = 2.555 \cdot 10^9,$$
$$d_1 = 2.593 \cdot 10^9,$$
$$d_2 = 2.635 \cdot 10^9,$$
$$d_3 = 2.680 \cdot 10^9,$$
$$d_4 = 2.728 \cdot 10^9,$$
$$d_5 = 2.780 \cdot 10^9.$$

Note that $J(a, r_0)$ is a sum of quadratic terms that measure the deviation between the output of the model and the observation data. It follows that if $J(a, r_0)$ is small, then (9.9)–(9.14) are approximately satisfied. We thus seek to minimize J:

$$\min_{a, r_0} J(a, r_0).$$

The first order necessary conditions for a minimum

$$\frac{\partial J}{\partial a} = 0,$$
$$\frac{\partial J}{\partial r_0} = 0,$$

yield a nonlinear 2×2 system of algebraic equations for a and r_0:

$$\sum_{t=0}^{t=5} (r_0 e^{at} - d_t) r_0 t e^{at} = 0, \tag{9.15}$$

$$\sum_{t=0}^{t=5}(r_0 e^{at} - d_t)e^{at} = 0. \tag{9.16}$$

In Exercise 9.2 we ask you to solve this system with Newton's method and to analyze the resulting model.

It is important to note that we suggest determining a and r_0 by minimizing the deviation between the output of the model and the observation data. We thus search for parameter values that yield a model that provide optimal fit to the available data. This approach is referred to as the *output least squares* method. The concept is easy to memorize, often straightforward to formulate, and can be applied to a wide range of practical problems. The function J is often referred to as an *objective function* or a *cost-functional*.

Please note that the standard output least squares form of the present problem is

$$\min_{a, r_0} \left[\frac{1}{2} \sum_{t=0}^{t=5}(r(t; a, r_0) - d_t)^2 \right]$$

subject to the constraints

$$r'(t) = ar(t) \quad \text{for } t > 0, \tag{9.17}$$
$$r(0) = r_0. \tag{9.18}$$

However, due to the formula (9.8) available for the solution of (9.17)–(9.18), it can be analyzed in the manner presented above. For most problems involving differential equations, such formulas for the solution are not known, and both the mathematical and numerical treatments of the output least squares form can be very difficult.

9.1.1 A Simpler Problem

Instead of seeking to compute both the growth rate a and the initial condition r_0, we might consider a somewhat simpler but less sophisticated approach. More specifically, one can use (9.9) to choose

$$r_0 = 2.555 \cdot 10^9$$

and estimate a by defining an objective function only involving the observation data from 1951 to 1955,

$$G(a) = \frac{1}{2} \sum_{t=1}^{t=5}(2.555 \cdot 10^9 e^{at} - d_t)^2. \tag{9.19}$$

The necessary condition

$$G'(a) = 0$$

for a minimum leads to the equation

$$\sum_{t=1}^{t=5}(2.555 \cdot 10^9 e^{at} - d_t)2.555 \cdot 10^9 te^{at} = 0, \qquad (9.20)$$

which must be solved to determine an optimal value for a, see Exercise 9.1.

In this case, the standard output least squares form is

$$\min_a \left[\frac{1}{2} \sum_{t=1}^{t=5}(r(t;a) - d_t)^2 \right]$$

subject to the constraints

$$r'(t) = ar(t) \quad \text{for } t > 0,$$
$$r(0) = 2.555 \cdot 10^9.$$

9.2 The Backward Diffusion Equation

The estimation of constant parameters in differential equations often leads to one or more algebraic equations that must be solved numerically. What about non-constant parameters? As you might have guessed, this is a far more subtle issue and can involve equations that are very difficult to solve. This topic, in its full complexity, is certainly far beyond the scope of this book. Nevertheless, in order to provide the reader with some basic understanding of the matter, we will consider a classic example, commonly referred to as the *backward diffusion equation*, in some detail.

Assume that a substance in an industrial process must have a prescribed temperature distribution, say, g(x), at time T in the future. Furthermore, the substance must be introduced/implanted into the process at time $t = 0$. (This could typically be the case in various molding processes or in steel casting). What should the temperature distribution $f(x)$ at time $t = 0$ be in order to ensure that the temperature is $g(x)$ at time T?

For the sake of simplicity, let us consider a medium with a constant diffusion coefficient $k(x) = 1$ for all x, occupying the unit interval. In mathematical terms, we may formulate our challenge as follows: Determine the initial condition $f = f(x)$ such that the solution $u = u(x, t; f)$ of

$$u_t = u_{xx} \quad \text{for } x \in (0, 1), \, t > 0, \qquad (9.21)$$
$$u(0, t) = u(1, t) = 0 \quad \text{for } t > 0, \qquad (9.22)$$
$$u(x, 0) = f(x) \quad \text{for } x \in (0, 1), \qquad (9.23)$$

is such that

$$u(x, T; f) = g(x) \quad \text{for all } x \in (0, 1).$$

In short, we want to use $g(x)$ and the diffusion equation to compute $f(x)$. Consequently, $g(x)$ is our observation data, and the output least squares formulation of the problem becomes

$$\min_{f} \left[\int_0^1 (u(x, T; f) - g(x))^2 \, dx \right] \tag{9.24}$$

subject to $u = u(x, t; f)$ satisfying (9.21)–(9.23).

At first glance one can get the impression that (9.24) is extremely hard to solve. However, we have put ourselves in the fortunate position such that the problem can be studied with Fourier analysis. (In most practical situations this is not the case and some sort of minimization algorithm must be employed.)

Recall that the solution $u(x, t; f)$ of (9.21)–(9.23) can be written in the form

$$u(x, t, f) = \sum_{k=1}^{\infty} c_k e^{-k^2 \pi^2 t} \sin(k \pi x),$$

where

$$f(x) = \sum_{k=1}^{\infty} c_k \sin(k \pi x) \quad \text{for } x \in (0, 1),$$

and (9.24) can therefore be expressed in terms of the Fourier coefficients,

$$\min_{c_1, c_2, \dots} \left[\int_0^1 \left(\sum_{k=1}^{\infty} c_k e^{-k^2 \pi^2 T} \sin(k \pi x) - g(x) \right)^2 \, dx \right]. \tag{9.25}$$

Next, we insert the Fourier sine expansion

$$g(x) = \sum_{k=1}^{\infty} d_k \sin(k \pi x) \quad \text{for } x \in (0, 1)$$

of g into (9.25) and obtain the following form of our problem:

$$\min_{c_1, c_2, \dots} \left[\int_0^1 \left(\sum_{k=1}^{\infty} c_k e^{-k^2 \pi^2 T} \sin(k \pi x) - \sum_{k=1}^{\infty} d_k \sin(k \pi x) \right)^2 \, dx \right]. \tag{9.26}$$

The Fourier coefficients of g can be computed by invoking formula (8.74) - keep in mind that g is the given observation data.

Since

$$\left[\int_0^1 \left(\sum_{k=1}^{\infty} c_k e^{-k^2 \pi^2 T} \sin(k \pi x) - \sum_{k=1}^{\infty} d_k \sin(k \pi x) \right)^2 dx \right] \geq 0$$

for all choices of $c_1, c_2, \ldots$, we can solve (9.26) by determining $c_1, c_2, \ldots$ such that

$$c_k e^{-k^2 \pi^2 T} \sin(k \pi x) = d_k \sin(k \pi x) \quad \text{for } k = 1, 2, \ldots,$$

which is satisfied if

$$c_k = e^{k^2 \pi^2 T} d_k \quad \text{for } k = 1, 2, \ldots.$$

The solution $f(x)$ of (9.24) is

$$f(x) = \sum_{k=1}^{\infty} e^{k^2 \pi^2 T} d_k \sin(k \pi x) \quad \text{for } x \in (0, 1),$$

where

$$g(x) = \sum_{k=1}^{\infty} d_k \sin(k \pi x) \quad \text{for } x \in (0, 1).$$

From a mathematical point of view, one might argue that the backward diffusion equation is a simple problem, since an analytical solution is obtainable. On the other hand, the problem itself has an undesirable property. More specifically, the distribution $f(x)$ at time $t = 0$ is determined by multiplying the Fourier coefficients of $g(x)$ by factors of the form $e^{k^2 \pi^2 T}$. These factors are very large, even for moderate k, e.g. with $T = 1$,

$$e^{\pi^2} \approx 1.93 \cdot 10^4,$$
$$e^{2^2 \pi^2} \approx 1.40 \cdot 10^{17},$$
$$e^{3^2 \pi^2} \approx 3.77 \cdot 10^{38}.$$

For example, if $T = 1$ and
$$g(x) = \sin(3 \pi x),$$
then the solution of the backward diffusion equation is

$$f(x) = e^{3^2 \pi^2} \sin(3 \pi x) \approx 3.77 \cdot 10^{38} \sin(3 \pi x).$$

This is quite amazing, or what?

Furthermore, if a very small amount of noise is added to g, say

$$\widehat{g}(x) = g(x) + 10^{-20} \sin(3\pi x) = (1 + 10^{-20}) \sin(3\pi x),$$

then the corresponding solution $\widehat{f}$ of (9.24) changes dramatically, i.e.

$$\widehat{f}(x) = (1 + 10^{-20})e^{3^2\pi^2} \sin(3\pi x) \approx f(x) + 3.77 \cdot 10^{18} \sin(3\pi x).$$

In fact,

$$\widehat{f}(x) - f(x) = 3.77 \cdot 10^{18} \sin(3\pi x),$$

even though

$$\widehat{g}(x) - g(x) = 10^{-20} \sin(3\pi x).$$

The problem is extremely unstable: Very small changes in the observation data g can lead to huge changes in the solution f of the problem.

One can therefore argue that it is almost impossible to estimate the temperature distribution backward in time by only using the present temperature and the diffusion equation. Further information is needed. This issue has led mathematicians to develop various techniques for incorporating a priori data, for example, that $f(x)$ should be almost constant. More precisely, a number of methods for approximating unstable problems with stable equations have been proposed, commonly referred to as *regularization techniques*.

Do the mathematical considerations presented above agree with our practical experiences with diffusion? For example, is it possible to track the temperature distribution backward in time in the room you are sitting? What kind of information do you need to do so? If you want to pursue a career in applied math, these are the kind of issues that you must consider carefully.

The estimation of spatially and/or temporally dependent parameters in differential equations often leads to unstable problems. It is an active research field called *inverse problems*. (You can think of the backward diffusion equation as the inverse of forecasting the future temperature.)

9.3 Estimating the Diffusion Coefficient

The examples discussed above are rather simple, since explicit formulas for the solutions of the involved differential equations are known. This is, of course, not always the case, and we will now briefly consider such a problem.

Assume that one wants to use surface measurements of the temperature to compute a possibly non-constant diffusion coefficient $k = k(x)$ inside a medium. With our notation, the output least squares formulation of this task is

$$\min_{k} \left[\int_0^T (u(0, t; k) - h_1(t))^2 \, dt + \int_0^T (u(1, t; k) - h_2(t))^2 \, dt \right]$$

subject to $u = u(x, t; k)$ satisfying

$$u_t = (ku_x)_x \quad \text{for } x \in (0, 1), \ t > 0,$$
$$k(0)u_x(0, t) = 0 \quad \text{for } t > 0,$$
$$k(1)u_x(1, t) = 0 \quad \text{for } t > 0,$$
$$u(x, 0) = f(x) \quad \text{for } x \in (0, 1).$$

Here, $h_1(t)$ and $h_2(t)$ represent the measurements made at the endpoints $x = 0$ and $x = 1$, respectively. We assume that the initial condition $f(x)$ is given and that $\partial u / \partial x = 0$ at the boundaries.

This is certainly a very difficult problem. To solve it, a number of mathematical and computational techniques developed throughout the last decades must be employed. This exceeds the ambitions of the present text, but we encourage the reader to carefully evaluate his or her understanding of the output least squares method by formulating such an approach for a problem involving, e.g., a system of ordinary differential equations.

9.4   Exercises

Exercise 9.1. (a) Use the bisection method to solve (9.20).
(b) Create a plot similar to that shown in Fig. 5.13. That is, visualize the actual total world population and the graph of the function $2.555 \cdot 10^9 e^{at}$, where a is the growth rate computed in (a), in the same figure.
(c) Compare the actual population in the year 2000 with that suggested/predicted by the model $2.555 \cdot 10^9 e^{at}$. How large is the error in the prediction?
(d) Solve (9.20) with Newton's method.

$\diamond$

Exercise 9.2. (a) Apply Newton's method to estimate the initial condition r_0 and the growth rate a by solving the nonlinear system of algebraic equations (9.15) and (9.16). You can use the population in the year 1950 and the growth rate estimated in Exercise 9.1 as the initial guess for the Newton iteration.
(b) Redo assignments (b) and (c) in Exercise 9.1 with the parameters estimated by solving the system (9.15) and (9.16).

$\diamond$

Exercise 9.3. The model for logistic growth (9.3) and (9.4) involves three parameters: the size of the initial population r_0, the growth rate a, and the carrying capacity R. We want to use Table 5.4, which contains the number of people on Earth for the period 1990–2000, to estimate these quantities. To this end, let $t = 0$ correspond to 1990, $t = 1$ correspond to 1991, and so on, and let $d_0, d_1, \ldots, d_{10}$ represent the total world population in 1990, 1991, ..., 2000, respectively.

Consider the following problem:

$$\min_{a,r_0,R} J(a, r_0, R),$$

where

$$J(a, r_0, R) = \frac{1}{2} \sum_{t=0}^{t=10} (r(t; a, r_0, R) - d_t)^2$$

and $r(t; a, r_0, R)$ denote the solution of (9.3)–(9.4).

(a) Use the analytical solution (2.21) of the logistic growth model to derive formulas for

$$\frac{\partial J}{\partial a}, \frac{\partial J}{\partial r_0} \text{ and } \frac{\partial J}{\partial R}.$$

(b) Show that

$$\frac{\partial J}{\partial a} = \frac{\partial J}{\partial r_0} = \frac{\partial J}{\partial R} = 0$$

for

$$a = 0,$$

$$r_0 = R = \frac{1}{11} \sum_{t=0}^{t=10} d_t.$$

(c) Implement Newton's method for the nonlinear system

$$\frac{\partial J}{\partial a} = 0,$$

$$\frac{\partial J}{\partial r_0} = 0,$$

$$\frac{\partial J}{\partial R} = 0.$$

Let $(\widetilde{a}, \widetilde{r}_0, \widetilde{R})$ denote the initial guess for the Newton iteration.

(d) Run Newton's method with

$$(\widetilde{a}, \widetilde{r}_0, \widetilde{R}) = \left(0, \frac{1}{11} \sum_{t=0}^{t=10} d_t, \frac{1}{11} \sum_{t=0}^{t=10} d_t \right).$$

(e) Run Newton's method with

$$(\widetilde{a}, \widetilde{r}_0, \widetilde{R}) = (0.05, 5.0, 11.0).$$

(f) Run Newton's method with

$$(\widetilde{a}, \widetilde{r_0}, \widetilde{R}) = (0.01, 5.0, 11.0).$$

(g) Run Newton's method with

$$(\widetilde{a}, \widetilde{r_0}, \widetilde{R}) = (0.8, 5.0, 11.0).$$

(h) Compare your results with those reported in Sect. 5.2.2. Is the initial guess important for the performance of Newton's method?

◇

Chapter 10
A Glimpse of Parallel Computing

It should not be too difficult to imagine that applications of scientific computing in the real world can require huge amounts of computation. This can be due to a combination of advanced mathematical models, sophisticated numerical algorithms, and high accuracy requirements. Such large-scale applications easily involve millions (or more) of data entities, and thousands (or more) of time steps in the case of time-dependent problems. All these will translate into a huge number of floating-point arithmetic operations and enormous data structures on a computer. However, a serial computer that has only one CPU will have trouble getting all these computations done quickly enough and/or fitting needed data into its memory. The remedy to this capacity problem is to use parallel computing, for which the present chapter aims to provide a gentle introduction.

10.1   Motivations for Parallel Computing

Let us start with motivating the use of multiple-processor computers by looking at the limitations of serial computers from the perspectives of computing speed and memory size.

10.1.1   From the Perspective of Speed

The development of computing hardware was well predicted by the famous *Moore's law* [22], which says that the number of transistors that can be put in an integrated circuit grows exponentially at a roughly fixed rate. Consequently, the speed of a top serial computer, in terms of floating-point operations per second (FLOPS), kept growing exponentially. This trend held from the 1950s until the beginning of the twenty-first century, but it then showed signs of flattening out. A future serial computer that can solve a problem that is too large for today's serial computer is unlikely. Combing the forces of many serial computers in some way, i.e., parallel computing, thus becomes a reasonable approach.

A. Tveito et al., *Elements of Scientific Computing*, Texts in Computational Science and Engineering 7, DOI 10.1007/978-3-642-11299-7_10,

The most prominent reason for adopting parallel computing is the need to finish a large computational task more quickly. To illustrate this point, let us consider a simplified three-dimensional diffusion equation:

$$\frac{\partial u}{\partial t} = \frac{\partial^2 u}{\partial x^2} + \frac{\partial^2 u}{\partial y^2} + \frac{\partial^2 u}{\partial z^2}, \tag{10.1}$$

for which a more general form was introduced in Sect. 7.2.

We want to develop an explicit numerical scheme that is based on finite differences for the above 3D equation. This will be done in the same fashion as for the 1D version described in Sect. 7.4. Suppose superscript ℓ denotes a discrete time level, subscripts i, j, k denote a spatial grid point, Δt denotes the time step size, and Δx, Δy, and Δz denote the spatial grid spacing. Then, the temporal derivative term in (10.1) is discretized as

$$\frac{\partial u}{\partial t} \approx \frac{u_{i,j,k}^{\ell+1} - u_{i,j,k}^{\ell}}{\Delta t}.$$

For the spatial derivative term in the x-direction, the finite difference approximation is

$$\frac{\partial^2 u}{\partial x^2} \approx \frac{u_{i-1,j,k}^{\ell} - 2u_{i,j,k}^{\ell} + u_{i+1,j,k}^{\ell}}{\Delta x^2}.$$

The other two spatial derivative terms $\partial^2 u/\partial y^2$ and $\partial^2 u/\partial z^2$ can be discretized similarly.

Combining all the finite differences together, the explicit numerical scheme for the 3D diffusion equation (10.1) arises as

$$\begin{aligned}
\frac{u_{i,j,k}^{\ell+1} - u_{i,j,k}^{\ell}}{\Delta t} &= \frac{u_{i-1,j,k}^{\ell} - 2u_{i,j,k}^{\ell} + u_{i+1,j,k}^{\ell}}{\Delta x^2} \\
&+ \frac{u_{i,j-1,k}^{\ell} - 2u_{i,j,k}^{\ell} + u_{i,j+1,k}^{\ell}}{\Delta y^2} \\
&+ \frac{u_{i,j,k-1}^{\ell} - 2u_{i,j,k}^{\ell} + u_{i,j,k+1}^{\ell}}{\Delta z^2}.
\end{aligned} \tag{10.2}$$

For simplicity we assume that the solution domain is the unit cube with a Dirichlet boundary condition, e.g., $u = 0$. The values of $u_{i,j,k}^0$ are prescribed by some initial condition $u(x, y, z, 0) = I(x, y, z)$. If the spatial grid spacing is the same in all three directions, i.e., $\Delta x = \Delta y = \Delta z = h = 1/n$, the explicit numerical scheme will have the following formula representing the main computational work per time step:

$$u_{i,j,k}^{\ell+1} = \alpha u_{i,j,k}^{\ell}$$
$$+ \beta \left(u_{i-1,j,k}^{\ell} + u_{i,j-1,k}^{\ell} + u_{i,j,k-1}^{\ell} + u_{i+1,j,k}^{\ell} + u_{i,j+1,k}^{\ell} + u_{i,j,k+1}^{\ell} \right)$$
$$(10.3)$$

for $1 \leq i, j, k \leq n - 1$, where $\alpha = 1 - 6\Delta t / h^2$ and $\beta = \Delta t / h^2$.

The above formula requires eight floating-point operations per inner grid point: two multiplications and six additions. That is, the total number of floating-point operations per time step is $8(n - 1)^3$. Recall from Sect. 7.4.5 that explicit numerical schemes often have a strict restriction on the maximum time step size, which is

$$\Delta t \leq \frac{1}{6} h^2$$

for this particular 3D case. The minimum number of time steps N needed for solving (10.1) between $t = 0$ and $t = 1$ is consequently $N \geq \frac{6}{h^2} = 6n^2$. Therefore, the total number of floating-point operations for the entire computation is

$$6n^2 \times 8(n - 1)^3 = 48n^2(n - 1)^3 \approx 48n^5.$$

If we have $n = 1,000$, then the entire computation requires 48×10^{15} floating-point operations. How much CPU time does it need to carry out these operations on a serial computer? Let us assume that an extremely fast serial computer has a peak performance of 48 GFLOPS, i.e., 48×10^9 FLOPS; then the total computation will require 10^6 s, i.e., 278 h. This may not sound like an alarmingly long time. However, the sustainable performance of numerical schemes of type (10.3), which are computer-memory intensive, is normally far below the theoretical peak performance. This is due to the increasing gap between the processor speed and memory speed on modern microprocessors, commonly referred to as the "memory wall" problem [21]. Moreover, our simple model equation (10.1) has not considered variable coefficients, difficult boundary conditions, or source terms. Therefore, it is fair to say that a realistic 3D diffusion problem can require a lot more than the above theoretical CPU usage, making a serial computer totally unfit for the explicit scheme to work on a $1,000 \times 1,000 \times 1,000$ mesh.

As another consideration, numerical simulators are frequently used as an experimental tool. Many different runs of the same simulator are typically needed, requiring the computing time of each simulation to be within e.g. an hour, or ideally minutes.

It should be mentioned that there exist more computationally efficient methods for solving (10.1) than the above explicit scheme. For example, a numerical method with no stability constraint can use dramatically fewer time steps, but with much more work per step. Nevertheless, the above simple example suffices to show that serial computers clearly have a limit in computing speed. The bad news is that the speed of a single CPU core is not expected to grow anymore in the future. Also, as will be shown in the following text, the memory limit of a serial computer is equally prohibitive for large-scale computations.

10.1.2 From the Perspective of Memory

The technical terms of megabyte (MB) and gigabyte (GB) should sound familiar to anyone who has used a computer. But how much data can be held in 1 MB or 1 GB of random access memory exactly? According to the standard binary definition, 1 MB is $1024^2 = 1,048,576$ bytes and 1 GB is $1024^3 = 1,073,741,824$ bytes. Scientific computations typically use double-precision numbers, each occupying eight bytes of memory on a digital computer. In other words, 1 MB can hold 131,072 double-precision numbers, and 1 GB can hold 134,217,728 double-precision numbers.

We mentioned in the above text that time savings are the most prominent reason for adopting parallel computing. But it is certainly not the only reason. Another equally important reason is to solve larger problems. Let us consider the example of an 8 GB memory, which is decently large for a single-CPU computer. By simple calculation, we know that 8 GB can hold 1024^3 double-precision values. For our simple numerical scheme (10.3), however, 8 GB is not enough for the case of $n = 1,000$, because two double-precision arrays of length $(n+1)^3$ are needed for storing u^ℓ and $u^{\ell+1}$.

Considering the fact that a mesh resolution of $1,000 \times 1,000 \times 1,000$ is insufficient for many problems, and that a typical simulation can require dozens or hundreds of large 3D arrays in the data structure, single-CPU computers are clearly far from being capable of very large-scale computations, with respect to both computing speed and memory size.

10.1.3 Parallel Computers

In fact, at the time of this writing, serial computers will soon become history. Every new PC now has more than one processor core, where each core is an independent processing unit capable of doing the tasks of a conventional processor. It is likely that all future computers will be parallel in some form. Knowledge about designing parallel algorithms and writing parallel codes thus becomes essential.

A parallel computer can be roughly defined as a computing system that allows multiple processors to work concurrently to solve one computational problem. The most common way to categorize modern parallel computers is by looking at the memory layout. A *shared-memory* system means that the processors have no private memory but can all access a single global memory. Symmetric multiprocessors (SMPs) were the earliest shared-memory machines, where the memory access time is uniform for all the processors. Later, shared-memory computers adopted the architecture of non-uniform memory access to incorporate more processors. The recent multicore chips can be considered a revival of SMP, equipped with some level of shared cache among the processor cores plus tighter coupling to the global memory.

In the category of *distributed-memory* systems, all processors have a private local memory that is inaccessible by others. The processors have some form of interconnection between them, ranging from dedicated networks with high throughput and low latency to the relatively slow Ethernet. The processors communicate with each other by explicitly sending and receiving messages, which are arrays of data values in a programming language. Two typical categories of distributed-memory systems are proprietary massively parallel computers and cost-effective PC clusters. For example, Ethernet-connected serial computers in a computer lab fall into the latter category. We refer to Fig. 10.1 for a schematic overview of the shared-memory and distributed-memory parallel architectures.

There are, of course, parallel systems that fall in between the two main categories. For example, a cluster of SMP machines is a *hybrid* system. A multicore-based PC cluster is, strictly speaking, also a hybrid system where memory is distributed among the PC nodes, while one or several multicore chips share the memory within each node. Furthermore, the cores inside a node can be inhomogeneous, e.g., general-purpose graphics processing units can be combined with regular CPU cores to accelerate parallel computations. For a review of the world's most powerful parallel computers, we refer the reader to the Top500 List [2].

Note also that parallel computing does not necessarily involve multiple processors. Actually, modern microprocessors have long exploited hardware parallelism within one processor, as in instruction pipelining, multiple execution units, and so on. Development of this compiler-automated parallelism is also part of the reason why single-CPU computing speed kept up with Moore's law for half a century. However, the present chapter will only address parallel computations that are enabled by using appropriate software on multiple processors.

10.2 More About Parallel Computing

A parallel computer provides the technical possibility, but whether or not parallel computing can be applied to a particular computational problem depends on the existence of parallelism and how it can be exploited in a form suitable for the parallel

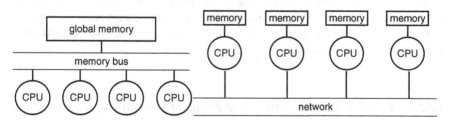

Fig. 10.1 A schematic layout of the shared-memory (*left*) and distributed-memory (*right*) parallel architectures

hardware. This section introduces the basic idea of parallel computing and its main ingredients.

10.2.1 Inspirations from Parallel Processing in Real Life

In everyday life, it is very common to see several people join forces to work on something together. Let us look at a scenario where two workers are set to paint a wall. Of course, each worker should first have a brush and a can of paint. Then, the two workers need to negotiate which area of the wall is to be painted by whom. When this work division is settled, each painter can start painting his or her assigned area, without having to interact with the other. The only assumption is that the two painters do not fight with each other when painting the boundary zone. In case more workers are available, the situation is similar except that the wall needs to be divided into more and smaller areas.

Compared with letting one worker paint the wall, employing several workers will normally speed up the project of wall painting. It has to be assumed, though, that the work division is reasonable, i.e., all the workers are able to finish their assigned areas using roughly the same amount of time. Otherwise the entire project of wall painting is not done until the slowest worker finishes his part. A second observation is that the initial negotiation of work division, which is not necessary for the single-worker case, should not cost too much time overhead. A third observation is that you cannot assign too many workers to paint a wall. That is, each worker should have a sufficiently large area to paint, without frequently coming over to others' areas and thereby slowing down each other. All in all, a speedup of the work for this very simple case is possible to achieve, but requires a large enough wall plus caution with respect to the work division.

There is, however, another important factor that is not addressed in the preceding example: Collaboration between the workers is often needed while the work is being carried out. Let us consider the example of bricklaying. Suppose a number of masons are hired to set up a brick wall. A possible means of work division is to let each mason have a vertical strip, inside of which she can lay the bricks layer-wise from bottom to top. For reasons of structural strength, the masons should preferably always work on the same layer of bricks. In addition, the positioning of bricks by neighboring masons should match up, without leaving large gaps between sections of bricks. The masons therefore need to collaborate, i.e., adopt a synchronized pace and exchange ideas frequently.

10.2.2 From Serial Computing to Parallel Computing

Not unlike the above cases of parallel processing in real life, parallel computing is only meaningful when an entire computational problem can be somehow divided evenly among the processors, which are to work concurrently, with coordination in the form of information exchange and synchronization. The purpose of information exchange is to provide private data to other processors that need them, whereas synchronization has the purpose of keeping the processors at same pace when desired. Both forms of coordination require the processors to communicate with each other. Moreover, for parallel computing to be beneficial, it is necessary that all the processors have a sufficient work load, and that the extra work caused by parallelization is limited.

Parallelism and Work Division

Parallelism in a computational problem arises from the existence of computations that can be carried out concurrently. The two most common forms of parallelism in scientific computing are *task parallelism* and *data parallelism*. Task parallelism arises from a set of standalone computational tasks that have a clear distinction between each other. Some of the tasks may need to follow the completion of other tasks. The possibility of parallel computing in this context arises from the concurrent execution of multiple tasks, possibly with some interaction between each other.

A simple example of task parallelism is parameter analysis: for example, a coefficient of a partial differential equation (PDE) is perturbed systematically to study its impact on the solution of the PDE. Here, each different value of the coefficient requires a new solution of the PDE, and all these different PDE-solving processes can proceed totally independently of each other. Therefore, one-PDE solving process, i.e., a computational task, can be assigned to one processor.

In the case that a certain dependency exists among (some of) the tasks, it is necessary to set up a dependency graph involving all the tasks. Parallel execution starts with the tasks that depend on nobody else and progressively includes more and more tasks that become ready to execute. When there are more waiting tasks than available processors, a dynamically updated job queue needs to be set up and possibly administered by a master processor. The worker processors can then each take one task off the queue, work on the assigned task, and come back for more until the queue is empty. Also, the worker processors may need to communicate with each other, if the assigned tasks require interaction.

When the number of tasks is small and the number of processors is large, an extra difficulty arises, because multiple processors have to share the work of one task. This is a more challenging case for parallel computing and requires further work division. The answer to this problem is data parallelism, which is a much more widespread source of parallelism in scientific computing than task parallelism. Data parallelism has many types and loosely means that identical or similar operations can be carried out concurrently on different sections of a data structure. Data parallelism can be considered as extreme task parallelism, because the operations applied on each unit of the data structure can be considered a small task. Data parallelism can thus produce, if needed, fine-grain parallelization, whereas task parallelism normally results in coarse-grain parallelization. The somewhat vague distinction between the two

forms of parallelism lies in the size and type of the tasks. It is also worth noticing
that some people restrict the concept of data parallelism to loop-level parallelism,
but we in this chapter we will refer data parallelism more broadly as parallelism
that arises from dividing some global data structure and the associated computing
operations. Unlike in the case of task parallelism, where the tasks can be large and
independent of each other, the different pieces in data parallelism are logically connected through an underlying global data structure. Work division related to data
parallelism, which is the first step of transforming a serial task into its parallel counterpart, can be non-trivial. So will be the interaction between the processors. Our
focus is therefore on data parallelism in the following text.

10.2.3 Example 1 of Data Parallelism

Let us consider a simple example of evaluating a uni-variable function $f(x)$ for a
set of x values. In a typical computer program the x values and evaluation results
are stored in one-dimensional arrays, and the computational work is implemented
as the following for-loop in C/C++ syntax:

```
for (i=0; i<n; i++)
   y[i] = f(x[i]);
```

An important observation is that the evaluations of f for different x values are
independent of each other. If each function evaluation $f(x)$ is considered as a separate task, this example can be categorized as task parallel. However, it is more
common to consider the present example as data parallel, because the same function is applied to an array of numerical values. Parallelism arises from dividing
the x values into subsets, and one subset is given to one processor. Suppose all
the evaluations of f are equally expensive and all processors are equally powerful.
Fairness thus requires that each processor get the same number of x values. (For
cases of different costs of the evaluations and inhomogeneous processors, we refer
to Exercises 10.2 and 10.3.)

Let P denote the number of processors and n the number of x values; then each
processor should take n/P values. In case n is not divisible by P, the following
formula provides a fair work division:

$$n_p = \left\lfloor \frac{n}{P} \right\rfloor + \begin{cases} 1 & \text{if } p < \text{mod}(n, P), \\ 0 & \text{else,} \end{cases} \tag{10.4}$$

where p denotes the processor id, which by convention starts from 0 and stops at
$P - 1$. In (10.4) the symbol $\lfloor \cdot \rfloor$ denotes the floor function that gives the same result
as integer division in a computer language, and $\text{mod}(n, P)$ denotes the remainder
of the integer division n/P. It is clear that the maximum difference between the
different n_p values is one, i.e., the fairest possible partitioning.

Example 10.1. If P is 6 and n is 63, then we have

$$\left\lfloor \frac{n}{P} \right\rfloor = \left\lfloor \frac{63}{6} \right\rfloor = 10 \quad \text{and} \quad \mod(n, P) = \mod(63, 6) = 3.$$

This means that processors with id $p = 0, 1, 2$ will be assigned with $n_p = 11$, whereas the other processors get $n_p = 10$. ∎

Finding the number of x values for each processor, e.g., by using (10.4), does not complete the work division yet. The next information is about which x values should be assigned to each processor. There can be many different solutions to this partitioning problem, and its impact on the resulting parallel performance is normally larger for distributed-memory systems than for the shared-memory counterparts. Often, letting each processor handle a contiguous piece of memory is performance friendly. In this spirit, the index set $\{0, 1, \ldots, n - 1\}$ corresponding to the array entries can be segmented into P pieces, where the start position for processor p is

$$i_{\text{start}, p} = p \left\lfloor \frac{n}{P} \right\rfloor + \min(p, \mod(n, P)). \tag{10.5}$$

Example 10.2. Following the above example, where $P = 6$ and $n = 63$, we can use (10.5) to find

$$i_{\text{start}, 0} = 0 \times \left\lfloor \frac{63}{6} \right\rfloor + \min(0, \mod(63, 6)) = 0,$$

$$i_{\text{start}, 1} = 1 \times \left\lfloor \frac{63}{6} \right\rfloor + \min(1, \mod(63, 6)) = 11,$$

$$i_{\text{start}, 2} = 2 \times \left\lfloor \frac{63}{6} \right\rfloor + \min(2, \mod(63, 6)) = 22,$$

$$i_{\text{start}, 3} = 3 \times \left\lfloor \frac{63}{6} \right\rfloor + \min(3, \mod(63, 6)) = 33,$$

$$i_{\text{start}, 4} = 4 \times \left\lfloor \frac{63}{6} \right\rfloor + \min(4, \mod(63, 6)) = 43,$$

$$i_{\text{start}, 5} = 5 \times \left\lfloor \frac{63}{6} \right\rfloor + \min(5, \mod(63, 6)) = 53.$$

∎

When the work division is ready, i.e., n_p and $i_{\text{start}, p}$ are computed, the parallelized computation can be implemented simply as follows:

```
for (i=i_start_p; i<i_start_p+n_p; i++)
   y[i] = f(x[i]);
```

The above code segment implicitly assumes that all processors can access the entire x and y arrays in memory, although each processor is only assigned to work

on a distinct segment. This assumption fits very well with a shared-memory architecture, i.e., arrays x and y are shared among all processors. Such shared-memory
programming closely resembles standard serial programming, but there are differences that require special attention. For example, the index integer i can not be
shared between processors, because each processor needs to use its own i index to
traverse an assigned segment of x and y.

In the case of distributed memory, the rule of the thumb is to avoid allocating
global data structures if possible. Therefore, each processor typically only allocates two local arrays, x_p and y_p, which are of length n_p and correspond to the
assigned piece of the global x and y arrays. Distributed-memory programming thus
has to consider more details, such as local data allocation and mapping between
local and global indices. However, the pure computing part on a distributed-memory
architecture is rather simple, as follows:

```
for (i=0; i<n_p; i++)
   y_p[i] = f(x_p[i]);
```

10.2.4 Example 2 of Data Parallelism

Parallelizing the previous example of function evaluation is very simple, because the
processors can work completely independently of each other. However, such embarrassingly parallel examples with no collaboration between the processors are rare in
scientific computing. Let us now look at another example where inter-processor
collaboration is needed.

The composite trapezoidal rule of numerical integration was derived in Chap. 1,
where the original formula was given as (1.16). Its purpose is to approximate the
integral $\int_a^b f(x)dx$ using $n + 1$ equally spaced samples. Before we discuss its
parallelization, let us first recall the computationally more efficient formula of this
numerical integration rule, which was given earlier as (6.2):

$$\int_a^b f(x)dx \approx h\left(\frac{1}{2}(f(a) + f(b)) + \sum_{i=1}^{n-1} f(a + ih)\right), \quad h = \frac{b-a}{n}. \quad (10.6)$$

Looking at (10.6), we see that it is quite similar to the previous example of
function evaluation. The difference is that the evaluated function values need to
be summed up, where the two end-points receive a half weight, in comparison with
the $n - 1$ inner points. An important observation is that summation of a large set of
values can be achieved by letting each processor sum up a distinct subset and then
adding up all the partial sums. Similar to the previous example, parallelization starts
with dividing the $n - 1$ inner-point function evaluations on P processors, which also
carry out a subsequent partial summation. More specifically, the sampling points
$x_1, x_2, \ldots, x_{n-1}$, where $x_i = a + ih$, are segmented into P equal pieces. Then
each processor can compute a partial sum in the form:

$$s_p = \sum_{i=i_{\text{start},p}}^{i_{\text{start},p}+n_p-1} f(x_i), \tag{10.7}$$

where, for this example, assuming equally expensive function evaluations and homogeneous processors, we have

$$n_p = \left\lfloor \frac{n-1}{P} \right\rfloor + \begin{cases} 1 & \text{if } p < \text{mod}(n-1, P), \\ 0 & \text{else,} \end{cases} \tag{10.8}$$

and

$$i_{\text{start},p} = 1 + p \left\lfloor \frac{n-1}{P} \right\rfloor + \min(p, \text{mod}(n-1, P)). \tag{10.9}$$

Example 10.3. Suppose $n = 200$ and $P = 3$; then we have from (10.8) $n_0 = 67$, $n_1 = n_2 = 66$. Following (10.9), we have $i_{\text{start},0} = 1$, $i_{\text{start},1} = 68$, and $i_{\text{start},2} = 134$. More specifically, processor 0 is responsible for inner points from x_1 until x_{67}, processor 1 works for inner points from x_{68} until x_{133}, and processor 2 is assigned to inner points from x_{134} until x_{199}. ∎

So parallel computing in connection with the composite trapezoidal rule arises from the fact that each processor can independently compute its s_p following (10.7). However, when the processors have finished the local computation of s_p, an additional computation of the following form is needed to complete the entire work:

$$\int_a^b f(x)dx \approx h \left(\frac{1}{2}(f(a) + f(b)) + \sum_{p=0}^{P-1} s_p \right). \tag{10.10}$$

Note that the local results s_p are so far only available on the different processors. Therefore, before we can calculate $\sum_{p=0}^{P-1} s_p$, the processors have to somehow share all the s_p values. There are two approaches. In the first approach, we designate one processor as the master and consider all the other processors as slaves. All the slave processors pass their s_p value to the master, which then computes the final result using (10.10). In case the slaves also want to know the final result, the master can send it to all the slaves. In the second approach, all processors have an equal role. The first action on each processor is to pass its own s_p value to all the other $P - 1$ processors, and get in return $P - 1$ different s_p values. The second action on each processor is then to independently carry out the computation of (10.10).

It is worth noting that the processors have to collaborate in calculating $\sum_{p=0}^{P-1} s_p$. In the first approach above, the communication is of the form all-to-one, followed possibly by a subsequent one-to-all communication. In the second approach, the communication is of the form all-to-all. No matter which approach, we can imagine that a considerable amount of communication programming is needed for summing up the different s_p values possessed by the P processors. Luckily, standard parallel programming libraries and languages have efficient built-in implementations for such collective communications with associated calculation (i.e., summation in this example) that are called *reduction operations*. Users thus do not have to program these from scratch. The time cost of these built-in reduction operations is typically on the order $\mathcal{O}(\log_2 P)$.

As an alternative to the above data-parallel approach to the example of the composite trapezoidal integration rule, let us now consider another approach that is more in the style of task parallelism. To this purpose, let us first note the following relation:

$$\int_a^b f(x)dx = \sum_{j=0}^{P-1} \int_{X_j}^{X_{j+1}} f(x)dx, \tag{10.11}$$

where $a = X_0 < X_1 < X_2 < \ldots < X_{P-1} < X_P = b$ is an increasing sequence of x-coordinates that coincide with a subset of the sampling points $\{x_i\}$. In other words, formula (10.11) divides the integral domain $[a, b]$ into P segments. The following new parallelization approach is motivated by the fact that a serial implementation of the composite trapezoidal rule probably already exists, such as Algorithm 6.2. The idea now is to let processor p call the serial trapezoidal(X_p, X_{p+1}, f, n_p) function as an independent task to approximate $\int_{X_p}^{X_{p+1}} f(x)dx$. Afterward, the global sum is obtained by a reduction operation involving all the P processors. We have to ensure that n_p this time gives a division of the n sampling intervals, instead of dividing the $n - 1$ inner points. (For computing the X_p values that match the P-way division of the n intervals, we refer the reader to Exercise 10.4.) The advantage is the reuse of an existing serial code, plus a more compact parallel implementation. The actual computation on each processor is done by a single call to the trapezoidal(X_p, X_{p+1}, f, n_p) function instead of a for-loop. The disadvantage is a small number of duplicated function evaluations and floating-point arithmetic operations, which arise because computation of $0.5f(X_p)$ is carried out on both processor $p - 1$ and processor p. In practical cases, where n is very large, the cost of the duplicated operations can be neglected.

10.2.5 Example 3 of Data Parallelism

The above example of computing the composite trapezoidal rule in parallel only requires one collective communication in the form of a reduction operation at the end of the computation. Let us now look at another example where inter-processor collaboration is more frequent.

Consider an explicit numerical scheme for solving the 1D diffusion problem (7.82)–(7.85) from Sect. 7.4. The finite difference formula for the $n - 1$ inner points is, as already given in (7.91),

$$u_i^{\ell+1} = u_i^\ell + \frac{\Delta t}{\Delta x^2}\left(u_{i-1}^\ell - 2u_i^\ell + u_{i+1}^\ell\right) + \Delta t f(x_i, t_\ell) \quad \text{for } i = 1, 2, \ldots, n - 1. \tag{10.12}$$

In addition, the Dirichlet boundary condition (7.83), i.e., $u(0, t) = D_0(t)$, is realized as

$$u_0^{\ell+1} = D_0(t_{\ell+1}),\tag{10.13}$$

whereas the Neumann boundary condition (7.84), i.e., $\partial u/\partial x(1, t) = N_1(t)$, is realized as

$$u_n^{\ell+1} = u_n^\ell + 2\frac{\Delta t}{\Delta x^2}\left(u_{n-1}^\ell - u_n^\ell + N_1(t_\ell)\Delta x\right) + \Delta t f(1, t_\ell),\tag{10.14}$$

see Sect. 7.4.3 for the details.

Parallelism in formula (10.12) is due to the fact that the computations on any two inner points $u_i^{\ell+1}$ and $u_j^{\ell+1}$ are independent of each other. More specifically, to compute the value of $u_i^{\ell+1}$, we rely on three nodal values from the previous time step: $u_{i-1}^\ell, u_i^\ell, u_{i+1}^\ell$, thus none of the other $u^{\ell+1}$ values. The $n-1$ inner points can actually be updated simultaneously for the same time level $\ell + 1$. Work division is the same as partitioning these inner points. If the index set $\{1, 2, \ldots, n-1\}$ is segmented into P contiguous pieces, the work division is equivalent to decomposing the solution domain into P subdomains, as is evident in Fig. 10.2. Actually, for many PDE problems, it is more natural and general to let domain decomposition give rise to the work and data division, so that each processor is assigned with a subdomain.

After a work division is decided, the local work of each processor at time step $\ell + 1$ is to compute $u_i^{\ell+1}$ using (10.12) for a subset of the i indices $\{1, 2, \ldots, n-1\}$. In addition, the processor that is responsible for x_0 has to update $u_0^{\ell+1}$ using (10.13), and similarly the processor responsible for x_n needs to compute $u_n^{\ell+1}$ using (10.14). It should be stressed again that concurrency in formula (10.12) assumes that the inner points on the same time level are updated. In other words, no processor should be allowed to proceed to the next time step before all the other processors have finished the current time step. Otherwise the computational results will be incorrect. Such a coordination among processors can typically be achieved by a built-in synchronization operation called *barrier*, which forces all processors to wait for the slowest one. On a shared-memory system, the barrier operation is the only needed inter-processor communication. All nodal values of u^ℓ are accessible by all processors on a shared-memory architecture, and therefore there is no need to communicate while each processor is computing its assigned portion of $u^{\ell+1}$.

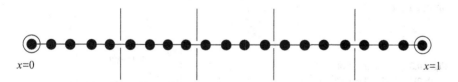

$x=0$ $x=1$

Fig. 10.2 An example of partitioning the 1D computational domain $x \in (0, 1)$, The *leftmost* and *rightmost* mesh points are marked for treatment of the physical boundary conditions, whereas the inner points are divided fairly among the processors

On a distributed-memory system, the parallelization is a bit more complex. We recall that a processor with only a local memory should avoid allocating global data structures if possible. So processor p should ideally only operate on two local arrays $\mathbf{u}_p^\ell$ and $\mathbf{u}_p^{\ell+1}$, both of length n_p, which contain, respectively, the assigned segments of the u_i^ℓ and $u_i^{\ell+1}$ values. The data segmentation can use the formulas (10.8) and (10.9). However, as indicated by (10.12), computing the leftmost and rightmost values of $\mathbf{u}_p^{\ell+1}$ requires one u^ℓ value each from the two neighboring subdomains. To avoid costly if-tests when computing the leftmost and rightmost values of $\mathbf{u}_p^{\ell+1}$, it is therefore more convenient to extend the two local arrays by one value at both ends. These two additional points are called *ghost points*, whose values participate in the owner subdomain's computations but are provided by the neighboring subdomains through communication. We refer the reader to Fig. 10.3 for an illustration. It should also be noted that on processor 0 the left ghost point coincides with the left physical boundary point $x = 0$, whereas on processor $P - 1$ the right ghost point coincides with the right physical boundary point $x = 1$.

Compared with a corresponding implementation on a shared-memory system, the implementation on a distributed-memory system is different in its inter-processor communications, in addition to using local data arrays. Here, each pair of neighboring subdomains has to explicitly exchange one nodal value per time step. For example, the value of $u_{p,1}^{\ell+1}$ that is computed on processor p needs to be sent to processor $p - 1$, which in return sends back the computed value on its rightmost inner point. A similar data exchange takes place between processors p and $p + 1$. The data exchanges need to be carried out before proceeding to the next time step. The resulting communications involve pairs of processors, commonly known as *one-to-one* communications. In this example, these one-to-one communications implicitly ensure that the computations on the neighboring processors are synchronized. There is therefore no need for a separate barrier operation, which is required in the shared-memory implementation. The complete numerical scheme suitable for a distributed-memory computer is given in Algorithm 10.1.

Remarks

To a beginner, whether or not parallelism exists in a computational problem can seem mysterious. A rule of the thumb is that there must be a sufficient amount of computational work and also that (parts of) the computations must not be dependent on each other.

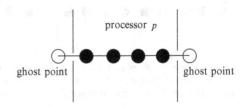

Fig. 10.3 An example of a subdomain that is assigned to processor p. The assigned mesh points consist of two ghost points and a set of inner points

Algorithm 10.1

Explicit Scheme for Solving the 1D Diffusion Problem (7.82)–(7.85) on a Distributed-Memory Computer.

Given $n + 1$ as the total number of global mesh points, $\Delta x = 1/n$, Δt as the time step size, m as the number of time steps, $\alpha = \Delta t/\Delta x^2$, and P as the total number of processors.

On processor p ($0 \le p \le P - 1$):

if $p < \mathrm{mod}(n - 1, P)$ then
$$n_p = \left\lfloor \tfrac{n-1}{P} \right\rfloor + 1$$
else
$$n_p = \left\lfloor \tfrac{n-1}{P} \right\rfloor$$
$$i_{\mathrm{start},p} = p\left\lceil \tfrac{n-1}{P} \right\rceil + \min(p, \mathrm{mod}(n - 1, P))$$
Allocate local arrays $\mathbf{u}_p^\ell$ and $\mathbf{u}_p^{\ell+1}$ both of length $n_p + 2$
Initial condition: $u_{p,i}^\ell = I((i_{\mathrm{start},p} + i)\Delta x)$, for $i = 0, 1, \ldots, n_p + 1$
for $\ell = 0, 1, \ldots, m - 1$

  for $i = 1, 2, \ldots, n_p$
$$u_{p,i}^{\ell+1} = u_{p,i}^\ell + \alpha \left(u_{p,i-1}^\ell - 2u_{p,i}^\ell + u_{p,i+1}^\ell \right)$$
$$+ \Delta t f((i_{\mathrm{start},p} + i)\Delta x, t_\ell)$$

  if $p = 0$ then
$$u_{p,0}^{\ell+1} = D_0(t_{\ell+1})$$
  else

    receive value from processor $p - 1$ into $u_{p,0}^{\ell+1}$
    send value of $u_{p,1}^{\ell+1}$ to processor $p - 1$
  if $p = P - 1$ then
$$u_{p,n_p+1}^{\ell+1} = u_{p,n_p+1}^\ell + 2\alpha \left(u_{p,n_p}^\ell - u_{p,n_p+1}^\ell + N_1(t_\ell)\Delta x \right)$$
$$+ \Delta t f(1, t_\ell)$$
  else

    send value of $u_{p,n_p}^{\ell+1}$ to processor $p + 1$
    receive value from processor $p + 1$ into $u_{p,n_p+1}^{\ell+1}$
  Data copy before next time step: $\mathbf{u}_p^\ell = \mathbf{u}_p^{\ell+1}$
end for

For instance, an addition operation between two scalar values is obviously not a subject for parallelization, whereas adding two long vectors is embarrassingly parallel. Solving a single ordinary differential equation (ODE), even if a huge number of time steps are employed, is normally serial by nature. This is because each time step contains too little work and the time steps have to be carried out one after another. For a system of ODEs, the situation with respect to parallel computing can be improved, especially when the number of involved ODEs is large. Parallelism can be found within each time step, either because the ODEs are handled separately in an explicit scheme, or because an implicit scheme solves a linear system that has some level of parallelism. Nevertheless, the number of ODEs in a system is normally not as many as the number of available processors, so exploiting such limited parallelism can be challenging. The best situation arises in an ODE–PDE coupled problem, because many numerical methods involve solving the ODE part for every spatial grid point. In this case, when the number of spatial grid points is large, the task of solving all these ODEs is again embarrassingly parallel.

Even when parallelism exists, work division can still be a challenge. Our previous examples are simple, because the work division there is essentially a one-dimensional partitioning applied to structured data entities. In the case of two- or three-dimensional unstructured computational meshes, partitioning not only has to make sure that each processor has approximately the same amount of work, but also has to keep the amount of resulting communication as low as possible. In the case of time-dependent problems that have a dynamically changing work load, dynamic load balancing can be another challenge.

10.2.6 Performance Analysis

The prominent reason for adopting parallel computing, time saving is also the main measure for evaluating the quality of a parallelized code. If we denote by $T(P)$ the computing time for solving a problem using P processors, the very important concept of *speedup* is thus defined as

$$S(P) = \frac{T(1)}{T(P)}, \tag{10.15}$$

where $T(1)$ should be the computing time used by a serial code, and not a parallelized code run on one processor. This is because a parallel implementation typically has some additional operations that are not needed in a serial implementation.

An ideal result of parallelization is that the entire serial code is parallelizable with a perfect work division and that the parallelization-induced communication cost is negligible. In such an ideal situation, we can expect that $T(P)$ is exactly one Pth of $T(1)$, thus $S(P) = P$. The ideal situation is rarely achievable, so we understandably would like the value of $S(P)$ to be close to P; the larger the better quality of the parallelization. An equivalent quality measure, called *parallel efficiency*, can be defined as

$$\eta(P) = \frac{S(P)}{P} = \frac{T(1)}{P \, T(P)}, \tag{10.16}$$

which will have a normalized value between 0 and 1.

Amdahl's Law

The first obstacle to perfect parallelization is that there can exist some bits of a serial code that are inherently sequential. Suppose a serial code has a fraction of size α

that is not parallelizable. Consequently, even if the remaining fraction $1 - \alpha$ is perfectly parallelized, there is a theoretical upper limit on the best possible achievable speedup. Amdahl's law [3] states the following:

$$S(P) \leq \frac{T(1)}{\left(\alpha + \dfrac{1-\alpha}{P}\right) T(1)} = \frac{1}{\alpha + \dfrac{1-\alpha}{P}} < \frac{1}{\alpha}. \tag{10.17}$$

For instance, if 1% of a serial code is not parallelizable, the best possible speedup cannot exceed $1/\alpha = 1/0.01 = 100$, no matter how many processors are used.

Amdahl's law was derived in the 1960s, a time when parallel computing was still in its infancy. The theory was in fact used as an argument against this new-born technology. An important observation is that Amdahl's law assumes a fixed problem size, no matter how many processors are used. It in effect says that it does not pay to use too many processors if a threshold of parallel efficiency is to be maintained (see Exercise 10.6). Amdahl's law indeed gives an unnecessarily pessimistic view, because the non-parallelizable fraction of a code typically decreases as the problem size increases. In other words, it pays off to use more processors when the problem becomes larger.

Gustafson–Barsis's Law

There is another theory, known as Gustafson–Barsis's law [16] that looks at the issue of scaled speedup. The basic idea is that P processors should be used to solve a problem that is P times the problem size on one processor. Suppose the absolute amount of serial computing time does not grow with problem size, and let α_P denote the fraction of serial computing time in $T(P)$; then $T(1)$ would have been $(\alpha_P + P(1 - \alpha_P)) T(P)$. The scaled speedup can be computed as follows:

$$S(P) = \frac{(\alpha_P + P(1 - \alpha_P)) T(P)}{T(P)} = \alpha_P + P(1 - \alpha_P) = P - \alpha_P(P - 1). \tag{10.18}$$

An important assumption of Gustafson–Barsis's law is that the fraction of serial computing time becomes increasingly small as the problem size increases, which is true for most parallel codes. The consequence is that it is possible to achieve a scaled speedup very close to P, provided the problem size is large enough.

Gustafson–Barsis's law can also be used to estimate speedup without knowing $T(1)$. For example, if $\alpha_P = 10\%$ is assumed for $P = 1,000$, then the scaled speedup of using 1,000 processors is, according to (10.18), $P - \alpha_P(P - 1) = 1000 - 0.1(1000 - 1) = 900.1$. This gives a corresponding parallel efficiency of 90%.

As indicated by Gustafson–Barsis's law, when a desired speedup value is not achievable due to Amdahl's law, it is time to consider enlarging the problem size. After all, solving larger problems is the other important reason for adopting parallel computing.

10.2.7   Overhead Related to Parallelization

Neither Amdahl's law nor Gustafson–Barsis's law has considered the impact of overhead on the achievable speedup. To a great extent, the attention on the fraction of non-parallelizable computation is over emphasized by both theories. In many large-scale problems, however, the actual fraction of non-parallelizable computation is virtually zero. The real obstacles to achieving high speedup values are different types of overhead that are associated with the parallelization.

First of all, a parallelized algorithm can introduce additional computations that were not present in the original serial algorithm. For example, work division often requires a few arithmetic operations, such as using (10.8) and (10.9). The cost of partitioning unstructured computational meshes, in particular, is often considerable. Second, when perfect work division is either impossible or too expensive to achieve, the resulting work load imbalance will decrease the actual speedup. Third, explicit synchronization between processors may be needed from time to time. Such operations can be costly. Fourth, communication overhead can surely not be overlooked in most parallel applications. For a collective communication involving all P processors, the cost is typically of the order $\mathcal{O}\left(\lceil \log_2 P \rceil\right)$, where $\lceil \cdot \rceil$ denotes the ceiling function which gives the smallest integer value that is equal to or larger than $\log_2 P$. For one-to-one communication on distributed-memory systems, an idealized model for the overhead of transferring a message with L bytes from one processor to another is as follows:

$$t_C(L) = \tau + \xi L, \tag{10.19}$$

where τ is the so-called *latency*, which represents the start-up time for communication, and ξ is the cost for transferring one byte of data. By the way, $1/\xi$ is often referred to as the *bandwidth* of the communication network. The parameter values of τ and ξ can vary greatly from system to system. The reason for saying that (10.19) is an idealized model is because it does not consider the possible situation of several processors competing for the network, which can reduce the network's effective capacity. Also, the actual curve of $t_C(L)$ is often of a staircase shape, instead of a straight line with constant slope. (For example, transferring one double-precision value can take the same time as transferring a small number of double-precision values in one message.) Although shared-memory systems seem to be free of one-to-one communications, there is related overhead behind the scene. For example, each processor typically has own private cache. We remark that cache is a small piece of fast memory that dynamically duplicates a small portion of the main memory. The speed of cache is much faster than that of the main memory, and the use of cache is meant to overcome the memory-speed bottleneck. A variable in the shared memory can thus risk being updated differently in several caches. Keeping the different caches coherent between each other can therefore incur costly operations in the parallel system.

It should be mentioned that there are also factors that may be speedup friendly. First, many parallel systems have the capability of carrying out communications at the same time as computations. This allows for the possibility of hiding the

communication overhead. However, to enable communication–computation overlap
can be a challenging programming task. Second, it sometimes happens that by using
many processors on a distributed-memory system, the subproblem per processor
suddenly acquires a much better utilization of the local cache, compared with solving the entire serial problem using one CPU and its local cache. This provides the
potential of superlinear speedup, i.e., $S(P) > P$, for particular problem sizes.

Example 10.4. Let us analyze the speedup of the parallel composite trapezoidal rule
(10.7)–(10.10). As we can see from (10.6), the serial algorithm involves $n + 1$ function evaluations plus $\mathcal{O}(n)$ additions and multiplications. If we assume that the
function evaluations are much more expensive than the floating-point operations,
the computational cost of the serial algorithm is

$$T(1, n) = (n + 1)t_f,$$

where t_f denotes the cost of one function evaluation. The computational cost of the
parallel algorithm is

$$T(P, n) = \left\lceil \frac{n - 1}{P} \right\rceil t_f + C \lceil \log_2 P \rceil + 2t_f, \tag{10.20}$$

where the first term corresponds to the cost of computing s_p using (10.7), the second
term corresponds to the cost of a reduction operation needed to sum all the s_p values,
and the last term corresponds to the cost of computing $f(a)$ and $f(b)$ in the end.
Therefore, the speedup will be of the following form:

$$S(P, n) = \frac{T(1, n)}{T(P, n)} = \frac{(n + 1)t_f}{\left\lceil \frac{n-1}{P} \right\rceil t_f + C \lceil \log_2 P \rceil + 2t_f}$$

$$= \frac{n + 1}{\left\lceil \frac{n-1}{P} \right\rceil + 2 + \tilde{C} \lceil \log_2 P \rceil}. \tag{10.21}$$

Of course, to compute the actual values of $S(P, n)$, we need to know the value of
the constant $\tilde{C} = C/t_f$. This can be achieved by measuring $T(P, n)$ for a number
of different choices of P and n and estimating the value of $\tilde{C}$ by the method of
least squares. Even without knowing the exact value of $\tilde{C}$ in (10.21), we can say
that the parallel efficiency $\eta(P)$ will get further away from 100% for increasing P.
This is because the $\tilde{C} \lceil \log_2 P \rceil$ term in the denominator of (10.21) increases with
P, meaning that the speedup will eventually saturate and decrease after passing a
threshold value of P, dependent on the actual size of n and $\tilde{C}$.

$\blacksquare$

Example 10.5. Table 10.1 shows some speedup results measured on a small cluster
of PCs, where the interconnect is 1 GB-Ethernet. The parallel program is written
using the message-passing interface (MPI [14, 26]) and implements the compos-

Table 10.1 Speedup results for computing the composite trapezoidal integration rule using an MPI program

P	1	2	4	8	16
$T(P)$	0.03909	0.02331	0.01599	0.008227	0.004398
$S(P)$	N/A	1.68	2.44	4.75	8.89
$\eta(P)$	N/A	0.84	0.61	0.59	0.56

ite trapezoidal integration rule, see Sect. 10.3.2. The integral to be approximated is $\int_0^1 \sin(\pi x)dx$ using $n = 10^6$ sampling intervals.

Due to the relatively small problem size and the slow interconnect, the speedup results in Table 10.1 are not very impressive. This example shows that communication overhead is an obstacle to good parallel efficiency, which typically deteriorates with increasing P.

■

In general, considering all the factors that have an impact on speedup, we should be prepared that a fix-sized problem often has an upper limit on the number of processors, beyond which speedup will decrease instead of increase.

10.3 Parallel Programming

So far we have discussed how serial scientific computations can be transformed into their parallel counterparts. The basic steps include parallelism identification, work division, and inter-processor collaboration. For the resulting computations to run on the parallel hardware, code implementation must be done accordingly. The hope of many people for an automatic tool that is able to analyze a serial code, find the parallelism, and insert parallelization commands has proved to be too ambitious. This observation is at least true for scientific codes of reasonable complexity. Therefore, some form of manual parallel programming is needed. There are currently three main forms of parallel programming: (1) recode in a specially designed parallel programming language, (2) annotate a serial code with compiler directives to provide the compiler with hints for parallelization tasks, and (3) restructure a serial code and insert calls to library functions for explicitly enforcing inter-processor collaboration.

The present section will consider the two latter options. More specifically, we will briefly explain the use of OpenMP [1] and MPI [14, 26], which target shared-memory and distributed-memory systems, respectively. Both are well-established standards of application programming interfaces for parallelization and thus provide code portability. For the newcomer to parallel programming, the most important thing is not the syntax details, which can be easily found from textbooks or online resources; rather, it is important to see that parallel programming requires a "mental picture" of multiple execution streams, which often need to communicate with each other by data exchange and/or synchronization.

10.3.1 OpenMP Programming

The OpenMP standard is applicable to shared-memory systems. It assumes that a parallel program has one or several parallel regions, where a parallel region is a piece of code that can be parallelized, e.g., a for-loop that implements the composite trapezoidal integration rule (10.6). Between the parallel regions the code is executed sequentially just like a serial program. Within each parallel region, however, a number of threads are spawned to execute concurrently. The programmer's responsibility is to insert OpenMP directives together with suitable clauses, so that an OpenMP-capable compiler can use these hints to automatically parallelize the annotated parallel regions. A great advantage is that a non-OpenMP-capable compiler will ignore the directives and treat the code as purely serial.

The OpenMP directive for constructing a parallel region is #pragma omp parallel in the C and C++ programming languages. The two most important OpenMP directives in C/C++ for parallelization are (1) #pragma omp for suitable for data parallelism associated with a for-loop, and (2) #pragma omp sections suitable for task parallelism. In the Fortran language, the OpenMP directives and clauses have slightly different names. For in-depth discussions about OpenMP programming, we refer the reader to [8] and [9].

Example 10.6. Let us show below an OpenMP parallelization of the composite trapezoidal rule (10.6), implemented in C/C++. Since the computational work in a serial implementation is typically contained in a for-loop, the OpenMP parallelization should let each thread carry out the work for one segment of the for-loop. This will result in a local s_p value on each thread, as described in (10.7). All the s_p values will then need to be added up by a reduction operation as mentioned in Section 10.2.2. Luckily, a programmer does not have to explicitly specify how the for-loop is to be divided, which is handled automatically by OpenMP behind the scene. The needed reduction operation is also incorporated in OpenMP's #pragma omp for directive:

```
h = (b-a)/n;
sum = 0.;

#pragma omp parallel for reduction(+:sum)
   for (i=1; i<=n-1; i++)
      sum += f(a+i*h);

sum += 0.5*(f(a)+f(b));
sum *= h;
```

In comparison with a serial implementation, the only difference is the line starting with #pragma omp parallel for, which is the combined OpenMP directive for parallelizing a single for-loop that constitutes an entire parallel region. The reduction(+:sum) clause on the same line enforces, at the end of the loop, a reduction operation of adding all the sum variables over all threads. Additional clauses can also be added. For example, schedule(static,chunksize) will result in loop iterations that are divided statically into chunks of size chunksize, and the chunks are assigned to the different threads cyclically. An appropriate choice of work division and scheduling, with a suitable value of chunksize, is important for the parallel performance.

∎

Example 10.7. Next, let us look at the most important section of an OpenMP implementation in C/C++ of the 1D diffusion problem (7.82)–(7.85), for which the explicit numerical method was given as (10.12)–(10.14).

```
u_prev = (double*)malloc((n+1)*sizeof(double));
u = (double*)malloc((n+1)*sizeof(double));
t = 0.;

#pragma omp parallel private(k)
  {
#pragma omp for
    for (i=0; i<=n; i++)          /* enforce initial condition */
      u_prev[i] = I(i*dx);

    for (k=1; k<=m; k++) {    /* time integration loop */
#pragma omp for schedule(static,100)
      for (i=1; i<n; i++)       /* computing inner points */
        u[i]=u_prev[i]+alpha*(u_prev[i-1]-2*u_prev[i]+u_prev[i+1])
             +dt*f(i*dx,t);

#pragma omp single
      {
      u[n] = u_prev[n]
             +2*alpha*(u_prev[n-1]-u_prev[n]+N1(t)*dx)
             +dt*f(1,t); /* right physical boundary condition */
      t += dt;
      u[0] = D0(t);      /* left physical boundary condition */
      }

#pragma omp for schedule(static,100)
      for (i=0; i<=n; i++)
        u_prev[i] = u[i];  /* data copy before next time step */
      }
  }
```

The above code contains considerably more instances of #pragma than in the previous example. Apart from parallelizing the for-loop for enforcing the initial condition, two for-loops, one for computing $u_i^{\ell+1}$ on the inner points and the other for copying array $\mathbf{u}^{\ell+1}$ to array $\mathbf{u}^{\ell}$, are also parallelized inside each time step. It can be seen that almost the entire code section is wrapped inside a large parallel region, indicated by #pragma omp parallel. This means that P threads are spawned at the entrance of the parallel region and stay alive throughout the region. This is why the OpenMP directive #pragma omp single is necessary to mark that only one thread does the work of incrementing the shared t variable and enforcing the two physical boundary conditions. Otherwise, letting all the threads repeat the same work will produce erroneous results.

Another possibility is to not use the large parallel region, but use #pragma omp parallel for in the three locations where #pragma omp for now stand. Such an approach will avoid using `#pragma omp single`, but will have to repeatedly spawn new threads and then terminate them, giving rise to more overhead in thread creation.

∎

OpenMP programming is quite simple, because the standard consists only of a small number of directives and clauses. Much of the parallelization work, such as work division and task scheduling, is hidden from the user. On the one side, this programming philosophy provides great user friendliness. On the other side, however, the user has very little control over which thread accesses which part of the shared memory. This limit is normally bad with respect to the performance on a modern processor architecture, which relies heavily on the use of caches. To fully utilize a cache, it is important that the data items that are read from memory to cache should be reused as much as possible. Data locality – either when data items that are located close by in memory participate in one computing operation, or when one data item is repeatedly used in consecutive operations – gives rise to good cache usage, but is hard to enforce in an OpenMP program.

10.3.2 MPI Programming

MPI is a standard specification for message-passing programming on distributed-memory computers. On shared-memory systems it is also common to have MPI installations that are implemented using efficient shared-memory intrinsics for passing messages. Thus MPI is the most portable approach to parallel programming. There are two parts of MPI: The first part contains more than 120 functions and constitutes the core of MPI [26], whereas the second part contains advanced extensions [14]. Here, we only intend to give a brief introduction to MPI programming. For more in-depth learning, we refer the reader to the standard MPI textbooks [15, 23].

A message in the context of MPI is simply an array of data elements of a predefined data type. The logical execution units in an MPI program are called *processes*, which are initiated by the `MPI_Init` call at the start of a program and terminated by the `MPI_Finalize` call at the end. The number of started MPI processes is usually the same as the number of available processors, although one processor can in principle be assigned with several MPI processes. All the started MPI processes constitute a so-called global MPI *communicator* named `MPI_Comm_World`, which can be subdivided into smaller communicators if needed. Almost all the MPI functions require an input argument of type MPI communicator, which together with a process rank (between 0 and $P - 1$) is used to specify a particular MPI process.

Compared with OpenMP, MPI programming is clearly more difficult, not only because of the large number of available functions, but also because one MPI call normally requires many arguments. For example, the simplest function in C for sending a message is

```
MPI_Send(void *buf, int count, MPI_Datatype datatype,
          int dest, int tag, MPI_Comm comm)
```

The first three arguments constitute the outgoing message, by specifying the initial memory address of the array, the number of data elements, and the data type. The `dest` argument gives the rank of the receiving process relative to the `comm` communicator, whereas `tag` is an integer argument used to label the particular message.

Correspondingly, the simplest MPI function for receiving a message is

```
MPI_Recv(void *buf, int count, MPI_Datatype datatype,
          int source, int tag, MPI_Comm comm, MPI_Status *status)
```

Compared with the `MPI_Send` function, the extra argument in the `MPI_Recv` function is a pointer to a so-called `MPI_Status` object, which can be used to check some details of a received message.

Below we will give two examples of MPI programming in the C language. Compared with Examples 10.6 and 10.7, we will see that a programmer is responsible for more details of the parallelization. Communications have to be enforced explicitly in MPI. Despite the extra programming effort, parallel MPI programs are usually good with respect to data locality, because the local data structure owned by each MPI process is small relative to the global data structure. In addition, the user has full control over work division, which can be of great value for performance enhancement.

For a beginner, it is important to realize that each MPI process executes the same MPI program. The distinction between the processes, which are spawned by some parallel runtime system, is through the unique process rank. The rank typically determines the work assignment for each process. Moreover, an `if`-test with respect to a particular process rank can allow the chosen process to perform different operations than the other processes.

Example 10.8. The following is the most important part of an MPI implementation of the composite trapezoidal integration rule (10.6):

```
#include <mpi.h>

/*
  code omitted for defining function f
*/

int main (int nargs, char** args)
{
    int P, my_id, n, n_p, i_start_p, i, remainder;
    double a, b, h, x, s_p, sum;

    MPI_Init (&nargs, &args);
    MPI_Comm_size(MPI_COMM_WORLD,&P);
    MPI_Comm_rank(MPI_COMM_WORLD,&my_id);

    n = 1000000; a = 0.; b = 1.;

    remainder = (n-1)%P;
```

```
n_p = (my_id<remainder) ? (n-1)/P+1 : (n-1)/P;
i_start_p = 1+my_id*((n-1)/P)+min(my_id,remainder);

h = (b-a)/n;
s_p = 0.; x = a+i_start_p*h;
for (i=0; i<n_p; i++) {
  s_p += f(x); x += h;
}

MPI_Allreduce (&s_p,&sum,1,MPI_DOUBLE,MPI_SUM,MPI_COMM_WORLD);

sum += 0.5*(f(a)+f(b));
sum *= h;

MPI_Finalize();
return 0;
}
```

The MPI functions MPI_Comm_size and MPI_Comm_rank give, respectively, the number of processes P and the unique id (between 0 and $P - 1$) of the current process. The work division, i.e., computation of n_p and $i_{start,p}$, follows the formulas (10.8) and (10.9). The for-loop implements the formula (10.7) for computing s_p. The MPI_Allreduce function is a collective reduction operation involving all the processes. Its effect is that all the local s_p values are added up and the final result is stored in the sum variable on all processes. An alternative is to use the MPI_Reduce function, which runs faster than MPI_Allreduce, but the final result will only be available on a chosen process. ∎

Example 10.9. Let us show below the most important part of an MPI implementation in C of Algorithm 10.1.

```
#include <mpi.h>
#include <malloc.h>

/*
  code omitted for defining functions I,f,D0,N1
*/

int main (int nargs, char** args)
{
  int P, my_id, n, n_p, i_start_p, i, k, m, remainder;
  int left_neighbor_id, right_neighbor_id;
  double dt, dx, alpha, t, *u_prev, *u;

  MPI_Init (&nargs, &args);
  MPI_Comm_size(MPI_COMM_WORLD,&P);
  MPI_Comm_rank(MPI_COMM_WORLD,&my_id);

  /*
    code omitted for choosing n and dt, computing dx,alpha etc.
  */

  remainder = (n-1)%P;
  n_p = (my_id<remainder) ? (n-1)/P+1 : (n-1)/P;
  i_start_p = my_id*((n-1)/P)+min(my_id,remainder);
```

```
u_prev = (double*)malloc((n_p+2)*sizeof(double));
u = (double*)malloc((n_p+2)*sizeof(double));

for (i=0; i<=n_p+1; i++)   /* enforce initial condition */
  u_prev[i] = I((i_start_p+i)*dx);

left_neighbor_id = (my_id==0) ? MPI_PROC_NULL : my_id-1;
right_neighbor_id = (my_id==P-1) ? MPI_PROC_NULL : my_id+1;

t = 0.;
for (k=1; k<=m; k++) {     /* time integration loop */
  for (i=1; i<=n_p; i++)   /* computing local inner points */
    u[i] = u_prev[i]+alpha*(u_prev[i-1]-2*u_prev[i]+u_prev[i+1])
           +dt*f((i_start_p+i)*dx,t);

  if (my_id==0)        /* left physical boundary condition */
    u[0] = D0(t+dt);

  if (my_id==P-1)      /* right physical boundary condition */
    u[n_p+1] = u_prev[n_p+1]
               +2*alpha*(u_prev[n_p]-u_prev[n_p+1]+N1(t)*dx)
               +dt*f(1,t);

  MPI_Sendrecv(&(u[1]),1,MPI_DOUBLE,left_neighbor_id,100,
       &(u[n_p+1]),1,MPI_DOUBLE,right_neighbor_id,100,
       MPI_COMM_WORLD,&status);
  MPI_Sendrecv(&(u[n_p]),1,MPI_DOUBLE,right_neighbor_id,200,
       &(u[0]),1,MPI_DOUBLE,left_neighbor_id,200,
       MPI_COMM_WORLD,&status);

  for (i=0; i<=n_p+1; i++)/* date copy before next time step */
    u_prev[i] = u[i];
  t += dt;
}

MPI_Finalize();
return 0;
}
```

We can see that the above code uses a slightly modified Algorithm 10.1, in that two calls to the MPI_Sendrecv function replace two pairs of MPI_Send and MPI_Recv. The modified code is thus more compact. A more important motivation for the modification is to avoid potential communication *deadlocks*, which will arise if a pair of neighboring processes both start with MPI_Recv before calling the matching MPI_Send command. These two MPI processes will be blocked, i.e., none will be able to return from its MPI_Recv call, which relies on its neighbor to have issued the MPI_Send call. A bullet-proof sequence of MPI_Send and MPI_Recv calls is that process *A* calls MPI_Send and MPI_Recv, whereas process *B* calls MPI_Recv and MPI_Send.

■

10.3.3 Concluding Remarks

Parallel programming is an evolving science, with new tools, languages, and paradigms steadily coming along. We should therefore bear this in mind and be prepared for future developments. There is no guarantee that MPI and OpenMP will always be the dominant standards for parallel programming. For example, the arrival of multicore chips has given rise to a serious debate on the applicability of these two approaches. The important principles of parallel computing will survive, no matter the standing of programming techniques.

Let us emphasize again that syntax details are not the most important stuff. A programmer needs to have a correct notion about the multiple execution streams that go side by side in a parallel program. A proper work division should give each execution stream a distinct set of operations to carry out, balanced among all the streams. The execution streams normally need to communicate with each other for the purpose of exchanging data and/or synchronizing pace. Although parallel programming libraries can provide a lot of ready-made communication functions, challenges will remain, such as the treatment of possible conflicts between the streams and performance-harming pitfalls. Interested readers are therefore referred to the rich literature on parallel programming and computing.

10.4 Exercises

Exercise 10.1. The peak performance of the world's most powerful parallel computer has undergone an exponential growth. Table 10.2 lists the history between 1993 and 2009, according to the Top500 List [2]. Suppose the growth can be described mathematically by the following formula:

$$G(t) = G_0 \cdot 2^{\frac{t-t_0}{t^*}},$$

where $G(t)$ denotes the development of the theoretical peak performance as a function of time, in terms of GFLOPS, and G_0 denotes the GFLOPS value at some initial time t_0. The purpose of this exercise is to estimate the value of t^*, i.e., the average length of time over which the peak performance doubles. The method of least squares should be used, where the value of t_0 can be chosen as year 1990. If the same growth continues, what will the peak performance of the most powerful parallel computer be in year 2020?   ◇

Exercise 10.2. We want to partition a set of n non-equal computational tasks among P homogeneous processors. Let us first suppose that half of the tasks cost twice the computing time of the other half's. How should a fair work division be? Next, suppose task 1 requires one unit of computing time, task 2 requires two time units, task 3 requires three time units, and so on. How should fair work division take place for this case?   ◇

Table 10.2 The development of the world's most powerful parallel computer according to the Top500 List [2]

Time	System model	CPU type	CPUs	GFLOPS
June 1993	TMC CM5	SuperSPARC I 32MHz	1024	131
June 1994	Intel Paragon	Intel 80860 50MHz	3680	184
June 1995	Fujitsu VPP	Fujitsu 105MHz	140	235.79
June 1996	Hitachi SR2201	HARP-1E 150MHz	1024	307.2
June 1997	Intel Paragon	Pentium Pro 200MHz	7264	1453
June 1998	Intel Paragon	Pentium Pro 200MHz	9152	1830.4
June 1999	Intel Paragon	Pentium Pro 333MHz	9472	3154
June 2000	Intel Paragon	Pentium Pro 333MHz	9632	3207
June 2001	IBM SP	POWER3 375MHz	8192	12288
June 2002	NEC SX6	NEC 1000MHz	5120	40960
June 2005	IBM BlueGene/L	PowerPC 440 700MHz	65536	183500
June 2006	IBM BlueGene/L	PowerPC 440 700MHz	131072	367000
June 2008	IBM BladeCenter Cluster	PowerXCell 8i 3200MHz	122400	1375776
June 2009	IBM BladeCenter Cluster	PowerXCell 8i 3200MHz	129600	1456704

Exercise 10.3. Suppose now that the P processors are not equal in computing speed, where half of them have double the speed of the others. How should n equally expensive computational tasks be divided?   ◇

Exercise 10.4. We said in Sect. 10.2.2 that parallelizing the composite trapezoidal rule can be done by two approaches. In the first approach the $n - 1$ inner points are divided among P homogeneous processors using (10.8) and (10.9). In the second approach the integral domain $[a, b]$ is divided into P segments, so that each processor applies the numerical integration rule on its assigned subdomain $[X_p, X_{p+1}]$. Derive the formula for calculating X_p, so that the n intervals are fairly divided among the processors. Also find out the exact difference (summed over all P processors) in the numbers of function evaluations and floating-point arithmetic operations between the two approaches.   ◇

Exercise 10.5. Derive the parallel algorithm of the composite Simpson's rule (6.4) using two approaches that are similar to the case of the composite trapezoidal rule.   ◇

Exercise 10.6. If $S^\star$ is a desired speedup value, how can we choose the number of processors P according to Amdahl's law (10.17)? Prove also that, if $\alpha > 0$, the parallel efficiency is a monotonically decreasing function of P as the result of Amdahl's law. If $\eta^\star$ is the acceptable threshold of parallel efficiency, what can be the maximum value of P?   ◇

Exercise 10.7. We recall that Gustafson–Barsis's law (10.18) uses α_P to denote the fraction of serial computing time in $T(P)$. What is the corresponding fraction α of non-parallelizable computing time in $T(1)$? If we know that the current speedup is $S(P) = S_1$ associated with $T(P)$ and the present value of α_P. By how much should the parallelizable computations be increased such that we can achieve an improved speedup S_2?   ◇

Exercise 10.8. The so-called "Ping-Pong" test is a well-used technique for measuring the values of latency τ and bandwidth $1/\beta$, which are used in (10.19). More specifically, a pair of processors keeps exchanging a message of length L_1 several times, using the MPI_Send and MPI_Recv commands. The measured time is then divided by twice the number of repetitions that the message was bounced between the two processors, giving rise to the actual cost of $t_C(L_1)$. Thereafter, the Ping-Pong test is repeated for different values of L, resulting in measurements of $t_C(L_2)$, $t_C(L_3)$, and so on. Write an MPI program that implements the above Ping-Pong test, and then uses the method of least squares to estimate the values of τ and $1/\beta$.   ◇

Exercise 10.9. Implement a serial code that performs a matrix-vector multiplication, where the matrix is dense and of dimension $n \times n$, and the vector is of length n. Parallelize the code using both OpenMP and MPI and compare the obtained speedup results.   ◇

Exercise 10.10. Carry out the performance analysis of Algorithm 10.1 in the same fashion as in Example 10.4. The one-to-one message transfer cost model (10.19) should be used in the analysis, while the actual values of τ and β are obtained from Exercise 10.8. Then, use the MPI program as described in Example 10.9 to verify the actual speedup results. Based on the time measurements, can you quantify how expensive it is to update one inner point $u_i^{\ell+1}$, relative to the value of τ?   ◇

10.5 Project: Parallel Solution of a 2D Diffusion Equation

Let us consider the following scaled diffusion equation in two space dimensions:

$$\frac{\partial u}{\partial t} = \frac{\partial^2 u}{\partial x^2} + \frac{\partial^2 u}{\partial y^2} + f(x,y,t), \quad (x,y) \in (0,1) \times (0,1), \ t > 0. \quad (10.22)$$

For the initial condition we have $u(x,y,0) = I(x,y)$, and the boundary conditions are as follows:

$$u(1,y,t) = D_1(y,t), \quad t > 0,$$
$$u(x,1,t) = D_2(x,t), \quad t > 0,$$

$$\frac{\partial}{\partial x}u(0,y,t) = N_3(y,t), \quad t > 0,$$
$$\frac{\partial}{\partial y}u(x,0,t) = N_4(x,t), \quad t > 0.$$

Following Sect. 7.4, we will use an explicit method based on finite differences for solving this two-dimensional diffusion equation. First, we introduce a uniform spatial mesh $(x_i = i\Delta x, y_j = j\Delta y)$, with $0 \le i \le n_x, 0 \le j \le n_y$ and $\Delta x = 1/n_x, \Delta y = 1/n_y$. Then, the approximate solution of u will be sought on the mesh points and at discrete time levels $t_\ell = \ell\Delta t, \ell > 0$.

(a) Show that an explicit numerical scheme for solving (10.22) on the inner mesh points at $t = t_{\ell+1}$ is of the following form:

$$u_{i,j}^{\ell+1} = \alpha u_{i,j}^\ell + \beta \left(u_{i-1,j}^\ell + u_{i+1,j}^\ell \right) + \gamma \left(u_{i,j-1}^\ell + u_{i,j+1}^\ell \right) + \Delta t f_{i,j}^\ell ,$$
$$(10.23)$$

where $1 \le i \le n_x - 1, 1 \le j \le n_y - 1$ and

$$\alpha = 1 - 2\left(\frac{\Delta t}{\Delta x^2} + \frac{\Delta t}{\Delta y^2} \right), \quad \beta = \frac{\Delta t}{\Delta x^2}, \quad \gamma = \frac{\Delta t}{\Delta y^2}.$$

(b) Derive the formulas for computing the numerical solutions on the four boundaries, following the discussions in Sects. 7.4.2 and 7.4.3.

(c) Implement the above explicit numerical scheme as a serial program in the C programming language, where the main computation given by (10.23) is implemented as a double-layered for-loop. For each of the four boundaries, a single-layered for-loop needs to be implemented according to (b).

(d) Parallelize the serial program using the OpenMP directive #pragma omp for. What are the obtained speedup results?

(e) Propose a two-dimensional domain decomposition for dividing the inner mesh points into $P_x \times P_y$ subdomains. More specifically, you should find a mapping $(i, j) \rightarrow (p_x, p_y)$ that can assign each inner mesh point (i, j) to a unique subdomain with id $(p_x, p_y) \in [0, P_x - 1] \times [0, P_y - 1]$.

(f) On each subdomain, the assigned mesh points should be expanded with one layer of additional points, which either lie on the actual physical boundaries or work as ghost points toward the neighboring subdomains. For a subdomain with id (p_x, p_y), which values of the local $u^{\ell+1}$ solution should be sent to which neighboring subdomains?

(g) Implement a new parallel program using MPI based on the above domain decomposition. The MPI_Sendrecv command can be used to enable inter-subdomain communication.

(h) An approach to hiding the communication overhead is to use so-called *non-blocking* communication commands in MPI. On a parallel system that is capable of carrying out communication tasks at the same time as computations, non-blocking communication calls can be used to initiate the communication (without waiting for its conclusion), followed immediately by computations. The simplest non-blocking MPI commands are MPI_Isend and MPI_Irecv, which upon return do not imply that the action of sending or receiving a message is completed. The standard MPI_Send and MPI_Recv commands are thus called blocking communication calls, meaning that the message is safely sent away or received upon the return of the calls.

To enable computation and communication overlap in the present example, we need to divide the local inner points into two sets. More specifically, the outermost layer of the local inner points on each subdomain should be computed first. Then, communication for transferring these new values between the subdomains can be initiated by the non-blocking MPI_Isend and MPI_Irecv commands. Thereafter, computation on the remaining local inner points can be carried out at the same time as communication is being handled. Finally, the blocking MPI_Wait command can be used to ensure that all the needed incoming messages have been received.

Construct a new MPI implementation based the above strategy of overlapping computation and communication.

References

1. The OpenMP API specification for parallel programming. http://openmp.org/wp/.
2. Top 500 supercomputer sites. http://www.top500.org/.
3. G. M. Amdahl. Validity of the single-processor approach to achieving large scale computing capabilities. In *AFIPS Conference Proceedings*, volume 30, pages 483–485. AFIPS Press, 1967.
4. T. M. Apostol. *Calculus; Volume 1*. John Wiley & Sons, Inc., 2nd edition, 1967.
5. R. B. Banks. *Growth and Diffusion Phenomena: Mathematical Frameworks and Applications*. Springer, 1994.
6. R. G. Bartle. *The Elements of Real Analysis*. John Wiley & Sons, Inc., 1964.
7. M. Braun. *Differential Equations and Their Applications*. Springer, 1975.
8. R. Chandra, L. Dagum, D. Kohr, D. Maydan, J. McDonald, and R. Menon. *Parallel Programming in OpenMP*. Morgan Kaufmann Publishers, Inc., 2000.
9. B. Chapman, G. Jost, and R. van der Pas. *Using OpenMP: Portable Shared Memory Parallel Programming*. MIT Press, 2007.
10. S. D. Conte and C. de Boor. *Elementary Numerical Analysis, an Algorithmic Approach*. McGraw-Hill, 1972.
11. R. Courant, H. Robbins, and I. Stewart. *What is Mathematics?* Oxford University Press, 1996.
12. P. J. Davis, R. Hersh, and E. A. Marchisotto. *The Mathematical Experience*. Birkhäuser, study edition, 1995.
13. C. H. Edwards and D. E. Penney. *Calculus and Analytic Geometry*. Prentice Hall, Inc., 1986.
14. W. Gropp, S. Huss-Lederman, A. Lumsdaine, E. Lusk, B. Nitzberg, W. Saphir, and M. Snir. *MPI–The Complete Reference: Volume 2, The MPI Extensions*. MIT Press, 1998.
15. W. Gropp, E. Lusk, and A. Skjellum. *Using MPI: Portable Parallel Programming with the Message-Passing Interface*. MIT Press, 2nd edition, 1999.
16. J. L. Gustafson. Reevaluating Amdahl's law. *Communications of the ACM*, 31(5):532–533, 1988.
17. R. Haberman. *Elementary Applied Partial Differential Equations with Fourier Series and Boundary Value Problems*. Prentice-Hall, Inc., 1983.
18. C. Johnson. *Numerical solution of partial differential equations by the finite element method*. Studentlitteratur, 1987.
19. H. P. Langtangen. *Python Scripting for Computational Science*, volume 3 of *Texts in Computational Science and Engineering*. Springer, 3rd edition, 2008.
20. H. P. Langtangen. *A Primer on Scientific Programming with Python*, volume 6 of *Texts in Computational Science and Engineering*. Springer, 2009.
21. S. A. McKee. Reflections on the memory wall. In *Proceedings of the 1st Conference on Computing Frontiers*, pages 162–167. ACM, 2004.
22. G. Moore. Cramming more components onto integrated circuits. *Electronics*, 38(8), 1965.
23. P. S. Pacheco. *Parallel Programming with MPI*. Morgan Kaufmann Publishers, Inc., 1997.
24. K. Rottmann. *Mathematische Formelsammlung*. Wissenschafts, 4th edition, 1991.
25. H. L. Royden. *Real Analysis*. MacMillan Publishing Co., Inc., 2nd edition, 1968.

26. M. Snir, S. Otto, S. Huss-Lederman, D. Walker, and J. Dongarra. *MPI–The Complete Reference: Volume 1, The MPI Core*. MIT Press, 2nd edition, 1998.
27. J. Stoer and R. Bilirsch. *Introduction to Numerical Analysis*. Springer, 1993.
28. A. Tveito and R. Winther. *Introduction to Partial Differential Equations – A Computational Approach*. Springer, 1998.
29. H. F. Weinberger. *A First Course in Partial Differential Equations*. John Wiley & Sons, Inc., 1965.
30. S. Wolfram. *A New Kind of Science*. Wolfram Media, Inc., 2002.

Index

Editorial Policy

§1. Textbooks on topics in the field of computational science and engineering will be considered. They should be written for courses in CSE education. Both graduate and undergraduate textbooks will be published in TCSE. Multidisciplinary topics and multidisciplinary teams of authors are especially welcome.

§2. Format: Only works in English will be considered. For evaluation purposes, manuscripts may be submitted in print or electronic form, in the latter case, preferably as pdf- or zipped ps- files. Authors are requested to use the LaTeX style files available from Springer at: http://www.springer.com/authors/book+authors?SGWID= 0-154102-12-417900-0 (for monographs, textbooks and similar)

Electronic material can be included if appropriate. Please contact the publisher.

§3. Those considering a book which might be suitable for the series are strongly advised to contact the publisher or the series editors at an early stage.

General Remarks

Careful preparation of manuscripts will help keep production time short and ensure a satisfactory appearance of the finished book.

The following terms and conditions hold:

Regarding free copies and royalties, the standard terms for Springer mathematics textbooks hold. Please write to martin.peters@springer.com for details.

Authors are entitled to purchase further copies of their book and other Springer books for their personal use, at a discount of 33,3 % directly from Springer-Verlag.

Series Editors

Texts in Computational Science and Engineering

1. H. P. Langtangen, *Computational Partial Differential Equations*. Numerical Methods and Diffpack Programming. 2nd Edition

2. A. Quarteroni, F. Saleri, P. Gervasio, *Scientific Computing with MATLAB and Octave*. 3rd Edition

3. H. P. Langtangen, *Python Scripting for Computational Science*. 3rd Edition

4. H. Gardner, G. Manduchi, *Design Patterns for e-Science*.

5. M. Griebel, S. Knapek, G. Zumbusch, *Numerical Simulation in Molecular Dynamics*.

6. H. P. Langtangen, *A Primer on Scientific Programming with Python*.

7. A. Tveito, H. P. Langtangen, B. F. Nielsen, X. Cai, *Elements of Scientific Computing*.

For further information on these books please have a look at our mathematics catalogue at the following URL: www.springer.com/series/5151

Monographs in Computational Science and Engineering

1. J. Sundnes, G.T. Lines, X. Cai, B.F. Nielsen, K.-A. Mardal, A. Tveito, *Computing the Electrical Activity in the Heart*.

For further information on this book, please have a look at our mathematics catalogue at the following URL: www.springer.com/series/7417

Lecture Notes in Computational Science and Engineering

1. D. Funaro, *Spectral Elements for Transport-Dominated Equations*.

2. H. P. Langtangen, *Computational Partial Differential Equations*. Numerical Methods and Diffpack Programming.

3. W. Hackbusch, G. Wittum (eds.), *Multigrid Methods V*.

4. P. Deuflhard, J. Hermans, B. Leimkuhler, A. E. Mark, S. Reich, R. D. Skeel (eds.), *Computational Molecular Dynamics: Challenges, Methods, Ideas*.

5. D. Kröner, M. Ohlberger, C. Rohde (eds.), *An Introduction to Recent Developments in Theory and Numerics for Conservation Laws*.

6. S. Turek, *Efficient Solvers for Incompressible Flow Problems*. An Algorithmic and Computational Approach.

7. R. von Schwerin, *Multi Body System SIMulation*. Numerical Methods, Algorithms, and Software.

8. H.-J. Bungartz, F. Durst, C. Zenger (eds.), *High Performance Scientific and Engineering Computing.*

9. T. J. Barth, H. Deconinck (eds.), *High-Order Methods for Computational Physics.*

10. H. P. Langtangen, A. M. Bruaset, E. Quak (eds.), *Advances in Software Tools for Scientific Computing.*

11. B. Cockburn, G. E. Karniadakis, C.-W. Shu (eds.), *Discontinuous Galerkin Methods.* Theory, Computation and Applications.

12. U. van Rienen, *Numerical Methods in Computational Electrodynamics.* Linear Systems in Practical Applications.

13. B. Engquist, L. Johnsson, M. Hammill, F. Short (eds.), *Simulation and Visualization on the Grid.*

14. E. Dick, K. Riemslagh, J. Vierendeels (eds.), *Multigrid Methods VI.*

15. A. Frommer, T. Lippert, B. Medeke, K. Schilling (eds.), *Numerical Challenges in Lattice Quantum Chromodynamics.*

16. J. Lang, *Adaptive Multilevel Solution of Nonlinear Parabolic PDE Systems.* Theory, Algorithm, and Applications.

17. B. I. Wohlmuth, *Discretization Methods and Iterative Solvers Based on Domain Decomposition.*

18. U. van Rienen, M. Günther, D. Hecht (eds.), *Scientific Computing in Electrical Engineering.*

19. I. Babuška, P. G. Ciarlet, T. Miyoshi (eds.), *Mathematical Modeling and Numerical Simulation in Continuum Mechanics.*

20. T. J. Barth, T. Chan, R. Haimes (eds.), *Multiscale and Multiresolution Methods.* Theory and Applications.

21. M. Breuer, F. Durst, C. Zenger (eds.), *High Performance Scientific and Engineering Computing.*

22. K. Urban, *Wavelets in Numerical Simulation.* Problem Adapted Construction and Applications.

23. L. F. Pavarino, A. Toselli (eds.), *Recent Developments in Domain Decomposition Methods.*

24. T. Schlick, H. H. Gan (eds.), *Computational Methods for Macromolecules: Challenges and Applications.*

25. T. J. Barth, H. Deconinck (eds.), *Error Estimation and Adaptive Discretization Methods in Computational Fluid Dynamics.*

26. M. Griebel, M. A. Schweitzer (eds.), *Meshfree Methods for Partial Differential Equations.*

27. S. Müller, *Adaptive Multiscale Schemes for Conservation Laws.*

28. C. Carstensen, S. Funken, W. Hackbusch, R. H. W. Hoppe, P. Monk (eds.), *Computational Electromagnetics.*

29. M. A. Schweitzer, *A Parallel Multilevel Partition of Unity Method for Elliptic Partial Differential Equations.*

30. T. Biegler, O. Ghattas, M. Heinkenschloss, B. van Bloemen Waanders (eds.), *Large-Scale PDE-Constrained Optimization.*

31. M. Ainsworth, P. Davies, D. Duncan, P. Martin, B. Rynne (eds.), *Topics in Computational Wave Propagation.* Direct and Inverse Problems.

32. H. Emmerich, B. Nestler, M. Schreckenberg (eds.), *Interface and Transport Dynamics.* Computational Modelling.

33. H. P. Langtangen, A. Tveito (eds.), *Advanced Topics in Computational Partial Differential Equations.* Numerical Methods and Diffpack Programming.

34. V. John, *Large Eddy Simulation of Turbulent Incompressible Flows*. Analytical and Numerical Results for a Class of LES Models.

35. E. Bänsch (ed.), *Challenges in Scientific Computing - CISC 2002*.

36. B. N. Khoromskij, G. Wittum, *Numerical Solution of Elliptic Differential Equations by Reduction to the Interface*.

37. A. Iske, *Multiresolution Methods in Scattered Data Modelling*.

38. S.-I. Niculescu, K. Gu (eds.), *Advances in Time-Delay Systems*.

39. S. Attinger, P. Koumoutsakos (eds.), *Multiscale Modelling and Simulation*.

40. R. Kornhuber, R. Hoppe, J. Périaux, O. Pironneau, O. Wildlund, J. Xu (eds.), *Domain Decomposition Methods in Science and Engineering*.

41. T. Plewa, T. Linde, V.G. Weirs (eds.), *Adaptive Mesh Refinement – Theory and Applications*.

42. A. Schmidt, K.G. Siebert, *Design of Adaptive Finite Element Software*. The Finite Element Toolbox ALBERTA.

43. M. Griebel, M.A. Schweitzer (eds.), *Meshfree Methods for Partial Differential Equations II*.

44. B. Engquist, P. Lötstedt, O. Runborg (eds.), *Multiscale Methods in Science and Engineering*.

45. P. Benner, V. Mehrmann, D.C. Sorensen (eds.), *Dimension Reduction of Large-Scale Systems*.

46. D. Kressner, *Numerical Methods for General and Structured Eigenvalue Problems*.

47. A. Boriçi, A. Frommer, B. Joó, A. Kennedy, B. Pendleton (eds.), *QCD and Numerical Analysis III*.

48. F. Graziani (ed.), *Computational Methods in Transport*.

49. B. Leimkuhler, C. Chipot, R. Elber, A. Laaksonen, A. Mark, T. Schlick, C. Schütte, R. Skeel (eds.), *New Algorithms for Macromolecular Simulation*.

50. M. Bücker, G. Corliss, P. Hovland, U. Naumann, B. Norris (eds.), *Automatic Differentiation: Applications, Theory, and Implementations*.

51. A.M. Bruaset, A. Tveito (eds.), *Numerical Solution of Partial Differential Equations on Parallel Computers*.

52. K.H. Hoffmann, A. Meyer (eds.), *Parallel Algorithms and Cluster Computing*.

53. H.-J. Bungartz, M. Schäfer (eds.), *Fluid-Structure Interaction*.

54. J. Behrens, *Adaptive Atmospheric Modeling*.

55. O. Widlund, D. Keyes (eds.), *Domain Decomposition Methods in Science and Engineering XVI*.

56. S. Kassinos, C. Langer, G. Iaccarino, P. Moin (eds.), *Complex Effects in Large Eddy Simulations*.

57. M. Griebel, M.A Schweitzer (eds.), *Meshfree Methods for Partial Differential Equations III*.

58. A.N. Gorban, B. Kégl, D.C. Wunsch, A. Zinovyev (eds.), *Principal Manifolds for Data Visualization and Dimension Reduction*.

59. H. Ammari (ed.), *Modeling and Computations in Electromagnetics: A Volume Dedicated to Jean-Claude Nédélec*.

60. U. Langer, M. Discacciati, D. Keyes, O. Widlund, W. Zulehner (eds.), *Domain Decomposition Methods in Science and Engineering XVII*.

61. T. Mathew, *Domain Decomposition Methods for the Numerical Solution of Partial Differential Equations*.

62. F. Graziani (ed.), *Computational Methods in Transport: Verification and Validation*.

63. M. Bebendorf, *Hierarchical Matrices. A Means to Efficiently Solve Elliptic Boundary Value Problems.*

64. C.H. Bischof, H.M. Bücker, P. Hovland, U. Naumann, J. Utke (eds.), *Advances in Automatic Differentiation.*

65. M. Griebel, M.A. Schweitzer (eds.), *Meshfree Methods for Partial Differential Equations IV.*

66. B. Engquist, P. Lötstedt, O. Runborg (eds.), *Multiscale Modeling and Simulation in Science.*

67. I.H. Tuncer, Ü. Gülcat, D.R. Emerson, K. Matsuno (eds.), *Parallel Computational Fluid Dynamics.*

68. S. Yip, T. Diaz de la Rubia (eds.), *Scientific Modeling and Simulations.*

69. A. Hegarty, N. Kopteva, E. O'Riordan, M. Stynes (eds.), *BAIL 2008 – Boundary and Interior Layers.*

70. M. Bercovier, M.J. Gander, R. Kornhuber, O. Widlund (eds.), *Domain Decomposition Methods in Science and Engineering XVIII.*

71. B. Koren, C. Vuik (eds.), *Advanced Computational Methods in Science and Engineering.*

72. M. Peters (ed.), *Computational Fluid Dynamics for Sport Simulation.*

73. H.-J. Bungartz, M. Mehl, M. Schäfer (eds.), *Fluid Structure Interaction II - Modelling, Simulation, Optimization.*

74. D. Tromeur-Dervout, G. Brenner, D.R. Emerson, J. Erhel (eds.), *Parallel Computational Fluid Dynamics 2008.*

75. A.N. Gorban, D. Roose (eds.), *Coping with Complexity: Model Reduction and Data Analysis.*

For further information on these books please have a look at our mathematics catalogue at the following URL: www.springer.com/series/3527